An Introduction to
INDUSTRIAL MYCOLOGY

Sixth Edition

GEORGE SMITH
M.Sc., F.R.I.C., M.I.Biol.

*Sometime Senior Lecturer, Department of Biochemistry,
London School of Hygiene and Tropical Medicine,
University of London*

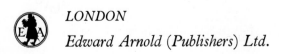
LONDON
Edward Arnold (Publishers) Ltd.

© *George Smith 1969*

First published	*1938*
Second edition	*1942*
Third edition	*1946*
Reprinted	*1948*
Fourth edition	*1954*
Fifth edition	*1960*
Sixth edition	*1969*

SBN: 7131 2208 0

261678

Printed in Great Britain by
R. & R. Clark Ltd., Edinburgh

An Introduction to

INDUSTRIAL MYCOLOGY

GEORGE SMITH
1895 – 1967

President of the British Mycological Society
1945

Foreword to the First Edition

by HAROLD RAISTRICK, SC.D., F.R.S.

Professor Emeritus of Biochemistry, University of London

With a Postscript, specially written for the Sixth Edition

It is a truism to say that were all types of micro-organisms suddenly to die out, human and animal life as we know it to-day would be impossible. Dead vegetable and animal remains are continuously being broken down by biochemical reactions brought about by micro-organisms, until their constituent elements are returned to the economy of nature as carbon dioxide, ammonia, nitrates, etc., to recommence the synthetic cycle for which the green plants are responsible. The part played by the lower fungi, or "moulds" as they are commonly known, in this chain of degradative processes is one of paramount importance.

Further, in all industries handling organic materials, e.g. those concerned with food production, leather, textiles, wood, pharmaceuticals, etc., the danger of spoilage through growth of moulds is one of which all those engaged in these industries are acutely aware. Thus the control and prevention of mould growth becomes a major problem to which, in the interests of increased efficiency, more and more attention is being given.

The harnessing of moulds for beneficial purposes, long established in some industries, as for example in the manufacture of Stilton, Gorgonzola and other types of cheese, has in recent years followed some very interesting lines. Thus citric acid, until this century obtained exclusively from the juice of citrus fruits, is now being made in thousands of tons per annum and in many different countries by growing chosen strains of the common black mould *Aspergillus niger* on sugar solutions, under carefully controlled conditions. The study of the biochemical changes, almost bewildering in their diversity, which can be brought about by moulds, is now a rapidly developing branch of biochemistry, and is attracting the attention of scientific workers in different parts of the world.

To those who are actively concerned in any of the industrial or scientific pursuits I have mentioned—and many more examples could be given if these were needed—it is scarcely necessary for me to point out the paramount importance of some knowledge of the moulds themselves, and the more detailed and accurate this knowledge is, the better. Thousands of different species of moulds have been described and their differences in response to a particular environment, their tolerance of adverse conditions, and their biochemical characteristics, are almost as varied as their numbers are great. There are in existence many admirable text-books on mycology which will probably meet the needs of the student who has had an adequate training in botany. But for those with little or no botanical knowledge, and particularly for those who are faced for the first time with an industrial problem of "mould" control, the lack of an adequate text-book setting forth in simple language the facts of the subject is a very real lack. To these and to any others who wish to acquire a first-hand knowledge of the common "moulds" I most warmly recommend this book written by my colleague, Mr. George Smith. The subject-matter of the book forms the basis of a course of lectures and practical work given by the author as part of the course to students working in this School for the post-graduate Academic Diploma in Bacteriology of the University of London. Readers of Mr. Smith's book will, I think, find it easy to read and stimulating to study, and will, I hope and believe, particularly appreciate, as I do, the really beautiful photomicrographs which form a very important part of the book.

H. RAISTRICK

LONDON SCHOOL OF HYGIENE
AND TROPICAL MEDICINE
July 1938

POSTSCRIPT

It is now thirty years since I wrote the Foreword to the first edition of this book and I have seen with pleasure the appearance of the further editions of it. I feel sure that this latest edition will be as much appreciated as the previous five. My association with Mr. George Smith extended over more than thirty years, not only as a colleague but also as a personal friend. It was with great sorrow that I heard of his death last year and I would like to take this opportunity of paying tribute to his valuable work in the field of Chemical Mycology.

H. RAISTRICK
July 1968

Preface to the First Edition

This book is intended to assist those who are commencing the study of "moulds" rather than of fungi in general. There is already an extensive literature of systematic mycology, plant pathology and medical mycology, but there has been up to the present no book in English, apart from highly specialized monographs, dealing particularly with the fungi which are of importance in industry.

Sufficient general mycology is included to enable the student to follow up the subject in the standard text-books. The major portion of the book, however, consists of descriptions and illustrations of most of the genera of moulds which are of regular occurrence in industrial products, with more detailed consideration of the genera which are of greatest importance. Chapters on laboratory methods are sufficiently detailed to enable those who have had no previous biological training, and who are unable to get personal instruction, to work from the beginning along the right lines.

Many, probably the majority, of those who are called upon to undertake the solution of problems connected with moulds in industry are chemists, most of whom have had no training in botany and who find it difficult to learn the special terminology of mycological literature. Throughout the book, therefore, I have endeavoured to explain all such terms and usages as are like to be unfamiliar to the non-botanical reader.

All the figures, except Fig. 101, are from original photomicrographs, this type of illustration being, in my opinion, more suitable than line drawings for the use of beginners. With few exceptions they are all at certain precise and selected magnifications and are readily comparable one with another. I am grateful for permission to include certain of my illustrations which have previously been published—Figs. 97, 104, 105, 107, 112, 116, 120, 122, 127 in *The Journal of the Textile Institute*; Figs. 8, 138, and 139 in *Transactions of the British Mycological Society*.

To the many colleagues and workers in other institutions who have supplied infected materials and cultures, or who have made practical suggestions, I tender my thanks. I am also greatly indebted to Messrs. Boardman & Baron, Ltd., of Great Harwood, for permission to make free use of the large number of photographs taken in their laboratories.

G. S.

July 1938

Preface to the Sixth Edition

In revising the text for this new edition I have endeavoured to take account of all important publications in the realm of Industrial Mycology. There are a number of changes in nomenclature, to bring this in line with modern opinion, and this has meant some rearrangement of the material. During recent years the tempo of work on soil mycology has increased so much that this subject has been given a separate chapter. Another field in which there has been considerable advance is in the taxonomy of the Actino-mycetes. The section on these organisms has, therefore, been rewritten and expanded.

There are six new photomicrographs, and a few deletions of illustrations which, for various reasons, were unsatisfactory.

It remains for me to tender my sincere thanks to all who have sent me material, reprints, criticisms, or appreciation, and particularly to Messrs. Edward Arnold (Publishers) Ltd, with whom I have had the happiest association since the script of the first edition was sent to them in 1938.

G. S.

HARROW

ADDENDUM

At the time my husband died he had wholly completed his revision of the text of this book and the script was virtually ready for submission.

I am indebted to my son, Dr. Geoffrey Howard Smith, for taking over the task of seeing this new edition through to publication.

Help given by Dr. G. C. Ainsworth and by two of my husband's former research students, Dr. Agnes H. S. Onions and Mr. John J. Elphick, is gratefully acknowledged. Mr. Elphick read through the entire script and made valuable suggestions for its improvement, notably in Chapters XI, XIV and XVII. Proof-correction and revision of the index were nobly undertaken by Dr. Onions and Dr. Ainsworth.

Mr. P. J. Price, Scientific Director of Edward Arnold (Publishers) Ltd., has throughout been most considerate and helpful, and I thank him for his cooperation.

NATHALIE SMITH

Harrow,
July 1968

Contents

Chapter I

Introduction

> We may rest assured that as green plants and animals
> disappear one by one from the face of the globe, some of the
> fungi will always be present to dispose of the last remains.
>
> B. O. Dodge, *Rep. 3rd Int. Congr. Microbiol.*, 1940

Mycology is concerned with the study of the Fungi, the term being derived
from the Greek word *mykes*, meaning a fungus. The Fungi were, until
comparatively recent times, regarded as members of the Plant Kingdom,
and certainly, in general aspect, the majority of them bear a superficial
resemblance to plants. Even at the present day nearly all the teaching in
mycology in this country is carried out as part of courses in botany in our
schools and universities, and the great majority of research workers in
mycology have been trained as botanists. The supposed relationship of the
Fungi to the various types of true plants (and some other groups of organisms) is usually set out somewhat as in Table 1.

TABLE I

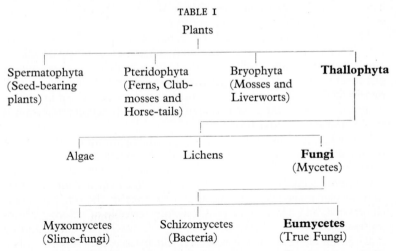

The Thallophyta (Gr. *thallos*, a young shoot; *phyton*, a plant) are plants
which show no differentiation into root, stem, leaf, etc., the vegetative
structure being known as a *thallus*. This may be unicellular, as in some
of the simplest fungi, or may show considerable specialization of structure
with corresponding specialization of function.

The Fungi are distinguished from the Algae in that they lack chlorophyll, the green colouring matter which enables plants to bring about photosynthesis, that is, the building up of complex organic compounds from carbon dioxide and water in the presence of sunlight.

The Lichens are compound organisms, consisting of algae and fungi in intimate association. Their study is a special branch of botany, since the alga-fungus association is so close that the Lichens may be classified into genera and species just as if they were single organisms, and many of the fungi are unknown apart from their algal associates. Those who are interested in this group should consult the excellent text by Hale (1967).

The Myxomycetes (Gr. *myxa*, slime; *myketes*, pl. of *mykes*) are a puzzling group of organisms. When they were first described they were regarded as Gasteromycetes (i.e. Basidiomycetes, including puffballs, stinkhorns, etc.), and were so classified by Persoon (1801) and Fries (1829); the latter, however, separating them as a distinct Suborder, Myxogastres. Link (1833) separated the group completely from the Gasteromycetes, and coined the term Myxomycetes. De Bary (1858, 1859) stated that these organisms are not fungi, but are to be regarded as amongst the lowest members of the animal kingdom, terming them Mycetozoa (Gr. *mykes* and *zoon*, an animal). Many subsequent students of the group have followed De Bary, but others maintain that the real place of the slime moulds is with the fungi (see Martin, 1960). No matter how they are to be classified their study does not come within the province of industrial mycology. There is a good monograph of the British Mycetozoa by Lister (1925), and with its aid identifications are not difficult.

The Bacteria, comprising in Table 1, one of the three classes of fungi, bear little resemblance to true fungi. Until modern developments in microscopic technique have advanced to the stage of enabling us to get a clear picture of their structure, their position in any scheme of classification will remain doubtful. In any case, their study is now a separate and important branch of science, with a technique of its own, and their consideration is outside the scope of this book.

The only organisms, listed in Table 1, coming within the province of the industrial mycologist are, therefore, the Eumycetes, or true fungi.

The modern view is that the Fungi do not belong to either of the Plant and Animal Kingdoms, but constitute a third, co-equal Kingdom. One fundamental difference from plants is that none of the fungi produces chlorophyll, and none, so far as is known, is capable of effecting photosynthesis. Plants, in general, utilize simple substances and build them up into substances of greater complexity. On balance they absorb carbon dioxide, using it to build up starch, cellulose, fats and the like, and liberate oxygen. Also their nitrogen requirements are met by the presence in the soil of simple inorganic salts, nitrates and ammonium salts. Animals differ fundamentally from plants in that they are entirely unable to utilize carbon dioxide as a source from which to build up tissue but, on the contrary, require oxygen and liberate carbon dioxide as a waste product. They cannot utilize inorganic compounds of nitrogen but require this element in the form of proteins, or at least the constituent amino-acids. Fungi resemble animals in that they require oxygen and invariably liberate carbon dioxide as a final metabolic product. On the other hand, very many species are able to utilize inorganic nitrogen, in this respect resembling plants.

According to Langeron (1945) a further fundamental difference between plants and animals on the one hand and fungi on the other is that the latter never form tissue, all structures, including the highly organized fruit-bodies of the larger fungi, consisting entirely of a system of tubes. Even when the tubes are apparently divided into individual cells by cross-walls, these, except when they are formed to cut off dead portions of the organisms, always have a central pore, through which both cytoplasm and nuclei can freely pass. Langeron maintains that fungi are thus essentially unicellular and the following classification is based on his arguments.

TABLE 2

Living Things

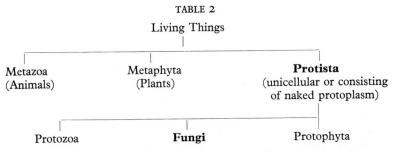

The term "protista", derived from Gr. *protistos*, superlative of *protos*, first, was coined by the German philosopher Haeckel.

The number of species of fungi is possibly about 250,000. This approximately equals the number of species of flowering plants, but in number of individuals the fungi surpass plants, for their range of habitat is wider and many species are of world-wide distribution. As is to be expected of such a large group of living things, the fungi show great differences in size, structure, and metabolic activities. Some, such as the Yeasts, grow as loose aggregates of single detached cells, whilst others, such as the mushrooms and toadstools, form large fruit-bodies of complicated structure, with elaborate mechanisms for propagation. Some of the larger fungi are prized by the epicure whilst others are shunned as amongst the deadliest of poisons. The majority of the known fungi live on dead organic matter, performing a useful service in returning to the soil nutrients originally extracted by plants, but there is a large group of species which are inimical to man's activities through their habit of parasitizing plants which are grown for food and clothing, and a smaller group which are parasitic on animals, including man himself. Many fungi attack manufactured products of all kinds, including foodstuffs, fabrics, leather, timber, cosmetics, pharmaceuticals, and even glass. On the other hand, a number of species are capable of synthesizing, under suitable conditions, substances useful to man, with an economy of effort which the chemist cannot emulate.

In mycology, as in other sciences, increased knowledge has resulted in complexity, and eventually the division of the science into a number of branches in which individual workers tend to specialize. What is usually termed pure mycology concerns the detailed structure, cytology, and modes of development of fungi. Field mycologists are interested in the fungi which are to be found in fields and woods, both the larger forms, known as mushrooms and toadstools, which grow on the ground or as parasites of forest trees, and the microscopic forms found on plant debris or as parasites of wild plants. The taxonomist studies structure with a view to classifying fungi, so as to show relationships and facilitate identifications by others. Although plant pathology is not a branch of mycology, since it is concerned with the study and prevention of all kinds of abnormalities in cultivated plants, the plant pathologist nevertheless must have a good knowledge of mycology, for many important diseases of plants are caused by fungi, and there are now a number of workers who specialize in the mycological side of the subject. Medical mycology deals with the fungi which cause disease in man. Another somewhat restricted branch of the science is the study of the wood-destroying fungi, both those which attack standing trees and those which rot felled and worked timber.

The field of industrial mycology includes both the harmful activities of fungi in rotting or spoiling industrial raw materials and manufactured goods, and the uses of fungi in industrial fermentations. The fungi concerned are commonly known as "moulds". They show considerable diversity of structure and are placed by the taxonomist in a number of widely separated groups. Although they are commonly visible to the naked eye, they all produce minute fruiting structures which cannot be studied without the aid of a microscope. One great advantage which industrial mycology has over some branches of the science is that the moulds are readily grown in the laboratory and can be studied quite independently of season or weather.

The various species of moulds differ in their responses to environmental conditions, in their abilities to attack various types of material, in toleration to antiseptics, and in synthetic activity. Hence little can be accomplished in the field of industrial mycology without a working knowledge of the moulds themselves, and the ability to recognize at least the more common species. Even in fermentation industries, which may use a single species of fungus, contaminants are likely to cause trouble unless they are recognized at an early stage. In addition, it is essential to be able to distinguish a highly active strain from other less useful strains of the same species, and to be able to recognize that species when searching for new and more active strains. A large proportion of this book is therefore devoted to descriptions and illustrations of most of the common species of moulds, and to the methods used in studying them for the purpose of identification. The later chapters deal with the activities of moulds, both harmful and beneficial, but only in outline and in such a way as to introduce to the reader the extensive literature on the various subjects.

REFERENCES

de Bary, A. (1858). Ueber die Myxomyceten. *Bot. Ztg*, **16**, 357–8, 361–4, 365–9.

de Bary, A. (1859). Die Mycetozoen. Ein Beitrag zur Kenntnis der niedersten Thiere. *Z. wiss. Zool.*, **10**, 88–175.

Fries, E. M. (1829). *Systema mycologicum*. Vol. III, 67–199.

Hale, M. E. (1967). *The Biology of Lichens*. London: Edward Arnold.

Langeron, M. (1945). *Précis de Mycologie*. Paris: Masson & Cie.

Link, D. H. (1833). *Handbuch zur Erkennung der nutzbarsten und am häufigsten vorkommenden Gewächse*. Vol. III, 405–24.

Lister, A. (1925). *A Monograph of the Mycetozoa*, 3rd Ed., revised by G. Lister. London: Brit. Museum Publ.

Martin, G. W. (1960). The systematic position of the Myxomycetes. *Mycologia*, **52**, 119–29.

Persoon, C. H. (1801). *Synopsis methodica fungorum*.

Chapter II

General Morphology and Classification

Ce ne sont pas les rêveries des phylogénistes qui nous donneront, en mycologie, la systematique idéale.

M. Langeron, 1945

When a mould begins to grow, whether it is on a natural substrate, on some manufactured material, or on a culture medium in the laboratory, there is an initial period during which nothing is visible to the naked eye. The time taken for visible growth to appear may vary from a few hours to many days, depending on a number of factors, the most important of which are the particular species of mould, the availability of any nutrient material present, and the relative humidity of the air. When the mould *colony* (as it is usually termed) has grown sufficiently to be readily seen, examination with a good hand-lens or low-power microscope will show the presence of a network of fine filaments. Each individual filament is termed a *hypha* (Gr. *hyphe*, a tissue), and the hyphae collectively are called *mycelium* (Gr. *mykes*, a fungus; *elos*, a wart). Some of the hyphae grow along the surface of the substrate, some may penetrate to a degree depending on the texture of the substrate, whilst others may stand above the surface, in some cases giving a hairy or fluffy appearance.

If the material on which the mould is growing is poor in nutritive value the mycelium spreads slowly and usually changes little in appearance, except that any aerial growth tends to collapse after a time. When, however, there is a reasonable amount of available food, and external conditions remain favourable, there is usually a gradual change in the appearance of the colony. Frequently there is a distinct change in colour, first manifested in the central and older parts of the mycelium. Microscopic examination at this stage will show the presence of fruiting structures, which are readily recognized as quite distinct from the ordinary hyphae. The individual reproductive bodies are called *spores* (Gr. *spora*, a seed). For example, the green mould usually found on leather (*Penicillium*) has more or less erect hyphae terminating in miniature broom-like structures, the individual bristles of the broom consisting of long chains of small roundish spores; the hairy-looking mould on bread (*Rhizopus*) bears small, round, dark heads on erect hyphae, and if a head is crushed on a microscope slide it will be found to be full of small oval spores; a mould commonly found

6

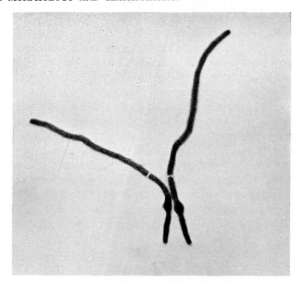

FIG. 1.—Germinating spores of *Penicillium notatum*. × 500.

FIG. 2.—Anastomosis. × 70.

on jam (*Aspergillus*) produces spores of two kinds, one kind arranged in radiating chains, arising from the tops of erect hyphae, the other enclosed in yellow spherical bodies which are big enough to be seen without a lens. Other moulds form spores in a variety of ways, but, however they are formed, and whatever their appearance, they are almost always readily recognizable as distinct from the mycelium.

The spores of a fungus are designed for dissemination and reproduction, just as seeds are produced as a means of propagation by green plants. A spore, however, differs from a seed in that it never contains a germ, or plant embryo. It may consist of a single cell, or be a compound structure of several cells, each consisting of a mass of protoplasm surrounded by a firm containing wall. In size spores of different species show considerable variation, ranging from little over 1μ, as in some species of *Penicillium*, to about 200μ, in greatest dimension, in certain species of *Helminthosporium*, but even the largest spores are light enough to be transported considerable distances by air currents, with the result that it is rare to find the atmosphere, either inside a building or outside, free from mould spores.

When a spore eventually alights it may remain dormant for a long period if conditions for growth are not favourable. As soon, however, as the relative humidity of the air becomes sufficiently high (the actual figure varies for different species), germination occurs. It should be noted that, although the term "germination" is always used in reference to spores of fungi, the process has nothing in common with the process of germination of a plant seed. The first stage in germination of a spore is the absorption of water, with a consequent increase in size of the spore. Then one or more **germ-tubes** emerge from points on the surface and rapidly elongate. Even a single-celled spore may give rise to more than one germ-tube, whilst in the case of many multicellular spores any cell may germinate separately. Fig. 1 shows two spores of *Penicillium notatum* which have been allowed to germinate on cellulose sheet and subsequently stained (for the method see Chapter XI). Note that each spore has put forth two germ-tubes and that these have issued successively and not at the same time. If a supply of suitable nutrients, as well as sufficient moisture, is available, the germ-tubes rapidly increase in length and soon branch repeatedly, so as to cover the surface of the substrate with a radiating network of hyphae. Frequently branches from one hypha fuse with other hyphae. This arrangement facilitates rapid transport of food material to the points where it is most required. The fusion of one hypha with another is termed **anastomosis** (Gr. *ana*, back; *stoma*, a mouth) and is illustrated in Fig. 2. When the mycelium has attained a certain size, that is, when the capacity for food intake has increased sufficiently, certain specialized hyphae begin to appear, and eventually these bear spores, ready to start the cycle over again. With very many moulds spores appear first on the older central parts of the mycelium, the fruiting area gradually spreading outwards, but always remaining some little distance short of the edge of the mycelial area.

CLASSIFICATION

The purpose of classification is twofold. On the one hand is the desire to exhibit natural relationships, and to show how all the different kinds of organisms have evolved. This discipline is known as Phylogeny. The second, and more practical, purpose of classification is to facilitate identifications. The more numerous a particular group of organisms is, the more necessary it is to have a scheme of classification which makes it as easy as possible to name any given species when found. From this point of view it does not matter whether or not the scheme is phylogenetically sound, but, of course, it is more satisfactory if the two aims of classification can be combined.

Mycelium may be colourless, brightly coloured, or dingy brown to almost black, the hyphae of different species may be of very varying average diameters, and characteristic pigments may be secreted and appear as nodules on the hyphae, or even as distinct crystals. However, the differences are seldom sufficient to allow of a fungus being recognized and named from a consideration of mycelial characteristics alone. The spores show much more variation, as between different species, and at the same time are more constant in size, shape, and colour for any one species. Hence all systems of classification are based chiefly on methods of spore production and the characteristics of the spores themselves.

Before discussing the characters of the various classes of fungi it is necessary to define a number of terms used in describing spores and fruiting structures. The term "spore" itself is used in two different senses. Most mycologists designate as spores all the types of specialized cells, formed on or within, or cut off from, special organs, and obviously designed as means of propagation and dissemination of the species, but some reserve the term for the product of what is the equivalent of a sexual process. The sexual nature of spore formation may be obvious, as in the Oomycetes and Zygomycetes, but in other cases it is obscure and has only been elucidated by cytological studies, although the presence of fruit-bodies or spore-bearing organs which regularly produce a definite number of spores, usually a multiple of two, is a fairly sure indication that a sexual process is involved.

Asexual spores are of various forms, are borne on very varied types of spore-bearing structures, and, unlike the sexual spores, are produced in indefinite numbers. Even in the orders of fungi which are classified primarily according to the characteristics of the sexual spores, the mode of occurrence and the size, shape, and markings of the accessory (asexual) spores are of considerable importance in classification, and, in the great group of Fungi Imperfecti, which includes most of the common moulds, are the only kinds of spores and, therefore, the only structures on which a system of classification can be based.

The following list includes most of the special terms which are in use to designate different kinds of spores and the organs which bear them. A further list of more specialized terms is given in Chapter VII.

Acervulus (dimin. of Lat. *acervus*, a heap). A structure characteristic of the Melanconiales (see p. 75). Fungi of this order, when growing on the host plant, break through the cuticle of the host and produce a cushion-like mass of conidiophores, sometimes accompanied by stiff, sterile hyphae known as setae.

Apothecium (Gr. *apo*, from, away from; *theke*, a receptacle). A cup-shaped or saucer-shaped structure, inside which are numerous, closely-packed, cylindrical or club-shaped asci. The term signifies that it is not a true receptacle, since it is wide open.

Ascus, pl. asci (Gr. *askos*, a skin bag). A structure characteristic of, and found only in, the class Ascomycetes. It is a thin-walled sac containing spores. The number of spores in each ascus is normally eight, but in certain cases is some other multiple of two. In the lower members of the Ascomycetes the ascus is rounded, with the spores packed tightly inside, whilst in the more highly developed species it is cylindrical or club-shaped, with the spores arranged in one or two rows. (See Fig. 35).

Basidium (Gr. *basis*, a pedestal; *eidos*, like). An organ characteristic of the class Basidiomycetes. In the larger Basidiomycetes, the mushrooms and toadstools, it is a short cell, wider at the top than at the base, and bearing four tiny projections (sterigmata) on each of which a spore is borne. It is in the basidium that the two nuclei of opposite "sex" fuse, the fusion being followed by two successive divisions to give four daughter nuclei, one passing into each developing spore.

Chlamydospore (Gr. *chlamys*, a cloak). A thick-walled resting spore, formed by swelling and thickening of single short cells in submerged vegetative hyphae, in aerial hyphae, or even in multiseptate conidia such as those of species of *Fusarium* (see Figs. 13, 63, 64).

Conidiophore (from *conidium* and Gr. *phoreo*, to bear or carry). A specialized hypha bearing conidia, either at the tip or, more rarely, along its length. In some genera the conidiophore terminates in a series of branches, arranged in a more or less definite pattern, on which the conidia are borne.

Conidium, pl. conidia (Gr. *konis*, dust; *eidos*, like). The term is often applied to a variety of types of asexual spores, but some authorities restrict the term to spores cut off successively from "phialides" (*q.v.*).

Coremium (Gr. *korema*, a besom). An erect, compact cluster of conidiophores. The latter may be of different lengths, the coremium being then more or less cylindrical and clothed with spores throughout the greater part of its length, or may be of approximately equal length, in which case the coremium has a distinct stalk and a spore-bearing head. (See Figs. 57, 70.)

Oidia. A term usually used only in the plural. (*Oidium* is the name of a genus of fungi.) In some fungi portions of the mycelium form close series of very short cells, which may remain roughly rectangular or may round

up to resemble a chain of beads. The mycelium breaks up at the septa
and the fragments, known as oidia, function as spores. (See Fig. 65)

Perithecium (Gr. *peri*, around; *theke*, a receptacle). A structure found
in the majority of Ascomycetes, and consisting of a receptacle in which
asci are produced, often in considerable numbers. The perithecium may
be roughly spherical and break up at maturity to liberate the spores, or,
more usually, flask-shaped and liberate spores through the neck. (See
Figs. 33, 100.)

Phialide (Gr. *phiale*, a vial; *eidos*, like). A highly specialized terminal cell
of a conidiophore, or, more rarely, the conidiophore itself, more or less
flask-shaped, which continually elongates at the tip, but from which
conidia are successively cut off, so that the length of the cell remains
sensibly constant. Phialides are often called "sterigmata" (this usage was
followed in earlier editions of this book), but the latter term should be
reserved for a very different type of structure, the spore-bearing projections on basidia.

Pycnidium (Gr. *pyknos*, compact). A fruit body resembling a perithecium, flask-shaped or somewhat irregular in shape, usually of a size
readily visible to the naked eye. Unlike perithecia, pycnidia do not contain asci but are lined with very short conidiophores. When crushed, a
pycnidium releases its spores as an irregular mass. (See Fig. 77.)

Sclerotium (Gr. *skleros*, hard). Neither a spore nor a spore-bearing
structure, but compacted masses of mycelium, often very hard, varying
in size (in mould fungi) from a fraction of a millimetre to several millimetres in diameter. Some of the higher fungi produce much larger
sclerotia.

Sporangiole. A small sporangium, containing few spores.

Sporangium (Gr. *angeion*, a vessel). A closed receptacle, usually round
or pear-shaped, borne on a stalk (the *sporangiophore*) and containing
an indefinite number of spores. Mould fungi which produce spores in
sporangia include the Mucorales (Chapter IV).

Sporodochium (Gr. *docheion*, a receptacle). A cushion-shaped, tightly
packed cluster of conidiophores. Really a type of coremium in which the
conidiophores are too short to form a stalk.

Sporophore (Gr. *phoreo*, to bear). Any spore-bearing structure. Sporangiophores and conidiophores are types of sporophores.

Sterigma, pl. sterigmata. See under "basidium".

Stroma, pl. stromata (Gr. *stroma*, a bed). As seen in laboratory cultures
is a gelatinous or leathery layer covering the surface of the culture
medium, and bearing spores on very short conidiophores or having
embedded in it perithecia or pycnidia.

Sympodium. A kind of false axis, in which the conidiophore or sporangiophore forms a terminal spore or sporangium, then puts out a
branch, which extends beyond the tip of the original sporophore, and
then produces a second terminal spore or sporangium. This process is

repeated, each new branch arising from the most recently formed branch. Fig. 14 shows a typical sympodium.

Sympodula. A term coined by Kendrick (1962) for a type of conidiophore which produces a terminal single conidium, then buds out from just below the conidium to produce a very short branch, on which a second spore is formed, and so on. The terminal portion of the conidiophore thus has the appearance of a very much longitudinally compressed sympodium. The conidia so produced are termed sympodioconidia. Sympodulae are formed in the genera *Helminthosporium* and *Curvularia* (q.v.).

Synnema, pl. synnemata (Gr. *syn*, together; *nema*, a thread). Much the same as a coremium. The term is also used for associations of conidiophores of less definite form than a true coremium.

There are some differences in detail between the schemes of classification propounded in the standard works on systematic mycology, but there is fairly general agreement as to the first great subdivisions of the Eumycetes. There are four classes, characterized as follows:

Class 1. PHYCOMYCETES. Mycelium in some primitive forms lacking; when present typically non-septate; hyphae usually of comparatively large diameter. Sexual reproduction by oospores or zygospores. Asexual reproduction by sporangia or modified sporangia.
    Sub-class OOMYCETIDAE. Sexual spores endogenous (i.e. formed within a closed receptacle); gametangia unlike, definitely recognizable as antheridium and oogonium.
    Sub-class ZYGOMYCETIDAE. Sexual spores exogenous (not enclosed); gametangia similar.
Class 2. ASCOMYCETES. Sexual spores endogenous, formed in asci; normal number of spores in ascus eight; asexual spores often present and very varied in form and disposition.
Class 3. BASIDIOMYCETES. Sexual spores exogenous, formed on basidia; normal number of spores per basidium four; asexual spores comparatively uncommon.
Class 4. FUNGI IMPERFECTI. No sexual spores known; reproduction exclusively by asexual spores.

The Phycomycetes (Gr. *phykos*, seaweed; hence, alga-like fungi) are distinguished from fungi of the other three classes by the character of the mycelium, which forms a system of continuous tubes without cross walls (septa), except in the fruiting organs and in some of the older hyphae. Fig. 3 shows vegetative hyphae of a species of *Rhizopus*, almost root-like in appearance and consisting of part only of a single enormously elongated and branched cell. Contrast this with Fig. 4, which is a photograph of germinated spores of an Ascomycete (*Aspergillus glaucus*), showing several septa formed in the primary germ-tubes. Fig. 5 shows the richly septate mycelium of one of the Fungi Imperfecti, *Alternaria tenuis*. If a young, vigorously growing culture of the very common mould *Rhizopus stolonifer*, preferably growing in a Petri dish, be examined with a low power of the

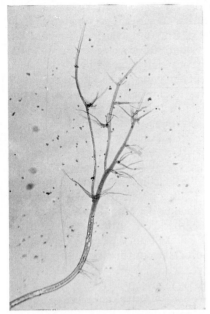

FIG. 3.—Non-septate mycelium—*Rhizo-pus stolonifer*. ×60.

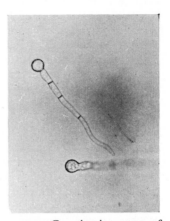

FIG. 4.—Germinating spores of *Aspergillus glaucus*, showing septation of primary hyphae. ×250.

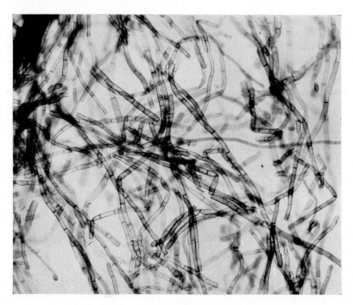

FIG. 5.—Septate mycelium—*Alternaria tenuis*. ×250.

microscope, it will be noticed that the cell contents are streaming along both the main hyphae and the secondary branches. This rapid movement is possible only because of the entire absence of septa. It is true that it is possible to detect, by patient observation, the movement of protoplasm along septate hyphae. This streaming, very slow compared with the rate in Phycomycetes, is possible because septa in living hyphae are not complete cell walls, but have small central pores. It is owing to the free connection between all parts of the mycelium that most of the Phycomycetes which can be grown in the laboratory spread with extreme rapidity.

The Oomycetidae. In this sub-class the sexual organs are readily distinguished as antheridium, the male organ (Gr. *antheros*, flowery), and oogonium, the female organ (Gr. *oon*, an egg; *gone*, seed, offspring). The oogonium, when ripe, shows most of the contained protoplasm aggregated into one or more oospheres (see Fig. 6). It is fertilized by the antheridium coming in contact with it or, in one family, the Monoblepharidaceae, by motile male cells liberated by the antheridium. After fertilization the oospheres round up and develop cell walls to form the oospores, which lie free within the oogonium (Fig. 7). Most Oomycetes also produce sporangia containing asexual spores.

Many species are aquatic or grow in very damp situations and, the better to aid rapid dissemination, produce motile spores known as zoospores, from their resemblance to the simplest forms of animal life. This sub-class includes a number of important plant parasites, such as *Phytophthora infestans*, the potato blight, and various species of the genus *Pythium*, causing damping-off diseases of seedlings. It also includes a few parasites of animal life, the best known being species of *Saprolegnia* on fish and insects in water. However, none of the species is likely to be encountered in industrial work and, therefore, their consideration will be limited to this brief account of their systematic position.

The Zygomycetidae (Gr. *zygon*, a yoke). In this sub-class the sexual organs are not distinguishable as antheridium and oogonium. They may be similar in size (the usual case) or markedly dissimilar as in *Zygorhynchus*, but they are essentially alike in nature and function. Whereas in the Oomycetes the oospores are formed within the oogonium, and are thus termed endogenous, the zygospore results from the fusion of two cells cut off from the sexual organs, and is thus exogenous. Two hyphae touch and at the point of contact develop protuberances, the progametangia, which gradually elongate and swell, remaining in contact themselves whilst pushing the parent hyphae further and further apart. Septa next appear in the progametangia, the cut-off portions which remain in contact being the gametangia. The latter fuse, the cell walls in contact disappearing, and the cell resulting from the fusion becomes rounded and, usually, thick-walled and dark-coloured. Fig. 8 shows six stages in zygospore formation. A slight variant of this process is found in *Piptocephalis*, a genus of moulds parasitic on other fungi belonging to the same order, the zygospores in this case

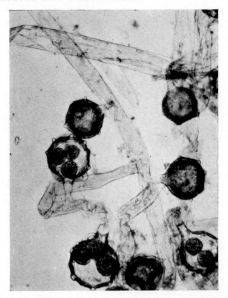

FIG. 6.—*Achlya* sp.—Oogonia with oospheres.
× 150.

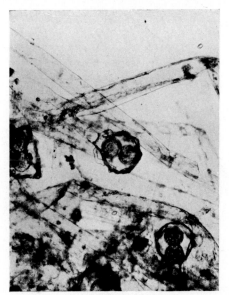

FIG. 7.—*Achlya* sp.—Oogonia with oospores.
× 150.

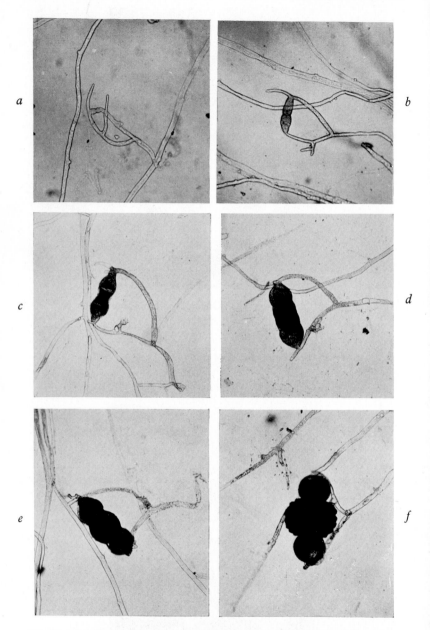

FIG. 8, *a-f*.—Six stages in zygospore formation—*Rhizopus sexualis*. × 100.

being formed in a bud put out from the cell which is produced by fusion of the gametangia. In some genera the zygospore is surrounded by stiff hairs or by a web of protecting hyphae, whilst in others it lies free amongst the aerial hyphae. In several genera which obviously belong here zygospores have never been observed.

Asexual spores are always non-motile and, in the best-known genera, are produced in round or pear-shaped sporangia borne on simple or

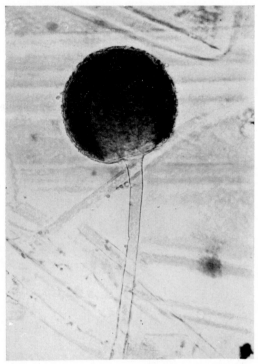

FIG. 9.—Sporangium of *Rhizopus stolonifer*. × 250. Somewhat immature, but showing the essential points of structure.

branched sporangiophores. Fig. 9 is a photograph of a sporangium of *Rhizopus stolonifer*. In a few genera, sporangia are lacking and asexual reproduction is by conidia.

The Zygomycetes include about 200 species, amongst which are a number of moulds of very common occurrence and considerable economic importance. They are considered in more detail in Chapter IV.

The Ascomycetes. In this class sexual spores are produced in asci (defined above). In most species each ascus contains eight spores. More rarely a larger or smaller multiple of two is the rule, and only in some of the

Yeasts is it usual to find asci with an odd number of spores. The group shows a gradual transition from primitive forms which produce single, naked, globose asci to species which build up elaborate fruit-bodies containing large numbers of club-shaped asci arranged in parallel series. The details of ascus formation are not readily observed and, up to the present, have been worked out for only a limited number of species. Full discussion of the class from this standpoint is to be found in several modern works on systematic mycology. Asexual reproduction is by conidia, which show great diversity of form and arrangement, or by oidia. The class comprises about 15,000 species and includes both saprophytes and parasites, microscopic species and large fleshy fungi, some of which are edible. The genera which are of importance to the industrial mycologist are considered further in Chapter V.

The Basidiomycetes. The spores which give the name to the group are the basidiospores, borne exogenously on special organs, the basidia. In the typical species each basidium bears four spores. In the higher Basidiomycetes, the mushrooms and toadstools, the basidia are found in serrated

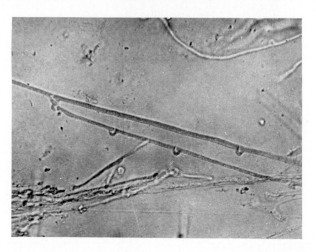

FIG. 10.—Clamp connections in mycelium of a Basidiomycete—
Polyporus sp. × 500.

ranks on the gills of the large fleshy sporophores and there are specialized arrangements for ensuring the widespread distribution of the spores. The lower Basidiomycetes, including the destructive parasites known as smuts or rusts, have a more complicated life cycle, and their position in a scheme of classification is not quite so obvious. Since most of these are obligate parasites, no attempt will be made to discuss their relationships and life histories. They have been extensively studied and are treated in detail in a number of text-books (see Chapter XVII). A fair number of species of the

Agaricales (the gill fungi) and the Polyporales (the fungi with pores instead of gills) can be grown in artificial culture and, by using special culture media, can even be induced to produce typical sporophores, but it is rarely that any of these are encountered in industrial work. Occasionally, however, when plating out various materials, particularly soils, Basidiomycetes appear on the culture plates. A few such species produce abundant conidia, but in most cases the growths are sterile. Whether forming conidia or not it is often easy to recognize that such species belong to the Basidiomycetes by the presence, in the mycelium, of what are known as **clamp connections**. These peculiar structures are found only in the Basidiomycetes. A little distance from the tip of a growing hypha a very short branch appears; this curls backwards and reunites with the parent hypha close to the point of emergence; in the meantime a septum is formed midway in the hypha and another at the base of the hook. These structures are readily detected, as slight bulges on the hyphae, under a low power of the microscope—when one knows what to look for they can be seen through the walls of a culture tube, using a $\frac{2}{3}$-inch objective—but a high power is required for resolution of details (Fig. 10).

The Fungi Imperfecti. This class comprises a very large number of fungi which produce neither ascospores nor basidiospores, but reproduce solely by means of conidia. (It should be noted that moulds which produce sporangia are not included in the Fungi Imperfecti, but in the Zygomycetidae, even though zygospores have never been seen.) It is safe to assume that, in many cases, their life cycles are incompletely known rather than incomplete. Most of them are probably Ascomycetes which produce the perfect stage (ascopores) only under special conditions which have not yet been discovered, or which have entirely lost the power of producing asci. It is now known with certainty in a few cases that heterothallism exists amongst the Ascomycetes (see below), and it is probable that this phenomenon accounts for many other cases of apparent loss of sexuality.

In the absence of perfect fructifications, it is necessary to classify the Fungi Imperfecti on characters of the asexual spores. According to the systematists, the genera thus set up are not true genera at all, since it is known, on the one hand, that a single genus of Ascomycetes may include species with very different types of imperfect fructifications and, on the other hand, that the conidial forms of species belonging to widely different genera may be similar. It follows, therefore, that a so-called "form-genus", comprising a number of species with the same type of conidial fructification, may be a purely artificial group of fungi which are really quite unrelated. However, until real relationships are discovered, it is necessary to have a convenient classification of the enormous number of Fungi Imperfecti, as otherwise identification of any given form would be a hopeless task.

The great majority of the common fungi which are popularly known as moulds or mildew belong to this group and will be dealt with at greater length in Chapters VII to X.

HETEROTHALLISM

It was for long a puzzle to mycologists why some species of the Mucorales regularly produced zygospores whilst others seemed to lose the power on continued cultivation and in still other species, which were obviously nearly related, zygospores were quite unknown. Various theories attempting to explain the matter on a nutritional basis were not supported by experiment, and it was not until 1904 that the riddle was solved by Blakeslee (1904, 1920). He showed that the sporangiospores from a strain of *Mucor* which was producing abundant zygospores gave rise, on germination, to two distinct strains of mycelium, called by Blakeslee + and − . Each strain, by itself, was incapable of forming zygospores, but if both strains were grown in such a way that the two mycelia could come in contact, zygospores were produced along the line of contact. It was thus obvious that transfers of considerable masses of spores from a culture containing zygospores would usually result in a mixed culture of the + and − strains and zygospores would again be produced, but that new cultures made from single spores would invariably lack zygospores and that sub-cultures made by transfer of a few spores would often result in a pure culture of one strain only. Such species, where sexuality resides in a whole thallus, are called *heterothallic*. In contrast to these the *homothallic* moulds produce zygospores on mycelia derived from single sporangiospores. Formation of gametangia is induced by contact of neighbouring hyphae, or even of branches from the same hypha. The illustrations of zygospore formation, Fig. 8, obviously depict a homothallic species. Both + and − strains of a number of heterothallic moulds are known and can be obtained from the various collections of type cultures, but, in a great many cases, only one strain of a species has so far been discovered and hence zygospores have never been observed.

Heterothallism is most frequently encountered in species belonging to the Mucorales, and it is to the numerous studies of such species that our knowledge of the phenomenon is chiefly due, but in comparatively recent times it has been shown that it is not confined to this order. Many Ascomycetes have been found to be heterothallic and in certain of the Basidiomycetes a still more complicated state of affairs exists, the four spores from a single basidium being all of different mating types.

REFERENCES

Blakeslee, A. F. (1904). Sexual reproduction in the Mucorineae. *Proc. Amer. Acad. Arts Sci.*, **40**, 205–319.
Blakeslee, A. F. (1920). Sexuality in the Mucors. *Science*, **51**, 375–82, 403–9.
Kendrick, W. B. (1962). The *Leptographium* complex. *Verticicladiella* Hughes. *Canad. J. Bot.*, **40**, 771–97.

Chapter III

Nomenclature

Il ne faut pas oublier qu'il y a dans les livres beaucoup
d'espèces . . . que dans la nature.

Konrad and Maublanc, *Les Agaricales*, 1948

In mycology, as in other branches of science, a stable and universally
accepted system of nomenclature is a necessity for progress. However,
although it is important, nomenclature is not studied for its own sake but
as an aid to the orderly classification of organisms which we call taxonomy.
Stability must therefore be combined with flexibility in order that nomenclature shall not impede advances in taxonomy.

In the early days of mycology there was no uniform method of naming
fungi. To refer to any particular species it was necessary to use cumbersome
descriptive phrases, as for example "Fungus campestris albus superne
inferne rubens", literally "the fungus of the fields, white on top and redden-
ing underneath", for the common mushroom. A great advance, not only
for mycology but for the whole of biology, was made when the great
Swedish botanist Linnaeus, in his *Species Plantarum* (1753), introduced the
use of binomials. In this system, soon to be universally adopted, the name
of any organism consists of two words, a Latin noun, which is the name of
the genus to which the organism belongs, followed by a specific epithet,
also Latin and agreeing with the noun according to the rules of Latin
grammar. The name is usually printed in italics, or underlined when
written, the generic name with a capital letter and the specific epithet not
capitalized. Each organism has its own name and the same name cannot
legitimately be applied to two different species.

In principle this method of naming living organisms is simple and
straightforward, but, in course of time, a number of difficulties have arisen.
So long as there were comparatively few mycologists, and only a scanty
literature of the science, it was not difficult for any one worker to become
acquainted with all the published descriptions of fungi, and to be in a
position to decide whether any species under examination was new to
science or had been described previously. However, the publication, in the
early part of last century, of a number of fungus floras, notably Persoon's
Synopsis Methodica Fungorum (1801) and Fries's *Systema Mycologicum*
(1821–32), together with great improvements in the microscope, stimulated
the search for fungi in all parts of the world, with a consequent enormous
increase in the volume of literature. Inevitably there was a good deal of

overlapping, many species being described under two or more distinct names. In addition, many illegitimate names were bestowed, in the sense that they had already been used for quite different organisms. Probably the greatest difficulty, however, has arisen from the fact that many of the descriptions in the older literature lack precision, the authors having been, of course, quite unaware of the enormous number of fungi yet to be discovered and of the necessity for detailed diagnosis. Many new names have been bestowed, in good faith, on fungi which were well known to the older mycologists. Subsequently, new interpretations of old publications have given rise to differences of opinion regarding the validity of many of these names, and the amount of attention which should be paid to the principle of priority. Saccardo, in his *Sylloge* (1882–1925), made an attempt to straighten out the confusion which, even at that time, had become serious, but eventually it was felt that only by international agreement was there any hope of putting nomenclature on a really sound basis.

INTERNATIONAL RULES

The nomenclatural confusion was by no means restricted to the realm of mycology, but obtained throughout botany. At a Botanical Congress held in Vienna in 1905 the matter was discussed at length, resulting in the publication, in 1906, of the *International Rules of Botanical Nomenclature*. The rules were accepted in Europe generally but not in America, where there was a much stricter interpretation of the law of priority. The rules were amended at the 2nd International Congress in 1910, and again at the 5th Congress in 1930, when truly international agreement was at last reached, and further modifications were made at the 6th Congress (1935), the 7th (1950), the 8th (1954) and the 9th (1959). The edition of the Rules as modified at the 10th Congress in 1964 appeared in 1966. It is not expected that finality has been reached, and there will certainly be a number of points for clarification at the next Congress.

Priority of names. The Rules lay down that each species of fungus shall bear one name only, and that this name shall be the earliest which was applied to the particular species. In the case of a fungus with more than one "state" the correct name is the first which was applied to the perfect state. Thus, many Ascomycetes produce not only typical perithecia but conidia in addition. In many such cases the connection between the two states was for long unsuspected, and each state, perithecial and conidial, or *status perfectus* and *status imperfectus*, received a separate name. As soon as it is definitely established that two names belong respectively to what are merely different states of the same fungus, the only correct name of the species is that of the perithecial or perfect state, even when the name of the conidial state is of earlier date. However, when a fungus commonly occurs as the conidial state alone, it is quite in order, and from the practical point of view often necessary, to continue to use a separate name for the

status imperfectus. For example, *Hypocrea rufa* Pers. ex Fries is an Ascomycete forming reddish stromata on dead wood. If the ascospores are
sown on laboratory culture media they give rise to typical cultures of the
imperfect fungus *Trichoderma viride* Pers. ex Fries, a mould which is of
common occurrence in the soil. The perithecial state is not produced in
cultures and, therefore, it would be useless for anyone seeing a culture of
Trichoderma for the first time to attempt to identify it from a key to the
Ascomycetes based on characteristics of perithecia and ascospores. In this
case the separate name is a convenience rather than a necessity, because
the connection with the perfect fungus *H. rufa* is quite definite, and there
is no logical reason why the latter name should not be used, instead of
T. viride, for the imperfect state. In other cases it is not possible to predict
the perfect state from an examination of the imperfect state, and a separate
name for the latter is a necessity.

Transfers. Owing to advances in our knowledge of fungi it often becomes necessary to transfer a species from one genus to another, or possibly
to make it the type of a new genus. Such changes are quite in order, but
with the proviso that the specific epithet shall remain unchanged, except
for inflection according to the rules of grammar. However often a species
is transferred, the correct specific epithet is always the first one which was
applied to that particular species. For example, Calmette in 1892 described
a mould isolated from Chinese rice as *Amylomyces rouxii*. Wehmer, 1900
decided that there was no justification in erecting a new genus for the fungus
and transferred it to *Mucor*, so that the correct citation is now *Mucor rouxii*
(Calmette) Wehmer. Again, Bainier in 1907 decided that *Penicillium brevicaule* Saccardo is not a true *Penicillium*, and made it the type of a new genus
Scopulariopsis, the correct citation being now *Scopulariopsis brevicaulis*
(Sacc.) Bainier.

Starting dates. One of the most important decisions embodied in the
Rules is the fixing of starting dates for the nomenclature of fungi. This
became necessary in order to give some degree of permanence to names of
fungi. If there were no definite starting-point it could easily happen that a
name which had been in use for many years, and which had gained common acceptance, would be rendered illegitimate by the discovery, in some
out of the way publication, of the earlier use of the same name for a different
organism, and there would never be any finality. The starting-point for the
Rusts, Smuts and Gasteromycetes (none of which concerns us here) is
Persoon's *Synopsis* (1801), and for all other fungi Fries's *Systema*. Until
the Congress of 1950 there was still some ambiguity, since some mycologists regarded 1821, the date of publication of the first volume of the
Systema, as the starting date for all species, whilst others, and these the
majority, were prepared to go back only to the date when any particular
species was described by Fries, the dates for the various parts of the
Systema (including a supplement, the *Elenchus*, which is to be regarded as
part of the *Systema*) being 1821, 1822, 1823, 1828, 1829, and 1832. It is

B

now decided that names used by Fries take precedence over synonyms or homonyms published by others during the period January 1, 1821, to December 31, 1832; also that names published by others during the period, and not included in the *Systema*, may be legitimate. Actually the wording of these rules is still somewhat ambiguous, probably because of lack of complete agreement between members of the Nomenclature Committee.

Hughes (1958) proposed that the starting-point for Hyphomycetes be taken as Persoon's *Synopsis* (December 31, 1801). This would involve a large number of changes of names, but it would have the advantage of legitimizing the names given by two assiduous collectors of Hyphomycetes, Persoon and Link. Hughes' proposal was rejected at both the 1959 and 1964 Congresses, so Fries's *Systema* is still the official starting-point for all the Fungi Imperfecti.

Provision is also made in the Rules for the "conservation" of generic names which have come into general use but which are not strictly legitimate. A few such names were, at the 1950 and 1954 Congresses, definitely recommended for conservation.

Authors' names. It is customary, when mentioning the name of a genus or species, to append the name (often abbreviated) of the author who published the first description and bestowed the name. This custom makes for precision and is of special value when the same name has been used inadvertently by different authors for different genera or species, as it prevents confusion until such time as the more latterly described fungus can be renamed. When a species is transferred from one genus to another, the name of the original describer is put in brackets and followed by the name of the worker who made the transfer. Two examples of this practice have already been given, *Mucor rouxii* (Calmette) Wehmer and *Scopulariopsis brevicaulis* (Saccardo) Bainier.

As stated above, nomenclature for most groups of fungi starts with Fries's *Systema*. However, when Fries accepted and used a name bestowed by some earlier author it is customary to cite the original author followed by "ex Fries". Two such examples have been given above, *Hypocrea rufa* Pers. ex Fries and *Trichoderma viride* Pers. ex Fries. Such citations are not obligatory and it is quite correct to write *Trichoderma viride* Fries.

Names of groups. A taxonomic group of any rank is referred to as a *taxon* (plural *taxa*). The naming of genera is limited only so far as that the names must be nouns of Latin form, that is, having endings as used for Latin nouns. Names of higher groups are usually derived from names of important genera within the groups, but have definite endings. Family names always end in -aceae (pronounced -ay'-se-ee). In modern usage the names of orders end in -ales (pronounced -ay'-lees), but in much of the older literature the ending -ineae is used. Names of classes end in -mycetes, usually pronounced as two syllables, though occasionally as three (-my-see' -tees). Names of classes, orders, and families are capitalized but are not printed in italics. In the larger classes of fungi, sub-classes, and sometimes sub-orders and sub-families, have been created to assist identifications and rational grouping, and these bear names ending respectively in -mycetidae, -ineae, and -oideae.

Examples: The class Phycomycetes is divided into two sub-classes, Oomycetidae and Zygomycetidae; one order of the latter is the Mucorales; the largest and most important family of the order is the Mucoraceae, and this includes the genus *Mucor*. The class Ascomycetes is divided into three sub-classes, of which one is the Pyrenomycetidae; one order of the latter is the Sphaeriales; one family of the order is the Sphaeriaceae, and a typical genus of this family is *Chaetomium*.

Many other matters are dealt with in the Rules, all aimed at removing ambiguity and ensuring that a name shall have a precise meaning which can be understood everywhere, but these cannot be summarized adequately. The Rules, as published in 1935, are reprinted, with a number of useful comments and illustrations, in a little book on Taxonomy and Nomenclature by Bisby (1945). The latest edition of the Rules, embodying the modifications introduced by the Edinburgh Congress of 1964, is published by Lanjouw *et al.* (1966). Those interested in questions of names should also consult the report of a discussion on nomenclature held by the British Mycological Society (1942).

MEANINGS OF NAMES

Names of genera and species should be, and often are, descriptive. A name may refer to method of spore production, characteristics of spores, colour or texture of colonies, odour, or any other distinctive feature. However, many names of species and of a fair number of genera have been bestowed in honour of well-known mycologists. Other names of species are derived from place-names, or refer to habitat, and specific epithets of many parasitic fungi are derived from the generic names of host plants.

There are two ways of using proper names as specific epithets. The more usual way is to use the genitive case of a latinized form of the name. In the case of the masculine form -i is added when the name of the person ends in a vowel or in -er; in all other cases -ii is added. The corresponding feminine ending is -ae. The other way is to use the adjectival ending -anus (m.), -ana (f.), -anum (n.). There are many cases in which epithets have been made in both ways from the same name. Thus, there are two species of *Penicillium* named in honour of Biourge, who monographed the genus, *P. biourgei* Arnaud and *P. biourgeianum* Zaleski. However, the present recommendation is that the use of both the genitive and adjectival forms of the same name is best avoided. It has been common practice in the past to write epithets derived from proper names with capital letters, but during recent years it has become more and more customary to write all specific epithets without capitals, and at the 1954 Congress this was a definite recommendation.

In the following chapters (and in Chapter II) derivations are given of all the special mycological terms which are used and of the names of all genera and species which are described. However, the meanings of one prefix and two suffixes which occur very frequently are usually omitted. The prefix *sporo-* or the suffix *-sporum* (or *-sporium*) obviously signifies that the descriptive Latin or Greek term refers to the spores, and the ending *-myces* simply means "fungus", from the Greek *mykes*. Throughout, the many Greek terms are written in Latin characters, for the benefit of readers who are unfamiliar with the Greek alphabet. All Latin adjectives are given as the masculine nominative singular, whatever the gender of the name of the fungus.

PRONUNCIATION

The pronunciation of names of fungi is, in this country, in accordance with the traditional method of pronouncing Latin, with the vowels mostly as in common English words. A terminal -i is long as in 'like'; a terminal -ii has the first -i short, as in 'tin', and the second long; 'ae' is pronounced 'ee'; 'g' and 'c' are soft before 'e', 'i', and 'y' and hard before other vowels, and so on. Endings such as *-genum* and *-ferum*, derived from Latin verbs, retain their original accents on the first of the two syllables. To give a few examples: *chrysogenum* is pronounced kry'-so-jee'-num, the 'y' and 'e' being both long and accented; *Sphaerotheca* is sfee'-ro-thee'-ka, whilst *Cephalothecium* is se'-fal-o-thee'-sium; fungus has a hard 'g', whilst the 'g' in fungi is soft and pronounced as 'j'; the 'c' is hard in ascus but soft in the plural asci. It is only fair to say that there is an increasing tendency, particularly amongst the younger school of mycologists, to modernize pronunciation by accenting the third syllable from the end. Thus *chrysogenum* becomes kry-soj'-enum, and *Sphaerotheca* sfeer-oth'-eka. This method of pronunciation takes all the meaning from many of the names, and there is much to be said for retaining a method which is in accordance with etymology, apart from the fact that knowing the meaning of a name is a great help towards remembering it.

Some knowledge of Latin is essential to any serious student of mycology, for not only is it useful to be able to appreciate the meanings of names and technical terms but most of the mycological classics are entirely in Latin, and it is now a rule, made by the International Botanical Congress in 1930, that every description of a new genus or species, in whatever language it is published, must be accompanied by an adequate diagnosis in Latin

REFERENCES

Bainier, G. (1907). *Bull. Soc. mycol. Fr.*, **23**, 99.
Bisby, G. R. (1945). *An introduction to the taxonomy and nomenclature of fungi.* Kew: Commonwealth Mycological Institute.

British Mycological Society (1942). Report of a discussion on mycological nomenclature. *Trans. Brit. mycol. Soc.*, **25**, 432–5.

Calmette, A. (1892). *Ann. Inst. Posteur*, **6**, 605.

Fries, E. M. (1821–32). *Systema mycologicum*.

Hughes, S. J. (1958). Revisiones Hypomycetum aliquot cum appendice de nominibus rejiciendis. *Canad. J. Bot.*, **36**, 727–836.

Lanjouw, J., Baehni, Ch., Robyns, W., Ross, R., Rousseau, J., Schopf, J. M., Schulze, G. M., Smith, A. C., De Vilmorin, R., and Stafleu, F. A. (1961). International Code of Botanical Nomenclature adopted by the Ninth International Botanical Congress, Montreal, 1959. *Reg. veg.*, Vol. 23. Utrecht: International Bureau for Plant Taxonomy and Nomenclature.

Linnaeus, C. (1753). *Species plantarum*.

Persoon, C. H. (1801). *Synopsis methodica fungorum*.

Saccardo, P. A. (1882–1925). *Sylloge fungorum omnium hucusque cognitorum*. Pavia, Italy.

Wahmer, C. (1900). *Zentralbt. f. Bakt.* II, **6**, 364.

Chapter IV

Zygomycetes

On se rend plus utile à la science en démolissant un genre
ou une espèce qu'en en créant cent nouveaux.

M. Langeron, 1945

The essential features of the Zygomycetes have already been outlined in
Chapter II and need not be further elaborated here. Most of the common
species are readily recognized as belonging to the group by their rapid rate
of growth and their characteristic appearance, colonies usually being loosely
floccose and of a grey or brownish-grey colour. Three genera, *Mucor*,
Rhizopus, and *Absidia*, include the great majority of the species which are
normally encountered in the laboratory, but members of several other
genera are found sufficiently frequently to justify descriptions being given
here. Zygospores have been found in comparatively few species, and hence
are not made the main basis of classification. However, although so many
species lack the perfect fruiting stage, there has never been any question of
placing these with the Fungi Imperfecti, owing to the very characteristic
type of mycelium and imperfect fruiting stage which distinguish the sub-
class.

The Zygomycetes are divided into two orders as follows:

Asexual spores occurring in sporangia .   .   .   MUCORALES
Asexual spores as conidia, forcibly shot away at
    maturity   .   .   .   .   ENTOMOPHTHORALES

The Entomophthorales are mentioned only for the sake of completeness,
since most of the species are obligate parasites on insects and have never
been grown in artificial culture. The largest and best-known genus is
Entomophthora, which now includes the species previously classed in *Em-
pusa* (an illegitimate name). *E. muscae* Cohn is of common occurrence in
the late summer, when it forms characteristic white haloes round the bodies
of dead flies.

Mucorales

The great majority of the species are saprophytic, occurring on a wide
variety of organic substrata, whilst the remainder are parasitic on other
members of the order. Typical colonies consist of coarse hyphae growing
loosely, white in the early stages of growth and becoming grey or brownish
with the production of fruiting structures. The usual mode of asexual re-

production is by spores produced in large numbers in globose sporangia, borne on sporangiophores which may be branched in various ways. In most genera the tip of the sporangiophore is swollen, the swollen end (the *columella* projecting into the sporangium. The columella may be of various shapes—globose, ovoid, hemispherical, etc. (Figs. 11, 12, 16)—and its characteristics are of importance in determination of species. The sporangial wall may be thin, when the spores are liberated by its rupture or dissolution, or may be cutinized and shot off, or broken off, in one piece. In a number of families various modifications occur of the typical globose, many-spored sporangium, the most important of these being described below.

A number of different classifications of the Mucorales have been proposed, but the total number of genera is not large and, in the great majority of cases, the placing of a particular isolate in its correct genus is a fairly simple matter whichever key is used for the purpose. The following simple key ignores division of the order into families and includes all the genera which are commonly found on industrial products. Many of the remainder live on dung, and, although of widespread occurrence, are unable to grow on ordinary culture media.

KEY[1] TO THE COMMON GENERA OF MUCORALES

1. Sporangia tubular, radiating from a vesicular
   swelling . . . . . . *Syncephalastrum*
   Sporangia globose or nearly so 2
2. Many-spored sporangia and few-spored sporangioles both present . . . . *Thamnidium*
   Sporangia alone produced 3
3. Sporangiophores stiff, dark-coloured, metallic in appearance . . . . . *Phycomyces*
   Sporangiophores otherwise 4
4. Rhizoids and stolons present 5
   Rhizoids and stolons absent 6
5. Sporangia large, globose; sporangiophores
   arising from points of attachment of rhizoids . . . . *Rhizopus*
   Sporangia small, pear-shaped; sporangiophores mainly as branches from the stolons *Absidia*
6. Homothallic; zygospores with very unequal
   suspensors . . . . . . *Zygorhynchus*
   Homo- or heterothallic; zygospores, when
   present, with approximately equal suspensors . . . . *Mucor*

[1] In the dichotomous system, used for this key and others later in the book (also used by many other authors), all the essential data required for identifications are arranged as *pairs* of contrasted characters, the pairs being numbered consecutively on the left. Each member of a pair leads, on the right of the page, either to the name of a genus or to another higher number, i.e. to a further pair of contrasted features.

Mucor Micheli ex Fries

(Lat. *mucor*, a fungus)

This is the largest genus of the order and includes a number of species which are of widespread occurrence and considerable importance. The criteria on which identifications are based are: the mode of branching, if any, and the dimensions of the sporangiophores; the character of the sporangial wall; the size and shape of the columellae; the dimensions and shape of the spores; the characteristics of zygospores and chlamydospores, if present; and general colony characteristics, such as colour and height of aerial growth. About 150 species are listed in Saccardo's *Sylloge*, but many of these are inadequately described and a considerable number of the names are synonyms of common species. Lendner (1908) describes 51 species, and his monograph was for many years the standard work on the Mucorales. His keys have been reproduced in a number of later publications but they are now out of date. Since the appearance of Lendner's book, Hagem (1910) has described 4 new species, Povah (1917) has added 6, and odd species have been described by a number of other workers. Until 1935, however, identification of species, even with all the relevant literature available, was difficult and often unsatisfactory. The reason is that in *Mucor*, as in many other genera of moulds, certain ubiquitous species exhibit minor differences between different strains. Taxonomists who have handled comparatively small numbers of strains have not sufficiently appreciated this fact, and hence have not always distinguished between true and spurious specific characters. Most original diagnoses have been drawn up in terms insufficiently broad to cover these strain differences, with the result that workers trying to identify species have but rarely found published descriptions which tallied exactly with their own data.

In a monograph on the Mucorales, Zycha (1935) considerably simplified the taxonomy of *Mucor* by taking into account this natural variation. The total number of species was reduced to 42, and of these, several, of which type cultures are not available, are accepted only provisionally. Zycha's monograph is still the only reasonably satisfactory treatment of the whole order of Mucorales. For a long time the book was out of print, and virtually unobtainable, but it has now been reprinted.

A more recent treatment of the Mucorales is by Naumov (1939). He includes a number of genera not recognized by Zycha and rejects others which Zycha considers as belonging to the order. The genus *Mucor* is divided into 13 sections, or sub-genera, and the total number of admitted species is 88. However, the majority of the 46 names not included in Zycha's list are regarded by the latter as synonyms. The whole of the diagnoses in Naumov's book are incorporated in the keys, and this makes the work much more difficult to use than Zycha's, in which short keys are supplemented by separate and complete diagnoses of all species.

At the present time C. W. Hesseltine, at Peoria, Illinois, is making an intensive study of the Mucorales. A number of papers have already appeared (Hesseltine, 1952, 1953, 1954, 1955, 1957; Hesseltine and Fennell, 1955; Hesseltine and Anderson, 1956; Hesseltine, Benjamin and Mehrotra, 1959; Hesseltine and Ellis, 1964; Ellis and Hesseltine, 1965) and it is probable that a monograph will eventually materialize.

M. mucedo Linn. ex Fr. (Lat. *mucedo*, mucus). This is the type species of the genus, having first been described in 1762. It occurs chiefly on dung, and can usually be found, when wanted, if some horse-dung is incubated for a few days in a very moist atmosphere. On its natural substrate it forms erect sporangiophores which may reach a height of 15 cm., with greyish sporangia encrusted with crystals said to be calcium oxalate. In culture the sporangiophores are usually shorter, and bear at the base a number of short branches terminating in small sporangia. The terminal sporangia are 100–200μ diam., with walls which disintegrate completely in fluid mounts; columellae pear-shaped to cylindrical, with orange-coloured contents; spores elliptical, 6–12μ long.

M. racemosus Fresenius (Lat. *racemosus*, branched). This is probably the most widely distributed of all the species of *Mucor*, and has frequently been described under other names. It is found on almost every kind of damp material. Colonies are grey or brownish-grey, of loose texture and normally

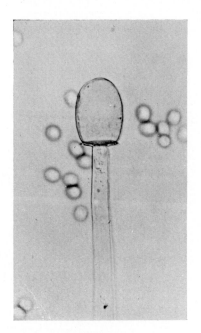 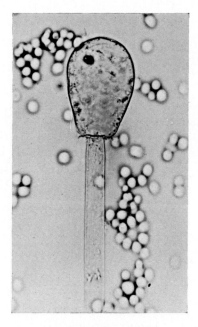

FIG. 11.—Oval columella with collarette—*Mucor racemosus*. × 500.

FIG. 12.—Pyriform columella with small collarette. × 500.

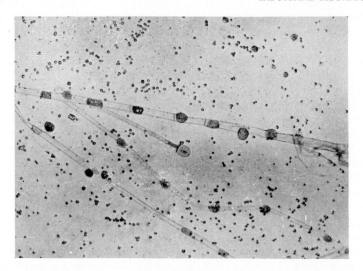

FIG. 13.—Chlamydospores in sporangiophore of *M. racemosus*. × 100.

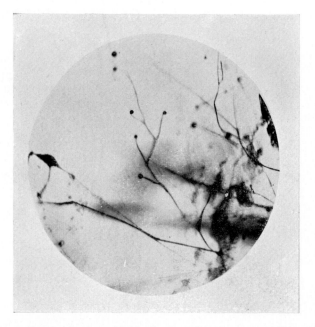

FIG. 14.—Sporangiophore of *M. plumbeus* showing cymose branching,
as seen in living culture (tube). × 50.

less than 1 cm. high; sporangiophores are simple at first, later becoming branched, with the branches irregularly arranged and very unequal in length; sporangia globose, very unequal in size but mostly small, 20–70μ diam., with walls which break in pieces when handled; columellae spheric or ovate, with collarette (a portion of the broken sporangial wall left *in situ*) (Fig. 11); spores mostly elliptical, 6–10 × 5–8μ. The most characteristic feature of the species is the abundant production of chlamydospores, which are formed in aerial hyphae, along the sporangiophores, and even in the columellae. They are of diverse shapes, colourless or yellow, smooth, and about 20μ diam. (Fig. 13). This species can grow submerged in liquid media and, like yeast, produces alcohol. It is heterothallic but, although numerous workers have tried to obtain zygospores by growing large numbers of strains in all possible pairs, success has very seldom been achieved.

M. plumbeus Bonorden (Lat. *plumbeus*, leaden). Two synonyms are often met with in the literature, *M. spinosus* van Tieghem (Lat. spiny) and *M. spinescens* Lendner (Lat. becoming spiny). Colonies at first white, then dull grey, and finally brownish-grey, most frequently only a few millimetres high but varying considerably in different strains; sporangiophores branched in sympodial cymes (see Fig. 14), about 1 mm. long and 10μ diam.; sporangia usually flattened, very even in size, about 65μ diam., and encrusted with fine needle-shaped crystals, thus appearing spiny (Fig. 15); columellae very characteristic, ovate or pear-shaped, with curious spiny projections at the top (Fig. 16); spores round, 7–8μ diam., occasionally somewhat smaller. Of very common occurrence.

M. hiemalis Wehmer (Lat. *hiemalis*, pertaining to winter). A common soil organism and hence found on numerous soil-contaminated products. Colonies yellowish or clear grey, 1–2 cm. high; sporangiophores simple or sympodially branched; sporangia 50–80μ diam., olive to greyish-brown when ripe, with diffluent walls; columellae spheric or ovate, with collarette, up to 50μ long; spores irregular in shape, but mostly roughly ovate, very characteristic, in spite of the variation in size, being a fairly even mixture of large and small forms; chlamydospores formed in the mycelium, but not so abundantly as in *M. racemosus*. The mould is heterothallic and both the + and – strains are fairly common. It is always interesting, when three or four strains have been isolated from different sources, to try growing them together in pairs.

M. ramannianus Möller (E. Ramann, German botanist). This is another fairly common soil organism, which macroscopically looks very little like a typical *Mucor*. Colonies at first pinkish to browning red, turning grey, only about 1 mm. high and almost velvety in appearance; sporangiophores mostly unbranched, 2–6μ diam.; sporangia reddish, small, 20–40μ diam., with diffluent walls; columellae globose, 5–10μ diam.; spores globose to short oval, 2–3μ long; chlamydospores numerous and of various shapes; giant cells formed in the submerged mycelium. This species is really a transition form between *Mucor* and *Mortierella*, a genus of the

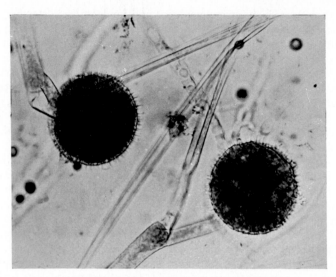

FIG. 15.—*M. plumbeus*—young sporangia showing spinescent walls.
× 500.

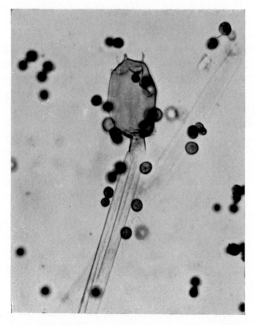

FIG. 16.—*M. plumbeus*—columella with terminal spines.
× 500.

Mucorales in which sporangia are small and completely lacking columellae. It has actually been transferred to *Mortierella* by Linnemann (1941).

M. piriformis Fischer (Lat. *piriformis*, pear-shaped). A fairly common species particularly associated with rotting fruits. Colonies dense cottony, 2–3 cm. high; sporangiophores mostly unbranched, up to 50μ diam.; sporangia globose and very large, 250–350μ diam., with diffluent walls; columellae definitely pear-shaped (giving the name to the species), 200–300μ long; spores elliptical, 5–13×4–8μ.

M. rouxii (Calmette) Wehmer (Emile Roux, French bacteriologist). Originally isolated from Chinese rice by Calmette and named by him *Amylomyces rouxii*. It was the first of a number of species of the Mucoraceae to be used for the manufacture of alcohol by the "Amylo process" (Calmette, 1892; Collette and Boidin, 1897, 1898). It grows, but does not usually form sporangia, on most ordinary culture media. On rice it grows normally, colonies being reddish in colour, with short, sparingly branched sporangiophores, small sporangia 20–30μ diam., elliptical spores about 5μ long, and abundant chlamydospores, yellow to brown and up to 100μ diam.

Zygorhynchus Vuillemin

(Gr. *zygon*, a yoke; *rhynchos*, a snout)

The genus is closely related to *Mucor*. It forms many-spored sporangia of similar structure to those of *Mucor*, but differs in that all the species are homothallic, and the suspensors of the zygospores are very markedly unequal in size. The genus has been recognized by all students of the Mucorales excepting only Lendner, who included the species in *Mucor*. Many species have been described, but Zycha recognized 6 only, the remaining names being reduced to synonymy. In a more recent detailed study of the genus Hesseltine, Benjamin and Mehrotra (1959) conclude that there are 6 species and one variety. All the species are typically soil fungi, being found only rarely on other substrata. Two species are isolated fairly frequently in this country, *Z. moelleri* being the more common of the two.

Z. moelleri Vuillemin (A. Möller, German mycologist). Grows well on all media, forming a loose felt only a few millimetres high; abundant zygospores formed within two to three days, with sporangia produced more sparingly and somewhat later; sporangiophores simple or irregularly branched; sporangia grey, usually slightly broader than long, mostly $48 \times 50\mu$; columellae definitely ovate but broader than high, averaging $25 \times 30\mu$; spores ovate, 4–$5 \times 3\mu$; zygospores formed on bifurcated hyphae, round, 35–50μ diam., dark brown to black and covered with short, thornlike projections.

Z. heterogamus Vuillemin (Gr. *hetero-*, uneven; *gamos*, marriage, union). Somewhat similar to the last species but with larger sporangia and zygospores. Sporangiophores branched, often more or less verticillate with two to four branches; sporangia globose, 50–60μ diam.; columellae globose; spores round, 2–3μ diam.; zygospores black, very rough, varying in size from 45μ to 150μ; chlamydospores formed in the mycelium.

Rhizopus Ehrenberg ex Corda

(Gr. *rhiza*, a root; *pous*, foot)

Species of this genus occur on all kinds of material and are common as aerial contaminants in the laboratory. On most culture media they grow with extreme rapidity, spreading widely by means of their stolons. They completely fill culture tubes and Petri dishes with dense cottony masses of mycelium, and can be a great nuisance from their habit of sporing along the line where the cover touches the edge of the dish, thus shedding spores outside. *Rhizopus* is readily distinguished from *Mucor* by the presence of stolons (runners), often several centimetres long, and of tufts of rhizoids (root-like hyphae) emerging from the points where the stolons touch the medium or the surface of the glass; also by a somewhat greyer colour and greater luxuriance of growth. In addition, the base of the columella is united with the sporangial wall, this type of structure being known as an apophysis (see Fig. 9 and also *Absidia*). Lendner (1908) describes 22 species but some of these are rare, and a number of others are, except for the specialist, best regarded as strains of *R. stolonifer*.

R. stolonifer (Ehrenberg ex Fr.) Linder (Lat. *stolo*, stolon; *fero*, to bear) (in previous editions of this book as *R. nigricans*.) A species of world-wide distribution and found on all kinds of mouldy material. It is frequently the first mould to appear on stale bread, and is often found on other foodstuffs. Colonies on most media spread very rapidly, completely filling tubes and dishes in a few days; stolons clearly differentiated, arising from and terminating in strong tufts of brown rhizoids (Figs. 18, 19); sporangiophores erect, arising in groups opposite the rhizoids, up to 2·5 mm. long and about 20μ diam.; sporangia globose, shining white at first, then turning black as the spores ripen, up to 200μ diam.; spores variously shaped—ovate, polygonal, or angular—striated, mostly 10–15μ in long axis.

Although growing so luxuriantly on most substrata, this species is unable to utilize nitrates and, therefore, will not grow on Czapek's solution (see Chapter XI). It grows on Czapek agar but only very sparsely, presumably utilizing some impurity in the agar as source of nitrogen.

R. oryzae Went and Prinsen-Geerlings (Lat. *oryzae*, of rice). Distinguished from the last species by its somewhat smaller spores (7–9μ) and by its ability to grow at 37–40° C. During the second world war it was used by British prisoners in Java to make soya beans digestible, so as to eke out the meagre diet allowed.

R. arrhizus Fischer (Gr. *a-*, lacking; *rhiza*, root). Grows much less rampantly than the two preceding species; rhizoids short, pale and ragged; sporangiophores frequently arising from hyphal swellings not provided

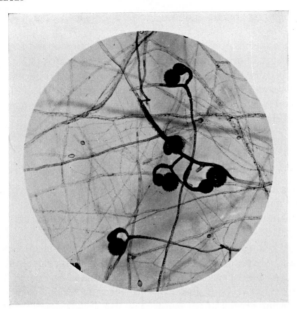

FIG. 17.—*Zygorhynchus moelleri*—zygospores (tube culture). × 100.

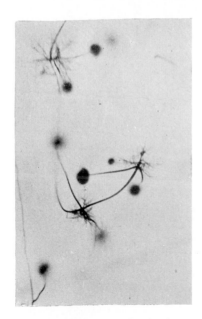

FIG. 18.—*Rhizopus stolonifer* showing rhizoids and stolons (Petri dish culture). × 25.

FIG. 19.—*R. stolonifer*—rhizoids and sporangium (tube culture). × 50.

with rhizoids; sporangia globose, 100–200μ diam.; spores 5–7μ long. Grows well at 37° C.

Two other species of *Rhizopus*, *R. japonicus* Vuillemin and *R. tonkinensis* Vuillemin, both of oriental origin, and both closely related to *R. oryzae*, are of interest because they successively replaced *Mucor rouxii* for the saccharification of starch in the Amylo process of manufacturing alcohol. They are of no importance in other connections, and are not met with in ordinary mycological work.

Absidia van Tieghem

(Gr. *apsis*, a loop; refers to the arched stolons)

This genus differs in several respects from *Rhizopus*. The rhizoids and stolons are not so clearly differentiated; the sporangiophores arise from the stolons and not from the points of attachment of the rhizoids; the sporangia are relatively small and pear-shaped; and, most characteristic feature of all, there is a well-marked "apophysis", i.e. a funnel-shaped base to the sporangium, where the walls of sporangium and columella are united (Fig. 20). The zygospores, when present, are surrounded by coarse hairy outgrowths from one or both suspensors, this feature serving to place at once the homothallic species of the genus (Fig. 21). Zycha describes 17 species,

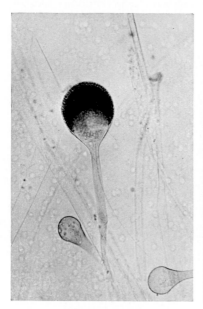

FIG. 20.—*Absidia ramosa*—sporangium, showing apophysis. × 250.

FIG. 21.—*Absidia spinosa*—formation of zygospores showing outgrowths from one suspensor. × 250.

whilst Naumov follows some earlier authors in splitting the genus into several, with a total of 29 species and 2 varieties. Hesseltine and Ellis (1964) and Ellis and Hesseltine (1965) have published an exhaustive study of the genus.

Some of the species are of common occurrence in the soil and sometimes appear in mixed cultures from mouldy material, or as aerial contaminants. A number are thermophilic, with optimum temperatures for growth close to 37° C, and some of these have been many times reported as pathogenic to various animals, including man. They can be a great nuisance in the bacteriological laboratory, for a single colony can almost cover a Petri dish in 24 hours at 37° C, and they thrive, better than most other moulds, on the usual bacteriological media.

A. corymbifera (Cohn) Sacc. and Trott. (Lat. *corymbus*, a cluster of flowers; *fero*, to bear); synonym *A. lichtheimii* (Lucet and Costantin) Lendner. Colonies at first white, then pale grey, up to 2cm. high; sporangiophores branched, often in whorls; rhizoids very sparingly produced; sporangia small and pear-shaped; columellae almost conical, with well-marked apophysis and often a short projection at the top; spores ovate, 2–3·5 × 3–4·5μ; thermophilic and pathogenic.

A. spinosa Lendner (Lat. *spinosus*, spiny). Colonies densely floccose, white then brownish grey; sporangiophores in clusters of 2–4, unbranched; columellae hemispherical, with one or more projections; a septum usually present in the sporangiophore, a little distance below the apophysis; spores cylindrical with slightly rounded ends, 4–5μ long; zygospores abundant, 50–80μ diam., with outgrowths from only one suspensor (Fig. 21).

A. ramosa (Lindt) Lendner (Lat. *ramosus*, branched). Colonies floccose, white then greyish; sporangiophores branched, often in whorls, without a septum below the sporangium; columellae pale bluish or violet, without projections; spores ovate or cylindrical, 4·8 × 2·8μ; zygospores not reported; temperature optimum 37° C; pathogenic to some animals.

Phycomyces Kunze ex Fries

(Gr. *phykos*, sea-weed; hence alga-like fungus)

Species of the genus are readily recognized by their very characteristic sporangiophores, very long, stiff, and with metallic sheen, looking like oxidized steel wire. Four species are known but only two are of any importance.

P. nitens (Agardh) Kunze ex Fr. (Lat. *nitens*, becoming shiny). Sporangiophores up to about 20 cm. high; sporangia black when ripe, about 500μ diam.; columellae cylindrical, constricted in the middle; spores elongated ovate, averaging 25 × 11μ. Found on dung, in soil, and occasionally in empty oil barrels.

P. blakesleeanus Burgeff (A. F. Blakeslee, American mycologist). This species was for many years distributed from various culture collec-

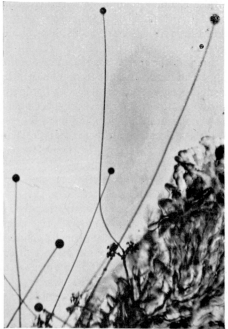

FIG. 22.—*Thamnidium elegans*
—young   sporangiophore
with clusters of sporangioles (Petri dish culture).
× 50.

FIG.   23.—*T.   elegans*—mature sporangiophores with terminal sporangia (from
culture on film of agar between two glass
plates).   × 25.

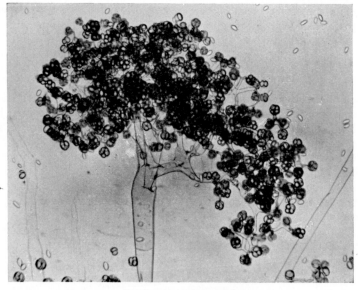

FIG. 24.—*T. elegans*—sporangioles.   × 200.

tions as *P. nitens*. Burgeff in 1925 showed that it differs from the latter, and renamed it. It is distinguished from *P. nitens* by its larger sporangia, up to 1000μ diam., and by its smaller spores, about 12 × 8μ. It grows well on wort agar and some media made from plant extracts, but does not thrive on synthetic media, since it needs to be supplied with thiamin. In test-tube cultures the sporangiophores curl round in order to accommodate their length to the confines of the tube, but in the open they are nearly straight and up to about 30 cm. long. Under suitable conditions they become much longer, since the fungus is strongly phototropic. An interesting experiment is to run a little rich medium, such as wort agar, to the bottom of a very tall vessel, sterilize, and inoculate with a long wire. Then wrap the vessel round with black paper to within about an inch of the plug, and stand in the light. By this means the sporangiophores can be induced to reach at least 2 feet in length. Another way of demonstrating phototropism in this species is illustrated in Fig. 166.

Thamnidium Link ex Wallroth
(Gr. *thamnos*, a bush; *eidos*, like)

This differs from the preceding genera in that the main sporangiophores bear lateral clusters of sporangioles as well as large terminal sporangia. The sporangioles resemble miniature sporangia, containing from two to a dozen or so spores, and are formed on richly branched outgrowths from near the base of the sporangiophore (Figs. 22, 23, 24).

T. elegans Link ex Fr. (Lat. *elegans*, elegant) is by no means a common mould. It has been reported most frequently as infecting meat in cold storage. Grows well on most culture media, greyish, granular, at first only a few millimetres high and with growth consisting chiefly of sporangioles, later producing sporangiophores up to 1 cm. or more long, bearing large terminal sporangia; sporangioles in dense clusters, borne on the ends of much branched hyphae; sporangia brown, 150–250μ diam.; spores in sporangioles indistinguishable from those in sporangia, ovate or bean-shaped, 8–12μ in long axis. Hesseltine and Anderson (1956) have reported the formation of zygospores when certain strains were mated (the mould being heterothallic). The zygotes are produced only on particular media, and only at low temperatures, 6–7° C.

Syncephalastrum Schröter
(Gr. *syn*, completely; *kephale*, a head; *aster*, a star)

The spores are formed in long tubular sporangia (termed by Zycha "part-sporangia") radiating from a swelling on the end of the sporangiophore. When mounted and viewed at a low magnification the heads bear a striking resemblance to heads of *Aspergillus* (Fig. 25). At a high magnification the chains of spores are seen to be enclosed in tubular membranes, and, if the development of the sporing structures is followed, it will be observed that

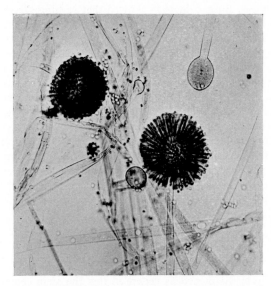

FIG. 25.—*Syncephalastrum racemosum*—spore heads.
× 250. Note the resemblance to *Aspergillus*.

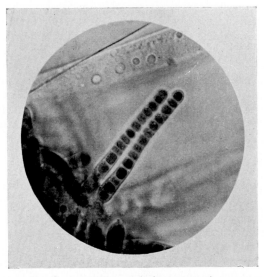

FIG. 26.—*S. racemosum*—tubular sporangia. × 1000.

the spores are formed exactly as in the normal type of sporangium. Fig. 26 shows clearly that the spores are formed within, and distinct from, the sporangial wall.

S. racemosum (Cohn) Schröter (Lat. *racemosus*, branched) is fairly widely distributed, but comes into this country chiefly on raw materials of tropical or sub-tropical origin. It grows luxuriantly on laboratory culture media, the colonies resembling in texture those of *Rhizopus stolonifer*, with black heads produced on short branches from aerial hyphae. The part-sporangia contain 5–10 spores, of somewhat irregular shape, $2\cdot5$–5μ diam.

REFERENCES

Calmette, A. (1892). *Ann. Inst. Pasteur*, **6**, 604.

Collette, A., and Boidin, A. (1897). German Patent 99253.

Collette, A., and Boidin, A. (1898). German Patent 100129.

Collette, A., and Boidin, A. (1898). English Patent 13053.

Ellis, J. J., and Hesseltine, C. W. (1965). The genus *Absidia*: globose-spored species. *Mycologia*, **57**, 149–97.

Hagem, O. (1910*a*). *Ibid.* (1910). 1–152.

Hesseltine, C. W. (1952). A survey of the Mucorales. *Trans. N.Y. Acad. Sci.*, *Ser. II*, **14**, 210–14.

Hesseltine, C. W. (1953). A revision of the Choanephoraceae. *Amer. Midl. Nat.*, **50**, 248–56.

Hesseltine, C. W. (1954). The section Genevensis of the genus *Mucor*. *Mycologia*, **46**, 358–66.

Hesseltine, C. W. (1955). Genera of Mucorales, with notes on their synonymy. *Mycologia*, **47**, 344–63.

Hesseltine, C. W. (1957). The genus *Syzygites* (Mucoraceae). *Lloydia*, **20**, 228–37.

Hesseltine, C. W., and Anderson, A. (1956). The genus *Thamnidium* and a study of the formation of its zygospores. *Amer. J. Bot.*, **43**, 696–703.

Hesseltine, C. W., Benjamin, C. R., and Mehrotra, B. S. (1959). The genus *Zygorhynchus*. *Mycologia*, **51**, 173–94.

Hesseltine, C. W., and Ellis, J. J. (1964). The genus *Absidia: Gongronella* and cylindrical-spores species of *Absidia*. *Mycologia*, **56**, 568–601.

Hesseltine, C. W., and Fennell, D. I. (1955). The genus *Circinella*. *Mycologia*, **47**, 193–212.

Lendner, A. (1908). Les Mucorinées de la Suisse. *Materiaux pour la flore cryptogamique Suisse*, **3**, 1–180.

Linnemann, G. (1941). Die Mucorineen-Gattung *Mortierella* Coemans. *Pflanzenforschung*, **23**, p. 29.

Naumov, N. A. (1939). *Clés des Mucorinées (Mucorales)*. Trans. from 2nd Russian Ed., with additional notes by the Author, by S. Buchet and I. Mouraviev. Paris: Paul Lechevalier.

Povah, A. H. W. (1917). A critical study of certain species of *Mucor*. *Bull. Torrey bot, Cl.*, **44**, 241–59, 287–313.

Zycha, H. (1935). *Kryptogamenflora der Mark Brandenburg*. Band VIa, Pilze II, *Mucorineae*. Leipzig: Gebrüder Borntraeger. Reprint 1963.

Chapter V

The Ascomycetes

The Ascomycetes constitute the largest of the classes of perfect fungi, the number of known species being approximately 15,000. As may be expected in such a large group there is considerable diversity of form and structure. At one end of the scale are the unicellular organisms commonly known as yeasts, and, at the other, species with extensive mycelium and large and elaborate fruiting structures, such as the truffles and morels.

The ascus, a structure which is peculiar to and which gives the name to the class, is a thin-walled receptacle enclosing the spores and usually rupturing at maturity. The ascus is distinguished from the sporangium of the Phycomycetes in the origin and method of formation of the spores, details of which are to be found in any text-book of General Mycology. A more obvious difference is that the sporangium contains an indefinite number of spores, whilst in the great majority of Ascomycetes the ascus invariably contains 8 spores. In a few cases, which are of no importance here, the regular number is a multiple of 8, or, more strictly, a higher multiple of 2. In a few primitive members of the class, notably in *Endomyces* and *Endomycopsis*, the number is 4, whilst in the yeasts the number ranges from 1 to 8.

In the simpler Ascomycetes the asci are globose or ovate, with the spores packed tightly together. They may be formed singly, in irregular loose clusters, or arranged irregularly in a more or less definite fruit-body (the **ascocarp**). The ascocarp, when present, is usually more or less globose, with a fairly firm wall, or occasionally is nothing more than a loose web of hyphae around the asci. In the higher members of the class the ascocarp is of more definite form, with club-shaped or cylindrical asci arranged in parallel series, often with elongated sterile cells (paraphyses) separating them. If the ascocarp is globose or flask-shaped and closed at maturity, except for a narrow passage through the neck (the ostiole) it is called a perithecium; if becoming wide open at maturity it is known as an apothecium. The primary classification of the Ascomycetes is based on the form of the ascocarp, as follows:

Ascocarp, if present, with no definite ostiole, usually
 with asci irregularly arranged . . . . PLECTOMYCETES
Ascocarp wide open at maturity; asci in parallel
 series . . . . DISCOMYCETES
Ascocarp flask-shaped when ripe; asci in parallel
 series . . . . PYRENOMYCETES

The Discomycetes (Gr. *diskos*, a dish) include about 4000 species, very many of them parasitic on plants, but some saprophytic and often found growing on soil or dung. A number of the larger forms are fleshy and a few, such as the morels, edible.

The Pyrenomycetes (Gr. *pyren*, a fruit stone) form the largest of the three sub-classes, with about 10,000 species. Many common and serious plant diseases are caused by members of this group, but there is also a very large number of saprophytic species, and a few which may be classed as moulds. The latter will be described below.

The Plectomycetes (Gr. *plektos*, twisted, plaited) are divided into three orders, Plectascales, Erysiphales, and Exoascales, of which the two latter include only obligate parasites and have asci arranged in parallel series like the Pyrenomycetes. The Plectascales include the majority of the Ascomycetes which are of industrial importance.

Byssochlamys Westling

(Gr. *byssos*, cotton; *chlamys*, a cloak)

This is a genus which is of interest to the systematic mycologist, as it produces clusters of 8-spored asci without any surrounding wall (peridium),

FIG. 27.—*Byssochlamys fulva*—conidiophores
(from slide culture). × 250.

and thus forms a link between the Endomycetaceae, with solitary 4-spored asci, and the Gymnascaceae and Aspergillaceae, with more or less definite perithecia containing 8-spored asci.

B. fulva Olliver and Smith (Lat. *fulvus*, tawny) is of considerable importance, since, for some years before it was isolated and described, it was the cause of serious spoilage of canned and bottled fruits, and it is still a

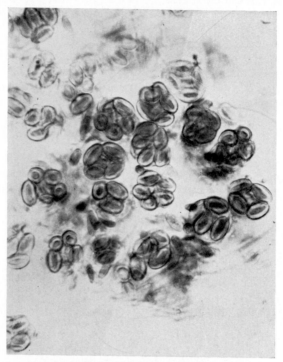

FIG. 28.—*B. fulva*—cluster of asci. × 1000.

potential source of trouble. The mature ascospores can withstand a temperature which is lethal to the spores of most fungi, and sometimes survive the commercial sterilizing process. The fungus is able to grow in an atmosphere containing very little oxygen, or even completely submerged in fluid, where the only oxygen available is the small amount dissolved. Spoilt fruit frequently shows no visible colonies of the mould and the cans are not blown as when the spoilage is due to anaerobic bacteria, the only evidence of the presence of the fungus being a general softening, or sometimes complete disintegration, of the fruit, such as would occur if cooking had been unduly prolonged.

The mould grows well on all ordinary media, with colonies of loose cottony texture, becoming fulvous (tan to sandy brown) as spores develop.

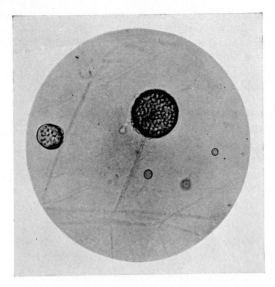

FIG. 29.—*Monascus purpureus*—stalked perithecium.
× 250.

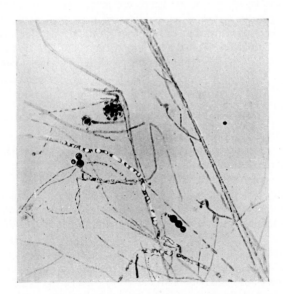

FIG. 30.—*M. purpureus*—young perithecium and
conidia. × 250.

The conidial fructification is of the *Paecilomyces* type (see Chapter X), with long chains of ovate spores, 4–9 × 2·3–2·5μ (Fig. 27). The asci develop rapidly in compact clusters, without any trace of peridium (Fig. 28), each ascus containing 8 ovate spores, 6–6·5 × 4·3–4·5μ.

This species is common in some soils, particularly in strawberry beds, with the result that spores are carried into the canning factories along with the early soft fruits.

B. nivea Westling (Lat. *niveus*, snow white) is the type species of the genus. Colonies are persistently white or very slightly brownish in age and the ascospores are somewhat smaller than those of *B. fulva*. It differs from the latter also in producing conidia much less abundantly and in forming very numerous macrospores (small terminal chlamydospores). This species is not common, having been found in English soil for the first time only comparatively recently.

Eurotium Link ex Fries

(Gr. *eurotiao*, to decay or become mouldy)

The genus is characterized by the production of globose perithecia, mostly bright yellow, without any neck or ostiole, breaking irregularly at maturity to liberate the spores. Asci are globose or ovate, very thin-walled, 8-spored, arranged irregularly in the perithecium, and the ascospores are more or less lenticular, splitting along a median longitudinal furrow at germination. The best-known species of the genus are more commonly known as the *Aspergillus glaucus* series. The conidial and ascosporic stages were long thought to be separate and distinct species, until De Bary in 1854 showed their real relationship. Those with and without asci are frequently considered as species of *Aspergillus*, the name of the imperfect state.

Monascus van Tieghem

(Gr. *monos*, single; hence forming a single ascus)

When the genus was first described it was thought that the perithecium consists of a single, many-spored ascus. It is now known that normal 8-spored asci are produced, but that these break up as the spores ripen, leaving the somewhat thin-walled perithecium full of loose spores.

M. purpureus Went (Lat. *purpureus*, purple) is not uncommon, particularly in dairy products. The author has isolated it a number of times from commercial crude lactose. It is also the organism which gives the characteristic colour to Chinese red rice. Colonies form a thin, spreading growth, with mycelium turning reddish or purplish, becoming greyish as conidia and perithecia develop, and with reverse dull purplish red. Perithecia are produced singly on stalks (Fig. 29), whilst the ovate conidia are formed in short chains (Fig. 30). Ascospores are ovoid, smooth, colourless, 5·5–6 × 3·5–4μ. Conidia are brown, ovate to barrel-shaped, 9–10·5 × 7–9μ.

Chaetomium Kunze ex Fries

(Gr. *chaitome*, a plume of hair)

This is the most common of the genera of Pyrenomycetes encountered in industrial work. The genus includes a fairly large number of species, all characterized by forming black perithecia with short necks and beset with long, dark-coloured, stiff hairs, which are variously straight, branched, or curled (Fig. 31). They are found chiefly on cellulosic materials and thrive particularly on paper. When kept in culture, a strip of filter paper, only partially immersed in the culture medium, usually ensures satisfactory production of perithecia.

The best monograph of the genus is embodied in two papers by the Canadian workers Skolko and Groves (1948, 1953). In the second of these there is a complete key to all the species of *Chaetomium* known at the time. References to their earlier paper (1948) are given for all species with dichotomously branched hairs, but all other species are described in full and illustrated. The key is lengthy and would be out of place here, and it would be invidious to pick out any species for special mention, since the worker on deterioration of cellulosic materials is likely to find any of 20 or more species.

Since 1953 a few new species have been described. Those who are interested in the genus can find the descriptions of these most easily from the *Index of Fungi* (see Chapter XVII).

A more recent monograph by Ames (1963) has a key to species which is by no means easy to use, and the general treatment of the genus is not so satisfactory as that by Skolko and Groves.

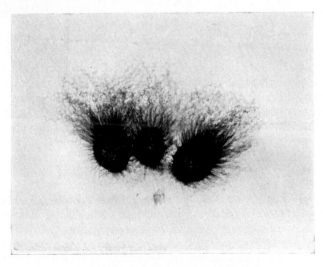

FIG. 31.—*Chaetomium globosum*—perithecia. × 50.

Neurospora Shear and Dodge

(Gr. *neuron*, nerve, vein; refers to the striate spores)

Species of this genus are the perfect states of the red bread-mould, *Monilia sitophila* (Montagne) Saccardo, and related species. All species except one are heterothallic, perithecia being obtained only when the mycelia of + and − strains come into contact (Shear and Dodge, 1927). A crushed

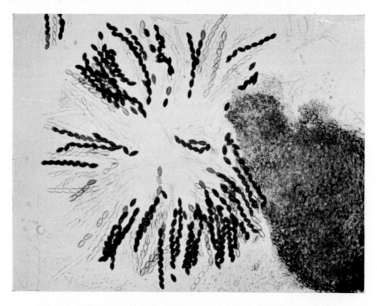

FIG. 32.—*Neurospora sitophila*—crushed perithecium. × 100.

perithecium is shown in Fig. 32. The ascospores are very resistant to heat and are frequently able to survive, in the interior of a loaf, the baking process. In addition, it is difficult to secure germination of the spores unless they are first heated to a fairly high temperature. This probably explains some of the outbreaks of "red mould disease" in sliced and wrapped bread.

Sordaria Cesati and de Notaris

(Lat. *sordes*, dirt, filth)

Species of this genus are common on dung; hence the name. As described in older mycological works the genus included a miscellany of Ascomycetes with dark, 1-celled spores, provided with appendages or surrounded by mucus. Moreau (1953) has adopted a more limited conception of the genus, restricting it to species having spores without appendages, but with a colourless outer membrane which swells in water. The species with appen-

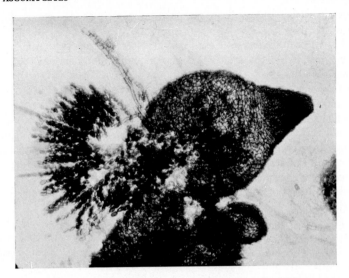

FIG. 33.—*Sordaria fimicola*—crushed perithecium. × 100.

FIG. 34.—*Pleurage unicaudata*—spores with appendages. × 500.

dages at one or both ends are regarded as belonging to the genus *Pleurage* Fries (see Fig. 34).

S. fimicola (Roberge) Cesati and de Notaris (Lat. *fimus*, dung, excrement; *colo*, to inhabit) is a fairly common mould, found on many substrates other than dung.

It spreads extremely rapidly on wort agar or other vegetable media, producing abundant perithecia, but grows very poorly or not at all on synthetic media, unless to these are added thiamin and biotin. The perithecia are brown and opaque, almost black in mass, and have short necks through which the spores are extruded at maturity. The spores are dark brown, ovate to biconvex, 17–24 × 11–13μ produced in single rows in long cylindrical asci. The spores have no tails and the amount of surrounding mucus is very small, detectable only on unripe spores. Fig. 33 shows a crushed perithecium.

The name *S. destruens* (Shear) Hawker, used previously, is no longer tenable. The original specimen of *Anthostomella destruens* Shear (= *Melanospora destruens* (Shear) Dodge; = *Sordaria destruens* (Shear) Hawker) has been examined by Cain, and shown to be a mixture of *S. fimicola* and another Ascomycete.

Fig. 34 shows spores of a species of *Pleurage*, some with a single appendage. The appendage tends to shrivel away as the spores age; hence the older spores in the photograph show no appendages.

Pleospora Rabenhorst

(Gr. *pleos*, full; refers to the many-celled spores)

Many species have been described but only one is of importance in the present connection.

P. herbarum (Pers. ex Fries) Rabenh. (Lat. *herbarum*, of plants) is found frequently on fruit of various kinds in storage. It forms black spots on the surface and may, on occasion, spread to the interior. The illustration is from an isolate obtained from the inside of a rotten lemon. The fungus grows rapidly on most culture media, forming a dense floccose mat of greyish mycelium, soon producing conidia. The conidial state is *Stemphylium botryosum* Wallroth. Perithecia take several weeks to mature and are black and fairly easily crushed. The 8-spored asci vary in length and contain the spores either in a single row (as shown here) or more or less in two rows. The golden-brown ascospores are very characteristic, having 7 cross septa and numerous longitudinal septa (Fig. 35). Conidia are produced fairly freely when the fungus is first isolated, but are seldom to be found in subsequent cultures.

Other Pyrenomycetes are occasionally isolated in the industrial laboratory. Some of these have pale-coloured walls through which the spores, and sometimes the asci, can be seen. Dark-coloured perithecia, with walls which are opaque under the microscope, can be distinguished from pycnidia of similar gross appearance by gently crushing them under a cover-glass, or, if too hard, by squashing them with a needle before putting on the cover-glass. If the perithecia are not completely mature it is nearly always possible, in this way, to find immature asci containing the normal 8 spores.

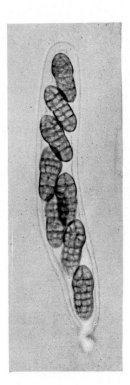

FIG. 35.—*Pleospora herbarum* —ascus, showing typical septation of spores. × 500.

REFERENCES

Ames, L. M. (1963). *A monograph of the Chaetomiaceae.* U.S. Army Res. and Devel. Series, No. 2.

De Bary, A. (1854). Entwickelung und Zusammenhang von *Aspergillus, glaucus* und *Eurotium. Bot. Ztg.,* **12,** 425.

Moreau, C. (1953). *Les genres Sordaria et Pleurage. Leurs affinities systématiques.* Paris: Lechevalier (*Encycl. mycol.,* Vol. 25).

Olliver, M., and Smith, G. (1933). *Byssochlamys fulva* sp. nov. *J. Bot., Lond.,* **71,** 196–7.

Shear, C. L., and Dodge, B. O. (1927). Life histories and heterothallism of the red bread-mould fungi of the *Monilia sitophila* group. *J. agric. Res.,* **34,** 1019–42.

Skolko, A. J., and Groves, J. W. (1948). Notes on seed-borne fungi. V. *Chaetomium* species with dichotomously branched hairs. *Canad. J. Res., Sect. C.*, **26**, 269–80.

Skolko, A. J., and Groves, J. W. (1953). Notes on seed-borne fungi. VII. *Chaetomium. Canad. J. Bot.*, **31**, 779–809.

Chapter VI

The Yeasts

The fungi known as yeasts are organisms of very great economic importance. Certain species are used in all parts of the world for the process of baking and for the production of alcoholic beverages by fermentation, for they secrete enzymes which convert sugars into alcohol and carbon dioxide. Others are responsible for the development of special flavours in certain wines, after the main fermentation is completed. Still others are encountered as contaminants in the fermentation industries, where their presence is undesirable because they reduce the yield of alcohol or produce unpleasant flavours. Some species flourish on substrates which contain high percentages of sugar, such as jams and honey, commodities which are usually considered to be safe from fungal attack. A few species are the causes of human diseases, the most common of which is thrush. Some of the yeasts are common accompaniments of moulds, frequently appearing in large numbers when mouldy products are plated out, and are obviously very widespread in nature.

During recent years a number of books have appeared dealing with various aspects of yeasts other than their taxonomy. Details of these are given in Chapter XVII.

General morphology. Young colonies of these organisms, when growing on solid media, have almost always a very characteristic appearance, being moist and somewhat slimy. The colour is usually whitish, cream, or pink, but a few species are otherwise coloured. In some species colonies change little with age, but others gradually become wrinkled and dryer in appearance. (It should be noted that some common saprophytic bacteria grow well on the usual mycological culture media and produce colonies which may readily be mistaken for those of yeasts. These may, of course, be distinguished by microscopical examination.)

The majority of the numerous species grow as loose aggregates of single cells, which may be globose, ovoid, more or less pear-shaped, or elongated and almost cylindrical. When growing very vigorously, and particularly in slide cultures, they often form chains of elongated cells, loosely attached, somewhat resembling mycelium and hence known as pseudo-mycelium. Some species form short lengths of true mycelium, often branched, which may break up finally into arthrospores (as in *Geotrichum candidum*, p. 107). There is, of course, no sharp line of demarcation between the yeasts and other fungi which form typical mycelium. The normal method of vegetative reproduction in all but a few of the industrially important species is by budding, and microscopic examination of specimens mounted in water, or fixed and stained, usually reveals all stages of this characteristic process (Figs. 36, 39, 40). When about to reproduce, the mature cell puts forth a nipple-like projection; as this grows a constriction develops between the future daughter cell and the parent; after a period of further growth, in which the projection rounds up to approximately the same shape as the parent cell, the constriction deepens and the new cell becomes detached.

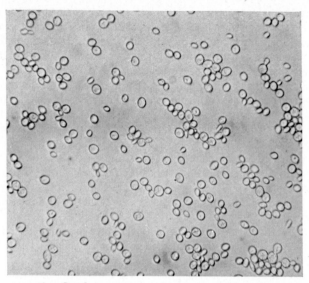

FIG. 36.—*Saccharomyces cerevisiae*—vegetative cells, some in various stages of budding. × 500.

A single cell may produce one, two, or several buds. In the process known as "bipolar budding", budding takes place only at one or both ends of an ellipsoidal or elongate cell (Fig. 39), whilst in "multipolar budding" one to several buds are produced, and these are not formed at any fixed points. A different process is found in the genus *Schizosaccharomyces*, and less regularly in some other genera, in which vegetative propagation is by binary fission, each mature cell dividing transversely into two approximately equal parts (Fig. 37).

The taxonomy of the yeasts is difficult and only an outline of modern ideas on the subject can be given here. The most authoritative monographs have come from the Dutch school, working mainly at the Centraalbureau voor Schimmelcultures, Yeast Division, Delft. In 1931 Stelling-Dekker produced the first of a trilogy of monographs, dealing with the sporing yeasts. The non-sporing yeasts were described in two volumes, by Lodder (1934) and by Diddens and Lodder (1942). More recently a new work has appeared, by Lodder and Kreger-van Rij (1952), dealing with all the yeasts in one volume, and making a number of important changes in the classification of these organisms.

The yeasts are classified into three families as follows:

1. ENDOMYCETACEAE. Includes all the species which used to be called "the True Yeasts", i.e. species which produce ascospores.
2. SPOROBOLOMYCETACEAE. Species belonging to this family produce "ballistospores", i.e. conidia which are forcibly shot off from the mother cells.
3. CRYPTOCOCCACEAE. Species producing neither ascospores nor ballistospores.

ENDOMYCETACEAE

The modes of production of ascospores have been much studied and are important diagnostic characters in all schemes of classification. In a broad sense it may be taken that any vegetative cell may, under suitable conditions, become an ascus, distinguished from a normal vegetative cell by the fact that it is an envelope containing one or more spores, each spore with its own definite limiting membrane. The number of spores in the ascus may be any number from one to eight.

Asci may be produced in three ways: by conjugation of two equal cells—isogamic conjugation; by conjugation of unequal cells—heterogamic conjugation; or from single cells without any trace of conjugation—parthenogenesis. Modern work has shown, however, that some of the so-called parthenogenetic species exhibit sexuality, in that the ascospores often fuse together in pairs before commencing to propagate by budding in the normal way. It has been found possible, in some cases, to separate single ascospores and to bring about the fusion of spores from different races of yeasts. The cells from the fusion often propagate to give normal colonies, and the production of hybrids by this method shows promise of giving new yeasts with unusual and interesting biochemical characteristics. A useful summary of work on this subject, though obviously not completely up to date, is given by Henrici (1941).

The criteria used for classification include a number of morphological characters, such as the presence of true mycelium or pseudo-mycelium, the formation of arthrospores, the shape of the cells, and the method of budding. Equally important are physiological characters, such as whether the action on various sugars is mainly fermentative or oxidative, the kinds of sugars fermented, the kinds of sugars utilized without fermentation, and whether nitrates can be utilized.

As in Stelling-Dekker's monograph the family includes some fungi which are not regarded as yeasts, but which resemble the yeasts in producing single naked asci. It is divided into 5 sub-families as follows:

Sub-family 1. EREMASCOIDEAE. Not yeasts. Vegetative growth entirely mycelial; spores hat-shaped, up to 8 per ascus; action on sugars mainly oxidative. There is only one genus, *Eremascus* Eidam, which is of academic interest only.

Sub-family 2. ENDOMYCETOIDEAE. Mycelium formed, breaking up into arthrospores or not; spores round, ovoid, or hat-shaped; oxidative or fermentative. There are 2 genera, *Endomyces* forming true mycelium and not regarded as yeasts, and *Schizosaccharomyces* with mycelium which breaks up into arthrospores, the latter also multiplying by binary fission, and asci formed from pairs of conjugating arthrospores.

Sub-family 3. SACCHAROMYCETOIDEAE. Mycelium and budding cells, or pseudo-mycelium and/or detached budding cells; propagation by fission and budding or by budding only; spores variously shaped, 1–4 per ascus; oxidative to predominantly fermentative. This is the largest of the sub-families and is further divided into 3 tribes.

Tribe A. ENDOMYCOPSEAE. Mycelium with blastospores and budding cells; propagation by fission and multipolar budding; ascospores hat-shaped, sickle-shaped, or round with a ridge; mainly oxidative but sometimes fermentative. There is one genus, *Endomycopsis*, which includes a number of species originally placed in *Endomyces*.

Tribe B. SACCHAROMYCETEAE. No mycelium but pseudo-mycelium and/or detached budding cells; reproduction by multipolar budding; ascospores variously shaped; oxidative to predominantly fermentative. This tribe includes most of the yeasts which are used commercially for the manufacture of potable liquors. There are 5 genera: *Saccharomyces*, *Pichia*, *Hansenula*, *Swanniomyces*, and *Debaryomyces*.

Tribe C. NADSONIEAE. No mycelium but budding cells and occasionally pseudo-mycelium; reproduction by bipolar budding on a more or less broad base; oxidative and fermentative. There are 3 genera: *Saccharomycodes*, *Hanseniaspora*, and *Nadsonia*, differing chiefly in the way the asci are formed.

Sub-family 4. NEMATOSPOROIDEAE. Mycelium and/or detached budding cells; reproduction by fission and/or multipolar budding; spores spindle-shaped or needle-shaped, with or without thread-like appendages; fermentative or oxidative. There are 3 genera: *Monosporella*, *Nematospora*, and *Coccidiascus*, none of which is of importance.

Sub-family 5. LIPOMYCETOIDEAE. Budding cells only, with no pseudo-mycelium; propagation by multilateral budding; asci formed as sac-like protuberances from vegetative cells; spores ovoid, amber-coloured, 4–16 per ascus; fermentation nil. There is one genus, *Lipomyces*.

Excluding *Eremascus* and *Endomyces*, which are regarded as true moulds and not yeasts, there are in all 13 genera. Several contain no industrially important species and will not be considered here. Some of the species have never even been cultivated, whilst others are of interest only from the systematic point of view. The descriptions given below of important genera and species are necessarily brief, and the reader who intends to make any kind of special study of these organisms is referred to the Lodder and Kreger-van Rij monograph, and to the special literature of the fermentation industries.

Schizosaccharomyces Lindner (Gr. *schizo*, to split). Three species are recognized, of which *S. octosporus* Beijerinck and *S. versatilis* Wickerham and Duprat both form normally 8 spores per ascus; both are rare. The third species, *S. pombe* Lindner, was originally isolated from African millet beer (Pombe is the local name for beer). It has also been found a number of times in molasses. Vegetative cells are cylindrical with rounded ends, mostly $6-16 \times 3-5\mu$ (Fig. 37); ascus formation often preceded by isogamous conjugation (Fig. 38); asci containing 1–4 spores, round and smooth, $2-2\cdot5\mu$ diam. This species has been shown by Underkofler, McPherson, and Fulmer (1937) to give very good yields of alcohol by fermentation of an unhydrolysed syrup prepared from Jerusalem artichokes.

Saccharomyces (Meyen) Reess (Gr. *sakchar*, sugar; hence, the sugar fungus). Cells round, ovate, or elongate, often forming pseudo-mycelium; reproduction by multipolar budding; isogamous or heterogamous conjugation may or may not precede ascus formation; spores 1–4 per ascus, usually round or ovate, seldom other shapes; spores may conjugate before starting growth; vigorous fermentation of glucose and usually of other sugars; nitrates not utilized. The genus *Zygosaccharomyces* was originally distinguished from *Saccharomyces* by ascus formation being preceded by conjugation. *Torulaspora* was distinguished by its formation of non-functional conjugation tubes in cells eventually becoming asci. Both these genera are now merged in *Saccharomyces*.

The genus includes 30 species and 3 varieties, separated in the key by their fermentation reactions and by their abilities to assimilate various sugars. It has been found that, without exception, any species which will ferment any other sugar will also ferment glucose, laevulose, and mannose (Kluyver's laws). In the diagnoses given below, therefore, either these three hexoses are omitted from the lists of sugars fermented, or, if glucose is mentioned, it implies the other two sugars. Ability to ferment raffinose 1/3 signifies that this trisaccharide is hydrolysed to laevulose and melibiose and only the former fermented.

S. cerevisiae Hansen (Lat. *cerevisia*, beer). This species is the typical "top-yeast" of the brewing industry (see below) and is also grown on a large scale for use in baking. Colonies on wort agar soft and moist, cream-coloured; cells round, ovate, or somewhat pyriform, in young cultures $4-14 \times 3-7\mu$, with a length/breadth ratio varying from 1 : 1 to 2 : 1; in wort a sediment, often a ring and occasionally a thin scum; ascospores round and smooth; galactose, saccharose, maltose, and raffinose 1/3 fermented; nitrates not utilized.

S. cerevisiae var. *ellipsoideus* (Hansen) Dekker (name refers to the cells being more ellipsoid than in the type). This species is commonly referred to as *S. ellipsoideus* and is the species used for the fermentation of wine. Its

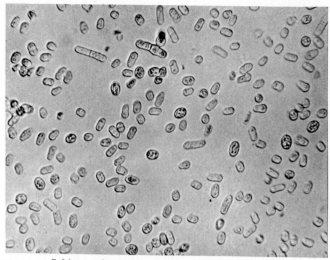

FIG. 37.—*Schizosaccharomyces pombe*—vegetative cells propagating by binary fission. × 500.

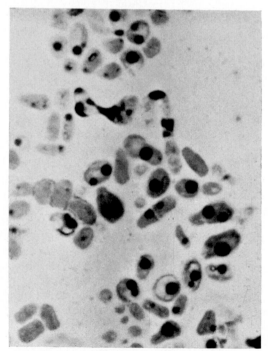

FIG. 38.—*S. pombe*—ascospores and isogamous conjugation. × 2000.

occurrence on grapes and in the soil of vineyards accounts for the success of vinous fermentation during countless ages before the nature of the process was understood. At the present time pure culture technique is used to some extent in the wine industry, and a number of strains with special trade names are in regular use for the production of different classes of wine. Other strains of this variety have been used in breweries, distilleries, and yeast factories.

Differs from *S. cerevisiae* in the average shape of the cells, which are seldom round but mostly long oval, with a length/breadth ratio of 2:1, and in its somewhat higher tolerance to alcohol.

S. carlsbergensis Hansen (of Carlsberg, the famous Danish brewery). The typical "bottom-yeast" used in brewing. Cells ovate, in young cultures 5–13 × 5–10μ, with an average length/breadth ratio of 2:1; differs from *S. cerevisiae* in its ability to ferment raffinose completely.

BREWING YEASTS. A number of strains with trade names have been regularly propagated since the introduction, by Hansen, of pure culture methods into the brewing industry. There are two main groups of these, known respectively as "top-yeasts" (*S. cerevisiae* and its variety *ellipsoideus*) and "bottom-yeasts" (*S. carlsbergensis*). The top-yeasts produce CO_2 vigorously and the cells have a tendency to become attached in clusters. They are carried to the surface during the fermentation by the rising gas, and eventually form a scum. These strains are used for brewing the typically English beers and ales. Bottom-yeasts are used for the manufacture of the light Continental beers typified by the well-known "lager". In these a prolonged fermentation is carried out at a comparatively low temperature and the yeast cells stay at the bottom of the liquor. Distillers use top-yeasts of high activity, and carry out the fermentation at a comparatively high temperature in order to obtain rapid production of alcohol in maximum yields.

BAKERS' YEAST. The pressed yeast used in bakeries, and in home baking, was at one time produced by breweries and distilleries, the mass of cells formed during the fermentation being separated by centrifuging, then washed and pressed for the market. Distillery yeast is more suitable than brewery yeast for baking, since the latter is dark in colour and somewhat bitter, due to the presence of extractive matter from the hops which are added to the wort prior to the fermentation. At the present time most of the yeast used in baking is manufactured specially for the purpose. In the manufacture of potable liquors the aim is to utilize to the full the biochemical activities of the yeast cells, with maximum production of alcohol and CO_2, and to keep down undue multiplication of the cells, which would result in loss of valuable fermentable material. In the manufacture of compressed yeast, on the other hand, the aim is to obtain maximum growth and minimum production of alcohol. This is achieved by vigorous aeration of the culture medium and by addition of nutrients in stages during the process, to keep pace with the increase in the number of cells. Full details of the manufacturing process are given by Walter (1940).

S. pastorianus Hansen (named after Louis Pasteur). Best known for its production of unpleasant flavours in beer, in which it is liable to occur as a contaminant. It is a bottom-yeast with cells ovate or, more usually, elongated, commonly described as sausage-shaped, $4–15 \times 3–7\mu$, often forming pseudo-mycelium; ascospores round and smooth; in wort a sediment and soon a scum; ferments saccharose, maltose, and raffinose $2/3$. It neither ferments nor utilizes galactose; hence hydrolyses raffinose and ferments only the glucose and laevulose produced.

S. fragilis Jörgensen (Lat. *fragilis*, tender, fragile). Synonym *S. kefyr*. This species and *S. lactis* are both used for the preparation of fermented liquors from milk. They are both bottom-yeasts capable of fermenting lactose, and producing therefrom much CO_2 and comparatively little alcohol. *S. fragilis* is chiefly associated with the manufacture of Kefyr and Koumyss. In the countries where these and similar drinks are prepared, pure culture methods are not used and the actual agent of fermentation is a mixture of various yeasts and bacteria, in which, however, *S. fragilis* predominates. Cells ovate to elongate, with very delicate walls which are easily ruptured, $6–10 \times 2\cdot5–5\cdot5\mu$, often forming chains; ferments galactose, saccharose, lactose, and raffinose $1/3$.

S. lactis Dombrowski (Lat. *lactis*, of milk). Associated with the manufacture of yogurt. It is distinguished from *S. fragilis* in having more robust cells, and in its ability to utilize maltose.

S. rouxii Boutroux (Emile Roux, French bacteriologist) and *S. mellis* (Fabian and Quinet) Lodder and Kreger-van Rij (Lat. *mellis*, on honey) are the commonest of the so-called "osmophilic yeasts", characterized by their ability to grow on, and ferment, materials containing very high percentages of sugar, such as will inhibit most other organisms. They are both very limited in their fermentative abilities. *S. mellis* ferments only glucose, laevulose, and mannose; *S. rouxii* ferments maltose in addition but often very weakly. Both grow well on an agar medium containing 60 per cent of glucose, and ferment vigorously a glucose solution of similar concentration. *S. rouxii* forms cells, in wort, which are round to ovate, $3\cdot5–8\cdot5 \times 2\cdot5–5\mu$, mostly detached. *S. mellis* has cells of similar shape and size which tend to clump together. These yeasts have been studied chiefly in connection with the spoilage of honey and jams. A useful summary of their characteristics is given by Henrici (1941).

Other species and varieties of the genus *Saccharomyces* are responsible for the alcoholic fermentation of cider and perry. For making home-brewed cider the yeasts which occur on the fruit are relied on exclusively, and usually a mixture of several species is present during the fermentation. Naturally the quality of the liquor produced varies considerably. The cider and perry produced by several large firms is made under carefully controlled conditions, including the use of pure cultures of selected strains of yeast. In the East, alcoholic liquors are made chiefly from rice and, whilst yeasts usually play some part in the fermentation, the main agents are moulds.

Pichia Hansen (P. Pichi, Italian mycologist) and *Hansenula* H. and P. Sydow (E. Chr. Hansen, Danish microbiologist). Both genera are best known in this country as contaminants in alcoholic liquors. They grow as dry, wrinkled pellicles on the surface of the liquid and their metabolism of sugars is more oxidative than fermentative. If fermentation occurs the main products are esters and not alcohol. Most species can utilize alcohol as a source of carbon and hence can constitute a serious nuisance in the fermentation industries. Species of both genera also occur in the "flor" of sherry and of certain wines of the Arbois district of France, this being a film of yeasts which is responsible for the development of the special flavours of these wines. A paper by Hohl and Cruess (1940) gives full information about these "flor" yeasts and their activities.

Species of both genera form cells of various shapes—round, ovate, and elongate cylindrical—with a strong tendency to form pseudo-mycelium; ascospores hemispherical, hat-shaped, or angular. Species of *Pichia* are mainly oxidative, ferment only slightly, and cannot utilize nitrates. Species of *Hansenula* ferment vigorously and can utilize nitrates.

Hanseniaspora Zikes (E. Chr. Hansen). The genus includes only one species, *H. valbyensis* Klöcker (Valby, a district of Copenhagen). Cells lemon-shaped to ovate, $4-10 \times 2-4\mu$, reproducing by bipolar budding, hence the common name "apiculate yeasts" (Fig. 39); ascospores hat-shaped, 1–4 per ascus; only glucose, laevulose, and mannose fermented. This species and its imperfect form, *Kloeckera apiculata*, are commonly found on grapes and predominate at the beginning of the wine fermentation. They are, however, sensitive to alcohol and are replaced by *S. cerevisiae* var. *ellipsoideus* as the fermentation proceeds.

Debaryomyces Lodder and Kreger-van Rij nom. conserv. (A. De Bary German mycologist). Cells round or short oval, propagating by multipolar budding; producing in liquid media a dry dull pellicle; ascus formation almost invariably preceded by conjugation, usually heterogamous, a cell conjugating with its own bud; spores round, usually with an oil-drop in the middle and sometimes finely warted. There are 5 species, mostly found on animal products—cheese, sausage, glue, rennet, etc. Many isolates studied by the Dutch workers were of human origin and are probably mildly pathogenic.

SPOROBOLOMYCETACEAE

This family includes only two genera, *Sporobolomyces* and *Bullera*, distinguished by the shapes of the ballistospores and by the presence or absence of carotinoid pigment.

Sporobolomyces Kluyver and van Niel (Gr. *bolis*, a missile; hence a fungus which throws its spores). Young colonies are red to salmon pink and moist, but later become powdery. If a Petri dish culture be incubated upside down a faint mirror image of the colony is produced in the lid of the dish. This mirror colony is composed of bean-shaped spores quite unlike ordinary yeast cells. These spores are formed on small projections put out by the ovate vegetative cells and are forcibly shot off by a curious mechanism which Buller (1933) has shown to be peculiar to the Basidiomycetes. Kluyver and van Niel, Buller, Derx and others have suggested that these organisms are actually primitive Basidiomycetes. However, Lohwag (1926) has given a number of reasons why this genus should not be regarded as belonging to the Basidiomycetes. Also there are a few fungi, definitely belonging to the Hypomycetes, which forcibly eject their spores by somewhat similar mechanisms (see Webster, 1952). Hence, the systematic position of this genus, and of *Bullera*, is still in doubt. Young colonies of *Sporobolomyces* are almost indistinguishable from colonies of *Rhodotorula* (see below) and it is possible, though not yet proven, that species of *Rhodotorula* are degenerate species of *Sporobolomyces* which have lost the power of producing ballistospores.

The genus is characterized by forming red to salmon pink colonies and kidney- or bean-shaped ballistospores. There are 7 recognized species.

Buller Derx (A. H. R. Buller, British mycologist). Colonies creamcoloured to pale yellowish, never reddish; ballistospores symmetrical, round to ovoid. Derx (1930), in describing the genus, stated that species of both genera of this family are of common occurrence on straw and on leaves of various plants, but their significance in nature has not been determined.

CRYPTOCOCCACEAE

This family belongs to the Fungi Imperfecti and is the only family in the order Cryptococcales. It includes all the yeasts which do not produce either ascospores or ballistospores. It probably includes many species which have lost the power of producing one or other of these types of spores, or for which methods of inducing spore formation are still unknown. A number of species of the family are now definitely regarded as imperfect forms of ascosporic yeasts, and, as indicated above, it is at least possible that all the species of the genus *Rhodotorula* are degenerated strains of *Sporobolomyces*.

There are 3 sub-families, characterized as follows:

Sub-family 1. *Cryptococcoideae*. Chiefly budding cells, but sometimes pseudo-mycelium or even true mycelium, but no arthrospores; propagation usually by budding, rarely by fission; no red or yellow carotinoid pigments produced. There are 7 recognized genera: *Cryptococcus, Torulopsis, Pityrosporum, Brettanomyces, Candida, Kloeckera,* and *Trigonopsis*.

Sub-family 2. *Trichosporoideae*. True mycelium, pseudo-mycelium, budding cells and arthrospores; no yellow or red carotinoid pigments; mostly strictly oxidative but occasionally fermentative. There is only one genus, *Trichosporon*, distinguished from sub-family 1 by the production of arthrospores.

Sub-family 3. *Rhodotoruloideae*. Chiefly budding cells, with occasional primitive pseudo-mycelium; reproduction by multi-lateral budding; yellow or red carotinoid pigment always produced; strictly oxidative. There is one genus, *Rhodotorula*.

Cryptococcus Kützing emend. Vuillemin (Gr. *kryptos*, hidden, secret; *kokkos*, seed or berry). Cells round, ovoid, or somewhat irregular in shape; reproduction by multipolar budding; cells surrounded by capsules, which usually contain a starch-like substance; ability to ferment sugars nil. There are 5 species and 3 varieties.

C. neoformans (Sanfelice) Vuillemin (Gr. *neo*, new, afresh; Lat. *formans*, forming, shaping). The name was bestowed because the fungus was supposed to produce tumours. It is highly pathogenic to man and animals, and hence is of interest chiefly to medical mycologists. The other 4 species are not common.

Torulopsis Berlese (Gr. *opsis*, like; hence like the genus *Torula*). Many of the so-called "wild yeasts" which sometimes cause trouble in breweries, and which appear in abundance on plating out many kinds of mouldy materials, belong to this genus. Cells are round or ovoid, reproducing by multilateral budding; capsules rare and never containing a starch-like compound; frequently fermentative.

There are 22 species and one variety, separated according to their fermentative abilities and ability to utilize nitrates. Although they are common associates of moulds little is known as to how far they contribute to "mouldiness", so that descriptions of individual species would be out of place here. The species previously known as *Torulopsis utilis* is now placed in the genus *Candida*.

Brettanomyces Kufferath and van Laer (*Brettano-* = British). This genus was erected to include a number of yeasts which have been isolated from English stouts and Belgian lambic beer, and which are stated to be responsible for the "after-fermentation" (Guilliermond 1920, Henrici 1941). The after-fermentation is a secondary fermentation which occurs in the bottles, residual oxygen being utilized and an appreciable amount of CO_2 produced. According to Hind (1940, p. 895), the development of the characteristic flavour of fully matured stock beers is preceded by a stage in which an objectionable odour is produced. Species of this genus are of fairly common occurrence in breweries and may constitute a source of trouble in beers which are not subjected to a prolonged maturing process. The yeasts differ markedly from *Saccharomyces cerevisiae*, which is responsible for the main fermentation. The cells are of various shapes, ovoid to almost globose, or elongated and pointed at both ends (so-called "ogive" cells). Under anaerobic conditions fermentation is very slow, taking several months to reach completion, but tolerance to alcohol is high so that eventually a greater percentage of alcohol is obtained than with *S. cerevisiae* alone. Under aerobic conditions they oxidize alcohol to acetic acid, eventually producing sufficient acid to kill the cultures unless the medium has had a considerable amount of calcium carbonate added to it.

There are 4 species and 2 varieties, distinguished by their reactions to various sugars. The taxonomy of the genus has been thoroughly studied by Custers (1940), and his classification is followed by Lodder and Kreger-van Rij.

Candida Berkhout (Lat. *candidus*, white). The genus contains 30 species plus 6 varieties, many of which are pathogenic to humans and/or animals. Others have been isolated from the surfaces of seeds and fruits, or from soil. Vegetative growth consists of budding cells and pseudo-mycelium, or true mycelium producing clusters of blastospores. A few of the species are of some industrial importance.

C. mycoderma (Reess) Lodder and Kreger-van Rij (Gr. *derma*, skin; hence the skin-forming fungus). The various species previously included in the genus *Mycoderma* are now regarded as a single, somewhat variable, species of *Candida*. It is the imperfect stage of *Pichia membranaefaciens* and, like the latter, forms dry wrinkled pellicles on the surface of liquid media. Cells are ovoid to more or less cylindrical, forming pseudo-mycelium; oxidative only, not fermentative.

C. utilis (Henneberg) Lodder and Kreger-van Rij (Lat. *utilis*, useful). This yeast was known as *Torula utilis* or *Torulopsis utilis*, or as "the food yeast". It is easy to grow in bulk and, since it contains a high percentage of protein, and also vitamins of the B group, it has been used for fodder and, to some extent, for human food (see Chapter XV). Cells normally ovate, $5-9 \times 3-5 \cdot 5\mu$, but in slide cultures tending to be cylindrical and forming pseudo-mycelium (hence the transfer to *Candida*); glucose, saccharose, and raffinose 1/3 fermented; nitrates utilized.

C. lipolytica (Harrison) Diddens and Lodder (Gr. *lipos*, fat; *lysis*, loosening; hence hydrolysing fats). Cells very variable, long oval to almost cylindrical and/or short oval; in slide culture well developed pseudo-mycelium and some true mycelium; liquefies gelatine rapidly and completely; grows on and hydrolyses fats; does not ferment sugars; found in butter and margarine.

C. pseudotropicalis (Castellani) Basgal var. *lactosa* (Hansen) Diddens and Lodder (*pseudotropicalis* = related to *C. tropicalis*). The parent species is a human pathogen. The variety is associated with buttermilk and Kefyr. It has previously been given a number of other names, including *Torula lactosa*, *Torula kefyr*, and *Mycotorula lactis*. Cells in liquid culture ovate to elongate, $7-15 \times 3 \cdot 5-6\mu$, forming a sediment and eventually a surface pellicle; pseudo-mycelium abundant in slide cultures; galactose, lactose, saccharose, and raffinose 1/3 fermented.

Kloeckera Janke (A. Klöcker, German microbiologist). Cells ovate to lemon-shaped; reproduction by bipolar budding. There are 8 species, of which all except one are rare.

K. apiculata (Reess emend. Klöcker) Janke (Lat. *apiculatus*, provided with little caps). This is the imperfect state of *Hanseniaspora valbyensis* and, like the latter, is found on grapes and is associated with the "flor" of sherry. Cells lemon-shaped or ovate, 5–8(10) × 2–4·5μ; in liquid media a sediment and thin surface ring; formation of pseudo-mycelium rare; only glucose, laevulose, and mannose fermented.

Rhodotorula Harrison (Gr. *rhodo-*, rosy red). The genus includes all the red, pink, and yellow yeasts which do not produce ballistospores. These are amongst the commonest of air-borne organisms, being frequently found as contaminants of cultures of moulds and bacteria. The genus includes 7 species and one variety. All are non-fermentative and hence are of no practical importance, being both useless and harmless so far as is known, except that they can be a nuisance in the culture laboratory. Their chief interest is to the systematist, since they are suspected of being imperfect forms of species of *Sporobolomyces*. A recent and useful paper on the taxonomy of the genus is by Hasegawa (1965).

METHODS OF IDENTIFICATION

Yeasts, unlike moulds, cannot usually be identified from morphological data alone. Biochemical tests are necessary and, indeed, are the basis of the published keys to many of the genera. In the descriptions of methods given below, a number of points of technique, and some of the culture media, are identical with those used in identifying moulds. These are fully dealt with in Chapter XI and are not elaborated here. On the other hand, many of the methods are used for yeasts alone and are described here in preference to including them with the more general methods.

The morphological data required are: appearance of cultures on wort agar and in liquid wort; shape and size of vegetative cells in wort and on slide cultures; method of propagation; mode of production of asci and size and shape of ascospores. Biochemical data include: ability to ferment various sugars; ability to utilize various carbohydrates for growth; ability to utilize nitrates as sole source of nitrogen; sometimes ability to utilize alcohol as source of carbon.

Appearance of cultures. Single colonies are grown in Petri dishes on wort agar and on wort gelatine, details of colour and texture being recorded over a period of up to 2 months. Cultures are made in liquid wort and development of sediment, superficial ring, or complete pellicle noted. Slide cultures are made, using wort agar as the medium, for microscopic examination.

Characteristics of cells. Slides are made from cultures in liquid wort after 24 hours' and 3 days' incubation. The shape and dimensions of the cells are noted, also the method of propagation, whether by multipolar budding, bipolar budding, or binary fission. Any tendency to form true mycelium or pseudo-mycelium is noted from examination of both the slides made from liquid cultures and of the slide cultures.

Production of ascospores. Some species, especially when newly isolated from a natural source, form spores readily. Other species, and most
isolates which have been kept for some time in artificial culture, may respond only after many different methods have been tried. The number of
methods which have been suggested for induction of spore formation is a
measure of the difficulties which most workers have experienced with these
organisms. The following is a selection of the more reliable of these
methods. In all the procedures which involve the use of special media,
transfer is made of a heavy inoculum from a vigorously growing culture on
wort agar. All these media are starvation media, in the sense that their
content of readily available source of energy is small.

1. The classical method, and one which often works, is to transfer to small
   sterile blocks of plaster of Paris moistened with water, very dilute
   wort, or a solution containing 2 per cent mannitol and 0·5 per cent
   K_2HPO_4.
2. Cultures are sown on sterile slices of carrot or potato, or on sterilized
   raisins.
3. Transfer is made to Gorodkowa's agar (glucose, 0·25 per cent; meat
   extract, 1·0 per cent; NaCl, 0·5 per cent; agar, 2 per cent) or to plain
   agar without any added nutrients, or to raisin agar.
4. Cultures on Gorodkowa's agar are irradiated with ultraviolet light, exposing different cultures for periods of 5, 10, and 15 minutes respectively.
5. Adams (1950) tested a number of methods and recommends the following: The yeast is subcultured twice on "presporulation medium",
   then transferred, from a 24-hour culture, to the "sporulation medium". Presporulation media are: (1) nutrient agar (Difco) plus 5 per
   cent dextrose and 0·5 per cent tartaric acid. (2) Commercial grape
   juice (containing about 5 per cent sugar) plus 2 per cent agar. The
   sporulation medium contains: dextrose, 0·04 per cent; anhydrous
   sodium acetate, 0·14 per cent; agar, 2 per cent.
6. The yeast is transferred to a sterilized 0·5 per cent solution of potato
   starch in water and incubated at 37° C (Almeida and Lacaz, 1940).
7. Bright, Dixon, and Whymper (1949) have shown that alcohol vapour
   and carbon dioxide have, separately or together, adverse effects on
   the sporulation of *Saccharomyces cerevisiae*. It is advisable, therefore,
   when using any of the methods described above, to incubate cultures
   in the open laboratory and not in an incubator along with other
   cultures.

When a satisfactory procedure has been found, it is necessary to make
microscopic preparations at frequent intervals, from cultures which are
actively sporing, in order to determine whether the process of spore formation is preceded by conjugation or is parthenogenetic.

Fermentation tests. These are usually carried out in special tubes sold for the purpose, the best-known being Einhorn tubes (illustrated in any catalogue of laboratory apparatus). Quite satisfactory tubes may, however, be made at little cost by anyone with a little skill in glass-blowing. A quantity of yeast extract is made by autoclaving 200 g of pressed yeast with 1 litre of tap-water, filtering twice, hot and cold, and finally making up to 1 litre. Two per cent solutions of the sugars to be tested are made up in this and sterilized. Each fermentation tube, previously sterilized, is charged with solution so that the closed limb is quite full and the bend is sealed so that no air can enter. A small quantity of the yeast taken with a needle from a culture on wort agar, is introduced through the open short limb and the tube is incubated at 25° C. If much gas is liberated this is definite evidence of fermentation, but, if only a small bubble collects, it should be tested with KOH to find whether it is CO_2 or air liberated from solution in the liquid medium.

Other tests. Two special media, one containing nitrate as sole source of nitrogen, the other containing alcohol as sole source of carbon, are used to test the availability of these substances for growth. The first medium contains: glucose 2 per cent; KNO_3, 0·1 per cent; KH_2PO_4, 0·1 per cent; $MgSO_4.7H_2O$, 0·05 per cent; washed agar, 2 per cent; in distilled water. The second contains: alcohol, 3 per cent; $(NH_4)_2SO_4$, 0·1 per cent; KH_2PO_4, 0·1 per cent; $MgSO_4.7H_2O$, 0·01 per cent; in tap-water.

For testing the ability of a yeast to utilize various sugars for growth, as distinct from fermenting them, and for testing the availability of various nitrogen compounds, the best, simplest, and most economical procedure is the auxanographic method of Beijerinck. The medium for testing sugars contains: $(NH_4)_2SO_4$, 0·5 per cent; KH_2PO_4, 0·1 per cent; $MgSO_4.7H_2O$, 0·05 per cent; washed agar, 2 per cent. About 2 ml. of a thick suspension of cells in sterile water is put in a Petri dish; 10 ml. of the above medium, previously melted and cooled to 40° C, is poured in and thoroughly mixed with the suspension; when the agar has completely set, very small amounts of the finely powdered sugars are placed at different spots on the surface, five or six sugars in one dish; finally the dish is incubated for a few days at 25° C. The sugars diffuse into the agar and, if utilized, stimulate into growth the cells lying within the diffusion zone. For testing compounds of nitrogen, the method is the same except that the medium contains 2 per cent glucose in place of the ammonium salt.

Microscopical methods. Cells from cultures in liquid media may be mounted in a drop of the culture fluid. The culture is shaken to give an even suspension, a small drop is placed on a slide and a cover-glass is well pressed down. Material from agar cultures may be mounted in lacto-phenol, but mounting in plain water gives better contrast and makes observation easier. Only a minute speck of material should be taken from the culture and the drop of water should be quite small. A useful method of demonstrating type of budding, and size and shape of cells, is to mount in a 5 per cent aqueous solution of nigrosin (water soluble). The cells are

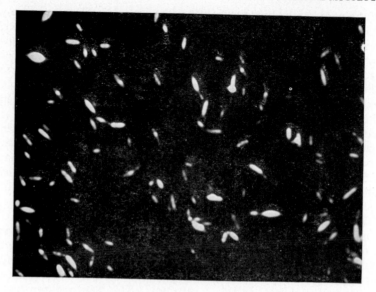

FIG. 39.—*Kloeckera apiculata*—relief stained. × 1000.

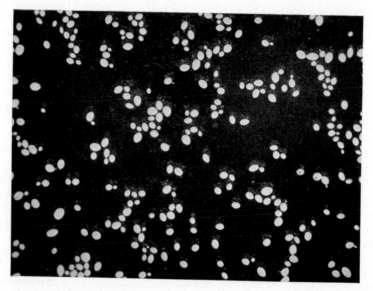

FIG. 40.—*Zygosaccharomyces* sp.—relief stained with nigrosin. × 500.

unstained and show as white on a blue-black ground (Figs. 39, 40). Such slides are not, of course, permanent, but usually remain in good condition long enough for a thorough examination. Their useful life can be considerably extended, and movement of the cells in the aqueous fluid largely prevented, by ringing them, immediately after mounting, with shellac cement.

Examination for the presence of ascospores is best carried out with stained films, since vacuoles are easily mistaken for spores in unstained cells. A simple method which is effective in most cases was described by Shimwell (1938) and modified by McClung (1943). A small sample is taken from the culture under examination and well mixed with a drop of water on a perfectly clean slide or cover-glass, spread into a thin film and allowed to dry in the air. The film is fixed by passing rapidly two or three times

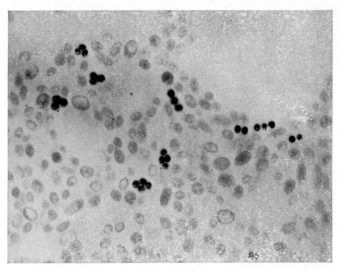

FIG. 41—*Saccharomyces* sp.—ascospores, stained film. × 1000.

through a bunsen flame, then flooded with malachite green (1 per cent solution in 1 per cent aqueous phenol, prepared without heating) and heated to steaming, but not boiled, over a small flame for 2 minutes. The slide is washed gently under the tap for one minute, then counterstained with 0·5 per cent aqueous safranin for 30 seconds. The ascospores are stained green, whilst the walls of the asci and the vegetative cells are pink. Fig. 41 is a photograph of a preparation made in this way, taken with a contrast filter for the green.

REFERENCES

Adams, A. M. (1950). A comparative study of ascospore formation by 43 yeast cultures. *Canad. J. Res., Sect, F.*, **28**, 413–16.

Bright, T. B., Dixon, P. A., and Whymper, J. W. T. (1949). Effect of ethyl alcohol and carbon dioxide on the sporulation of bakers' yeast. *Nature, Lond.*, **164**, 544.

Buller, A. H. R. (1933). *Researches on Fungi*. Vol. V. London: Longmans, Green & Co.

Custers, M. T. J. (1940). Onder zoekingen over het gistgeslacht *Brettanomyces*. Diss., Delft.

de Almeida, F., and Lacaz, C. da Silva (1940). Nova técnica para a demonstração rápida dos ascosporos. *Folio clin. biol. S. Paulo*, **12**, 129–30.

Derx, H. G. (1930). Étude sur les Sporobolomycètes. *Ann. mycol., Berl.*, **28**, 1–23.

Diddens, H. A., and Lodder, J. (1942). *Die anaskosporogenen Hefen*. Zweite Hälfte. Amsterdam: N.V. Noord-Hollandsche Uitgevers Maatschappij.

Guilliermond, A. (1920). *The yeasts*. Trans. by F. W. Tanner. New York: John Wiley & Sons.

Hasegawa, T. (1965). A report on the taxonomy of the red to orange *Rhodotorula*. *Annu. Rep. Inst. Ferm., Osaka*, No. 2 (1965).

Henrici, A. T. (1941). The yeasts: genetics, cytology, variation, classification and identification. *Bact. Rev.*, **5**, 97–179.

Hind, H. L. (1940). *Brewing*. Vol. II. London: Chapman & Hall.

Hohl, L. H., and Cruess, W. V. (1940). Observations on certain film-forming yeasts. *Zbl. Bakt.*, Abt. II, **101**, 65–78.

Lodder, J. (1934). *Die anaskosporogenen Hefen*. Erste Hälfte. Amsterdam: N.V. Hollandsche Uitgevers Maatschappij.

Lodder, J., and Kreger-van Rij, N. J. W. (1952). *The Yeasts. A taxonomic study*. Amsterdam: North-Holland Publ. Co.

Lohwag, H. (1926). *Sporobolomyces*, kein Basidiomyzet. *Ann. mycol., Berl.*, **24**, 194–202.

McClung, L. S. (1943). On the staining of yeast spores. *Science*, **98**, 159–60.

Shimwell, J. L. (1938). A simple staining method for the detection of ascospores in yeasts. *J. Inst. Brew.*, **44**, 474.

Stelling-Dekker, N. M. (1931). *Die sporogenen Hefen*. *Verh. Kon. Akad. Wetensch. Amsterdam, Afd. Natuurkunde*, **28**, 1–547.

Underkofler, L. A., McPherson, W. H., and Fulmer, E. I. (1937). Alcoholic fermentation of Jerusalem artichokes. *Industr. Engng. Chem.*, **29**, 1160–4.

Walter, F. G. (1940). *The manufacture of compressed yeast*. London: Chapman & Hall.

Webster, J. (1952). Spore projection in the Hyphomycete *Nigrospora sphaerica*. *New Phytol.*, **51**, 229–35.

Chapter VII

Fungi Imperfecti

> We cannot deny that we have not yet found out what is the best view to take about the numerous fungi which so far appear to be merely conidiophorous, and what criterion is to be trusted, so that we may safely arrange them in the most suitable place in the system; how, for example, we can discriminate without any risk of error between the simpler conidial apparatus of species of *Hypomyces* and *Hypocrea* (Pyrenomycetes), wonderfully constructed of superposed verticils, and the true species of *Verticillium*, if any there be, belonging to the Mucedinei.
>
> L. R. and C. Tulasne, *Selecta fungorum carpologia,*
> trans. W. B. Grove. Vol. I.

As already stated in Chapter II, the Class of fungi known as the Fungi Imperfecti[1] includes all species (except those belonging to the Phycomycetes) which have no perfect fruiting state, but reproduce solely by means of asexual spores or by fragmentation of mycelium.

In addition, it is common practice to include amongst the Fungi Imperfecti a number of species which are known to be merely the conidial stages (states) of perfect fungi, mostly of Ascomycetes. Many species of *Fusarium*, for example, are imperfect states of species of *Nectria*, *Calonectria*, *Gibberella*, and *Hypomyces*; the common mould *Trichoderma viride* is the conidial form of *Hypocrea rufa*; species of the *Aspergillus glaucus* group belong with *Eurotium*; several species of *Penicillium* produce ascocarps; and numerous other similar connections are known. From time to time fresh cases are brought to light and it is very probable that a large number of fungi which are at present classed with the Fungi Imperfecti have life cycles which are incompletely known rather than incomplete. A few purely conidial forms have been proved to be haplont strains of heterothallic Ascomycetes—*Monilia sitophila* and related species are examples—and it may be that there are more similar cases as yet undiscovered, awaiting some chance observations, or systematic experiment on a large scale, for their elucidation.

Strictly speaking, the discovery of the perfect state of any fungus known hitherto only in its imperfect form, or of the connection of an imperfect form with an already known perfect species, ought to result in the separate

---

[1] Saccardo uses the term Deuteromycetae (Gr. *deuteros*, second, inferior) instead of Fungi Imperfecti, whilst a number of French mycologists prefer the term Adelomycetes (Gr. *adelos*, uncertain).

name of the conidial form being dropped. The retention of such names, and the inclusion of the species in the Fungi Imperfecti, is simply a matter of convenience. Classification is intended to facilitate identifications just as much as it is designed to exhibit natural relationships. Many species, of which the perfect stages are known, form asci only under very special conditions of culture, or only when the fungi are parasitic, conidial fructifications alone being produced in ordinary laboratory cultures or on industrial materials. Such species would be very difficult to identify if they had to be sought in a classification based solely on characteristics of sexual spores and fruit-bodies, and it is much more satisfactory to classify them as Fungi Imperfecti. There are even some species, notably the *Aspergillus glaucus* group, the *A. nidulans* group, and ascosporic species of *Penicillium*, which are conveniently classified along with forms of similar conidial morphology, in spite of the fact that perithecia are formed readily and abundantly under almost any conditions of culture.

It should be understood that the terms "Order", "Family", and "genus", as used in classifying the Fungi Imperfecti, are terms of convenience and have not the same significance as they have in the Ascomycetes and Basidiomycetes. Species with similar cultural characteristics, and with conidial fructifications of the same type, and which are therefore placed in the same genus of Fungi Imperfecti, may produce, under suitable conditions, perithecia of quite different types, and hence belong to different genera of Ascomycetes. For example, *Aspergillus glaucus* (= *Eurotium herbariorum*) has soft yellow perithecia, whilst *A. nidulans* forms perithecia which are dark-coloured and brittle, and has been made the type of the genus *Diplostephanus* by Langeron; three different types of perithecia are produced by species of *Penicillium* and, in a rational classification, these species would be placed in three different genera; species of the genus *Botrytis* are known to be imperfect states of Ascomycetes belonging to several different genera, and the same thing obtains, as stated above, for species of *Fusarium*. In addition, it is not unusual for the different species included in a single genus of Ascomycetes to have conidial states which are sufficiently varied to be classed in different genera of Fungi Imperfecti. For example, the conidial states of species of the genus *Hypomyces* are placed in the imperfect genera *Fusarium*, *Tubercularia*, *Dactylium*, and *Diplosporium*.

On account of these facts many mycologists use the term "form genus" for a group of imperfect species with similar conidial fructifications, that is, for what is here termed a "genus" of the Fungi Imperfecti. It naturally follows that the grouping of genera into Families, and these into Orders, is purely artificial and, in many cases, takes no account of real relationships. Nevertheless, such grouping is necessary, since the number of imperfect fungi is so enormous, in order to facilitate identifications. In this and the following chapters the terms "genus", "Family", and "Order" are used without qualification, but they are not to be regarded as having the same significance as when applied to perfect fungi.

SYSTEMS OF CLASSIFICATION

The number of species of Fungi Imperfecti is very large, exceeding the number of Ascomycetes, and, therefore, the construction of a satisfactory scheme of classification is a matter of great difficulty. The first complete scheme to be evolved was that of Saccardo (1880, 1884, 1886), and it still remains the only one which has been fully worked out, so as to include nearly all known genera. From several points of view it is irrational and difficult to use, but it is essential for the student to be familiar with it until such time as one of the more logical schemes which have been suggested has been worked out so as to include at least the majority of the genera.

Saccardo's classification. The Fungi Imperfecti are divided into three Orders as follows (Saccardo's terminology has been modified as regards ordinal endings):

Conidiophores produced inside flask-shaped receptacles (pycnidia)	SPHAEROPSIDALES
Conidiophores occurring in a saucer-shaped receptacle, or forming a tuberculate mass, breaking through the surface of the substratum . .	MELANCONIALES
Conidiophores free, arising from the surface of the substrate, or from aerial mycelium . .	HYPHOMYCETALES

The three Orders correspond roughly with the three Sub-classes of Ascomycetes, Sphaeropsidales with Pyrenomycetes, Melanconiales with Discomycetes and Hyphomycetales with Plectomycetes. In view of the relative unimportance to the industrial mycologist of the first two Orders, the few genera which are likely to be met with are described and keyed along with the Hyphomycetales in the next chapter.

SPHAEROPSIDALES

(Gr. *opsis*, like; hence like the Spaeriales, an Order of the Pyrenomycetes)

The pycnidium, which is the fruiting structure characteristic of the Order, is a more or less flask-shaped body, superficially resembling a perithecium. However, instead of containing asci, it has the inside lined with very short conidiophores. When the pycnidium is squashed the spores are liberated in irregular masses.

Most of the species are parasites of plants, or saprophytes found on decaying plant material, but a few are found fairly frequently on industrial products and are of some practical importance. Almost all the species encountered belong to the one genus *Phoma* (p. 119).

MELANCONIALES

(Gr. *melano*-, black; *konis*, dust)

As in the case of the previous Order, most of the species are either parasitic or occur as saprophytes on fallen parts of plants. In their natural habitats they form pustular growths, consisting either of a saucer-shaped receptacle, somewhat resembling a pycnidium opened out flat, or of a dense cluster of short conidiophores breaking through the epidermis of the host, without any definite basal mycelial layer, and resembling a sporodochium. In some schemes of classification the Melanconiales are grouped along with the Tuberculariaceae (see below), on the grounds that the distinctions made by Saccardo have no real justification. It has often been found that one and the same species can produce fructifications of either type, depending on host or cultural conditions.

On the other hand, Grove (1935–7) unites the Sphaeropsidales and Melanconiales into one group, which he terms Coelomycetes (Gr. *koilos*, hollow), and this classification is used by the British Mycological Society in their lists of British Fungi.

The species which are likely to be isolated in the laboratory produce scattered, or more or less confluent, black pustules, consisting of bundles of very short conidiophores bearing slimy masses of spores. Only one genus, *Pestalotia* (p. 116), is of importance here.

HYPHOMYCETALES

(Consisting of hyphae, i.e. not forming definite receptacles)

This Order includes most of the common species of moulds. Saccardo divides the Order into four Families as follows:

1. Conidiophores detached, not aggregated into fascicles (bundles).

   (a) Hyphae and conidia colourless or in pale
       or bright colours   .   .   .   MUCEDINACEAE
   (b) Mycelium, spores, or both, dark brown to
       black   .   .   .   .   .   DEMATIACEAE

2. Conidiophores mostly aggregated in fascicles.

   (a) Conidiophores short, forming cushion-
       shaped aggregates (sporodochia), often
       waxy or gelatinous   .   .   .   TUBERCULARIACEAE
   (b) Conidiophores long, forming coremia   .   STILBACEAE

Each family is subdivided into a number of sections according to the septation of the spores.

Spores 1-celled	*Amerosporae*
Spores 2-celled	*Didymosporae*
Spores with 2 or more cross septa . . .	*Phragmosporae*
Spores with both cross and longitudinal septa . .	*Dictyosporae*
Spores forked or star-shaped	*Staurosporae*
Spores spirally curved, septate	*Helicosporae*

Some authors of taxonomic works, who have adopted Saccardo's classification, have used somewhat different terms for the various spore sections;

for pale-coloured spores, *Hyalosporae, Hyalodidymae, Hyalophragmiae,* etc., and for dark spores, *Phaeosporae, Phaeodidymae,* etc.

Each of the spore sections is then split into two groups:

1. Conidiophores short or obsolete, hardly distinguishable from the mycelium   .   .   .   .   .   *Micronemeae*
2. Conidiophores distinct from the mycelium   .   .   *Macronemeae*

Within these groups and sections the various Sub-families and genera are classified according to the arrangement of spores on the conidiophores, whether solitary or in heads, produced directly on the conidiophores or on specialized branches, arranged in clusters or in chains, and so on.

In addition to the multitude of species falling within the above scheme there are a number of fungi which are, so far as is known, permanently sterile. They produce mycelium and, in many cases, sclerotia, but no true

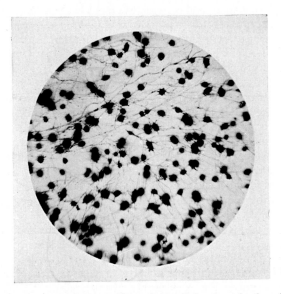

FIG. 42.—*Mycelium sterilium*—irregularly shaped sclerotia
in hyphal web (Petri dish culture). × 20.

reproductive structures or conidia. These have no place in any of the Families or groups founded on spore characteristics and are usually classed apart as *Mycelia sterilia*. Fig. 42 is a photograph of such a species which has been fairly frequently isolated. It forms a web of greyish hyphae, rapidly covering the surface of the medium, with numerous black sclerotia, which are irregularly placed and of various shapes. Purely mycelial forms occur fairly often in miscellaneous cultures, but it should not be hastily assumed that these belong with the *Mycelia sterilia*. Many of them are parasitic fungi, Ascomycetes and Basidiomycetes, which do not form spores on the usual artificial media. In addition, some species of *Fusarium* produce nothing but mycelium when first isolated, and can be recognized only from experience or after cultivation under conditions which will induce normal production of spores.

Saccardo's system is, of course, a key pure and simple, and not in any sense an attempt at a rational classification. This of itself is not a serious matter, since the main purpose of the scheme is to facilitate identifications. There are, however, a number of graver faults which diminish the usefulness of the scheme.

1. A number of the characteristics used as criteria are not as clear cut, in the majority of cases, as such a scheme demands, with the result that it is often necessary to search in more than one section of the key before a fungus can be identified. In particular, the distinctions between Mucedinaceae and Dematiaceae, and between Micronemeae and Macronemeae, are very vague. It must be admitted, of course, that this criticism is bound to apply to some extent to any scheme of classification, since there are no sharp dividing-lines in nature, but Saccardo's scheme is in many ways unnecessarily arbitrary.

2. Since it is obviously impossible for any one man to be directly acquainted with all the known fungi, numerous genera were, of necessity, known to Saccardo only from descriptions, many of which were totally inadequate for characterization. Some of these genera have been intensively studied and more completely defined by modern workers, but there are still a number of common fungi which have never been adequately described, and which are extremely difficult to track down from Saccardo's *Sylloge*, or from keys founded on it.

3. Saccardo used the term "conidium" very loosely, applying it to a number of totally different reproductive structures.

4. Very little account is taken of the fact that many moulds produce conidiophores or spores of more than one kind, one or other predominating under varying environmental conditions. Often the differences are so great as to justify placing a single species in two or more different genera, or even in different Families. For example, the common mould *Penicillium expansum* often forms almost velvety colonies devoid of coremia, and is placed in the Mucedinaceae. On apples, however, the same species usually produces large coremia, and hence should be placed in the Stilbaceae (where actually it is to be found under the name *Coremium glaucum*). The genus *Aspergillus* is placed in the Mucedinaceae, but *A. niger* has black spores and the beginner could hardly be blamed if he sought a description of it amongst the Dematiaceae.

Vuillemin's classification. Of a number of attempts to devise a system of classification more logical than Saccardo's, Vuillemin's system (1910, 1912) is the most important. It is based on differences in methods of spore formation instead of on the characteristics of the spores themselves. It has been worked out most fully for a group of fungi which do not fit at all well into Saccardo's scheme, the pathogenic fungi, and is in general use by medical mycologists. The system has much to commend it and, although we are still a long way from having a complete key to the Hyphomycetales based on Vuillemin's ideas, it is receiving increasing attention from modern authors on mycological subjects.

Vuillemin defines two main types of spores:

1. Thallospores—formed by transformation of pre-existing elements of the thallus (i.e. of vegetative hyphae) and, in general, not very readily detached from the thallus.
2. Conidiospores—external to the thallus, arising on the thallus as newly formed elements, terminal or lateral on the hyphae and essentially deciduous. They are separated from the hyphae which bear them by the action of the fungus itself.

Thallospores are of five different types:

1. Arthrospores (Gr. *arthron*, a joint)—formed by fragmentation of mycelium. A typical example is *Geotrichum candidum* (p. 107).
2. Blastospores (Gr. *blastos*, a bud)—arise as buds from pre-existing cells. They may drop off and themselves form new buds freely, or may remain *in situ* and bud repeatedly to form pseudo-mycelium. Examples are *Torulopsis* (non-sporing yeasts) and *Aureobasidium* (p. 90).
3. Dictyospores (Gr. *dictyon*, a network)—multicellular spores, having both cross and longitudinal septa. All the fungi producing thallospores of this type belong to Saccardo's Family Dematiaceae. A typical genus is *Alternaria* (p. 90).
4. Chlamydospores—large spores, terminal or intercalary, with thick walls and dense contents. Found in all groups of fungi.
5. Aleuriospores (Gr. *aleuron*, flour, farina. The spores, when liberated, form a powdery mass)—in method of formation resemble small chlamydospores, but are always terminal on the hyphae which bear them and resemble conidia in situation, colour, and form. They are not true conidia because they are not shed naturally, and are formed by condensation of protoplasm (like true chlamydospores). They are mostly isolated, rarely forming chains, and are attached to the parent hypha by a surface equal to that of a section of the hypha. Examples are *Trichothecium* (p. 131) and *Sepedonium* (an orange-coloured mould parasitic on agarics).

Conidiospores. The fungi which Vuillemin places in the Family Conidiosporae are divided into three groups:

1. No definite conidiophores. Conidia arise at any points on the hyphae, being borne on small projections or denticules. Example, *Sporothrix* (p. 123).

2. Conidiophores definite but not terminated by phialides. *Botrytis* (p. 93).

3. Spores borne on phialides. The phialide is a terminal cell of a hypha or branch, more or less swollen at the base and tapering to a tip of small diameter. From the extremity, or from the interior, the conidia arise. They are always thin-walled and separated as soon as formed. Examples are *Penicillium, Aspergillus, Cephalosporium,* and *Verticillium.*

Mason, in a series of publications (1933, 1937, 1941), has discussed the merits of various schemes of classification and has clarified the meanings of a number of terms used by Vuillemin and others. He has suggested a few new terms, for use in the classification of spore types, the most important of which is "radula spores" (Lat. *radula,* a scraper). These are true conidia, borne on tiny peg-like projections from a cell. The little pegs somewhat resemble the sterigmata on the basidia of Basidiomycetes, but are indefinite in number. Many of the fungi included in Vuillemin's first two groups of the Conidiosporae produce radula spores. The common mould *Botrytis cinerea* shows this type of spore production particularly well.

Mason's main thesis, however, is that, since the conidium is obviously a means of propagation of the fungus, the mechanisms designed for dispersal of conidia are of the greatest importance for classification. The great majority of species can be broadly divided into two groups, one including all species with dry spores which are normally dispersed by wind, and the other including the species with slimy spores, normally dispersed by water. The idea has not been worked out to the extent of providing a key to all the known genera, but it has been adopted by Wakefield and Bisby (1941) for a list of Hyphomycetes recorded for Britain. The two main divisions are designated Gloiosporae, with slimy spores, and Xerosporae, with dry spores. Separation into Mucedinaceae and Dematiaceae is abandoned, but Saccardo's spore sections are retained. It is frankly recognized that in this, as in any other arrangement, there are a number of intermediate forms which defy exact classification. However, the great majority of common moulds may readily be assigned to one or other of the two groups, and this scheme of classification is adopted in the key to the genera of moulds in the next chapter.

Hughes (1953) has published a scheme for classification of the Hyphomycetes which may be regarded as an extension of the conceptions of Vuillemin and of Mason, the emphasis being on methods of spore-production. The genera which are considered are divided into 8 main sections, and it must be admitted that some of Hughes' conceptions are far from easy to grasp. Also, the examples chosen to illustrate the various sections include very many fungi which are specialized saprophytes on decaying parts of plants, and many common moulds are omitted.

Tubaki (1958, 1963) has somewhat modified Hughes' grouping, and has added a ninth section. However, in a later paper he has adopted what is really a slight extension of the schemes of Vuillemin and Mason.

The most important of the groups from the point of view of the industrial worker is Section III, in which the conidia are similar to terminal chlamydospores (i.e. aleuriospores). Hughes has shown that the spore-bearing cell, after producing the first spore, may elongate *through* the scar, to produce a second spore, and so on, the spores eventually forming chains. The successive scars on the conidiophore show as annellations; hence Hughes proposes the term "annellophore" for this type of spore-producing cell. Annellophores are produced by two genera described in the next chapter, *Scopulariopsis* and *Doratomyces*. In these genera the spore-bearing cells were previously regarded as phialides, but it has long been recognized that their shape differs from that of typical phialides, and they are far more variable in length.

Annellations are not easy to see, even with the best optical equipment. However, if conidia are found to be truncate (i.e. appearing as if cut off at the base) they are likely to be aleuriospores. If, in addition, the spores occur in chains, and particularly if the length of the spore-bearing cell is very variable, there is every probability that the latter are annellophores.

ACTINOMYCETALES
(Gr. *aktis, aktinos*, a ray. From the growth habit. Commonly called Ray-fungi.)

This group of organisms is a puzzling one for the taxonomist. At one end of the series are species which are anaerobic, or microaerophilic, which are parasites of warm-blooded animals, including man, and which, in their colony characteristics, staining reactions, and microscopic appearance, are obviously closely related to the bacteria. At the other end the species are strictly aerobic, form true mycelium and aerial chains of spores, and are just as obviously related to the fungi. Many of the latter are common in soil, playing an important part in maintaining its fertility, whilst a few species are agents of plant diseases, usually known as "scab".

An interesting paper by Hesseltine (1960) discusses at length the relationships of the actinomycetes. He states: "Although related both to fungi and to bacteria, especially the propionic bacteria, the actinomycetes would appear to be a separate phylogenetic line." Also "From a practical standpoint we are certain, however, that actinomycetes must be isolated, grown and studied more as if they were fungi than bacteria."

The modern classification which is most widely used is due to Waksman and Henrici (1943, 1948) and Waksman (1967). The Order is divided into 4 Families as follows:

I. Mycelium rudimentary or absent; no spores formed
Family 1. Mycobacteriaceae.

II. True mycelium produced.
   A. Spores formed, but not in sporangia.
      (1) Spores formed by fragmentation of mycelium
           Family 2. Actinomycetaceae.
      (2) Vegetative mycelium mostly remains undivided
           Family 3. Streptomycetaceae.
   B. Spores in sporangia
           Family 4. Actinoplanaceae.

The only Family which the student of moulds needs to consider is the Streptomycetaceae. It includes 3 genera, thus:

I. Conidia produced in aerial hyphae, in chains .   *Streptomyces*
II. Conidia produced terminally and singly on short
    sporophores
    A. No growth at 50–65°C   .   .   .   .   *Micromonospora*
    B. Growth at 50–65°C .   .   .   .   .   *Thermoactinomyces*

Most of the species which appear when samples of soil are plated out belong to the genus *Streptomyces*. More rarely species of *Micromonospora* are isolated.

The genus *Streptomyces* has assumed great importance in recent times, because all the successful antibiotics, except penicillin, have been obtained from species of this genus. As a result, thousands of strains are being isolated in various laboratories, and screened for production of new antibiotics.

Species of *Streptomyces* are mostly sensitive to acid. A few colonies usually appear when soil is plated on Czapek agar made up with KH_2PO_4, which has a pH of approximately 4·5. If the isolation of species of *Streptomyces* is the primary object, it is better to use neutral Czapek, made up with K_2HPO_4. In many laboratories various antibacterial substances, and in some cases anti-mould substances also, are added to the isolation medium.

Streptomyces colonies grow slowly, and are usually tough and leathery at first, becoming powdery as spores are formed. They are variously coloured, white, grey, buff, lilac, yellow, greenish, red, brown, or almost black. Some species produce variously coloured soluble pigments, which diffuse into the medium, and such have diagnostic value when identifying species, but, for maximal production of such pigments, special media are required. Most of the common species have a distinct and pronounced odour, not unlike that of damp soil, but often more sour and unpleasant.

The aerial chains of spores may be straight, flexuous, loosely coiled, or in tight spirals, and may be unbranched, or branched in verticils or irregularly. Fig. 43 shows a species which produces spirals, as seen in a living culture. Fig. 44 is taken from a slide-culture, mounted in lactophenol.

It is natural that, in view of the increasing economic importance of these organisms, a good deal of attention has been paid to principles and methods of classification. This is all the more necessary because the tendency has been, when a new antibiotic has been isolated, to give the organism a new

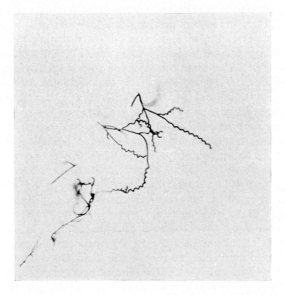

FIG. 43.—*Streptomyces* sp.—conidiophores with spiral
chains of conidia (Petri dish culture). × 200.

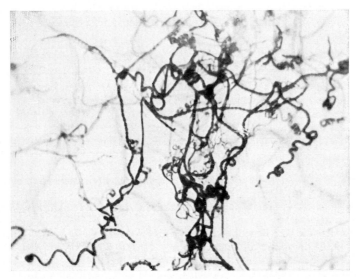

FIG. 44.—*Streptomyces* sp. with spiral conidiophores. × 1000.

name. The classification in Bergey's Manual, by Waksman and Henrici, is based very largely on biochemical characteristics, just as that of the bacteria is, but a number of investigators have advanced the view that the only satisfactory classification will have to be based mainly, or entirely, on morphology, as is the classification of the true fungi (Hesseltine *et al.*, 1954; Pridham *et al.*, 1958; Ettlinger *et al.*, 1958a and b; Round Table Conference, 1959; Bissett, 1959). Until fairly recently the number of morphological criteria used in descriptions of *Streptomyces* species was not sufficient for such an ideal classification. However, recent studies have shown that there is considerable variety in mode of branching of the sporophores, and electron microscope studies have shown that the spores of different species vary, not only in shape, but also in surface ornamentation, being respectively smooth, spiny, or hairy. Ettlinger *et al.* (1958b) give a key to species based on morphological criteria plus a single physiological characteristic, melanin production.

At a Round Table Conference on *Streptomyces*, held in Stockholm in Aug. 1958, in connection with the 7th International Congress of Microbiology, most of the leading workers on the Actinomycetales made contributions, which were published in the Proceedings of the Congress in 1959.

Co-operative research continues in an effort to sort out the confusion which exists in the taxonomy and nomenclature of the genus *Streptomyces*, and more than 40 laboratories are collaborating on an international basis. One of the most recent publications of this group is that of Shirling and Gottlieb (1968).

In the meantime Hütter (1961, 1962a and b) has published a series of interesting papers on the taxonomy of the group, and has used, almost exclusively, morphological characteristics to separate species.

REFERENCES

Bissett, K. A. (1959). The morphology and natural relationships of the saprophytic Actinomycetes. *Progr. industr. Microbiol.*, **1**, 31–43.

Ettlinger, L., Corbaz, R., and Hütter, R. (1958a). Zur Arteinteilung der Gattung *Streptomyces* Waksman et Henrici. *Experientia*, **14**, 334–5.

Ettlinger, L., Corbaz, R., and Hütter, R. (1958b). Zur Systematik der Actinomyceten, 4. Eine Arteinteilung der Gattung *Streptomyces* Waksman et Henrici. *Arch. Mikribiol.*, **31**, 326–58.

Grove, W. B. (1935–7). *British stem- and leaf-fungi (Coelomycetes)*. 2 vols. Camb. Univ. Press.

Hesseltine, C. W. (1960). Relationships of the Actinomycetales. *Mycologia*, **52**, 460–74.

Hesseltine, C. W., Benedict, R. G., and Pridham, T. G. (1954). Useful criteria for species differentiation in the genus *Streptomyces*. *Ann. N.Y. Acad. Sci.*, **60**, 136–51.

Hughes, S. J. (1953). Conidiophores, conidia and classification. *Canad. J. Bot.*, **31**, 577–659.

Hütter, R. (1961). Zur Systematik der Actinomyceten, 5. Die Art *Streptomyces albus* (Rossi-Doria emend. Kranisky) Waksman et Henrici, 1943. *Arch. Mikrobiol.* **38**, 367–83.

Hütter, R. (1962*a*). *Ibid.*, 7. Streptomyceten mit blauen, blaugrünen und grünen Luftmycel. *Arch. Mikrobiol.*, **43**, 23–49.

Hütter, R. (1962*b*). *Ibid.*, 8. Quirlbildene Streptomyceten. *Arch. Mikrobiol.*, **43**, 365–91.

Mason, E. W. (1933). Annotated account of fungi received at the Imperial Mycological Institute. List II (Fascicle 2). *Mycol. Pap.*, No. 3.

Mason, E. W. (1937). *Ibid.* List II (Fascicle 3—General part). *Mycol. Pap.*, No. 4.

Mason, E. W. (1941). *Ibid.* List II (Fascicle 3—Special part). *Mycol. Pap.*, No. 5.

Pridham, T. G., Hesseltine, C. W., and Benedict, R. G. (1958). A guide to the classification of Streptomycetes, according to selected groups. *Appl. Microbiol.*, **6**, 52–79.

Saccardo, P. A. (1880). Conspectus gererum fungorum italiae inferiorum nempe ad Sphaeropsideas, Melanconieas et Hyphomycetas pertinentium, systemate sporologico dispositorum. *Michelia*, **2**, 1–38.

Saccardo, P. A. (1884). *Sylloge fungorum omnium hucusque cognitorum*. Vol. 3. Pavia, Italy.

Saccardo, P. A. (1886). *Ibid.*, Vol. 4.

Shirling, E. B., and Gottlieb, D. (1968). Co-operative description of type cultures of Streptomyces. II Species descriptions from first study. *Internat. J. Syst. Bact.*, **18**,69-189.

Tubaki, K. (1958). Studies on the Japanese Hyphomycetes. V. Leaf and stem group, with a discussion of the classification of Hyphomycetes and their perfect stages. *J. Hattori bot. Lab.*, No. 20, 142–244.

Tubaki, K. (1963). Taxonomic study of Hyphomycetes. *Ann. Rep. Inst. Ferm. Osaka* (1961–62), 25–54.

Vuillemin, P. (1910). Matériaux pour une classification rationelle des Fungi Imperfecti. *C.R. Acad. Sci., Paris*, **150**, 882–4.

Vuillemin, P. (1912). *Les champignons. Essai de classification*. Paris: O. Doin et Fils.

Wakefield, E. M. and Bisby, G. R. (1941). List of Hyphomycetes recorded for Britain. *Trans. Brit. mycol. Soc.*, **25**, 49–126.

Waksman, S. A. (1967). *The Actinomycetes, a summary of current knowledge*. New York: The Ronald Press Co.

Waksman, S. A. and Henrici, A. T. (1943). The nomenclature and classification of the Actinomycetes. *J. Bact.*, **46**, 337–41.

Waksman, S. A., and Henrici, A. T. (1948). Streptomycetaceae, in *Bergey's Manual of determinative bacteriology*. 6th Ed. London: Baillière, Tindall & Cox.

Chapter VIII

Hyphomycetales

You will find a fungus and determine its characteristics. You turn to the books and decide on its genus. Then you look for the species. And you look and you look, and after a while you find it! In another genus!

J. J. Davis, *fide* Gilman, *Mycologia*, 1953.

As indicated in Chapter VII, the Order of Hyphomycetales includes all the Fungi Imperfecti which do not form pycnidia or dark-coloured stromata bearing short conidiophores, but which produce conidiophores irregularly on any part of the mycelium. The great majority of the moulds which are of industrial importance belong here.

The total number of generic names to be found in the literature is over 600 (with some 4000 species), but it is probable that many of these will sooner or later be shown to be synonyms. Many genera have been founded on insufficient data, obtained from study of single specimens in their natural habitats. In most of such cases descriptions have been hopelessly inadequate and have not taken into account the whole life histories of the fungi. The result has been that some of the commoner moulds are known by tradition and are difficult to recognize from published descriptions, whilst some genera exist only as names and it is impossible at the present time to know to what fungi these should refer.

The simple key given here takes into account only the more common forms, such as are fairly often found in industrial work. For identifications of fungi which are not included, and any student of moulds is likely to find rare species on occasion, the reader should consult one of the works of reference cited in Chapter XVII.

The key is based primarily on Mason's scheme of classification (see p. 80), as it is considered that this scheme is easier to use than is Saccardo's, with its division into four Families, based on characteristics which are seldom sharply defined. In particular the distinction between Mucedinaceae and Dematiaceae is often puzzling to the beginner, because some species of the former have dark spores, whilst many of the Dematiaceae have pale or hyaline mycelium, and many others are of a colour which is difficult to class as pale or dark. As stated earlier, two genera belonging respectively to the Sphaeropsidales and Melanconiales are here included amongst the Hyphomycetales. The key also includes, for the convenience of the reader, the Classes of fungi which have been considered in previous chapters. The descriptions of individual genera, which follow the key, are placed in alphabetical order for the sake of easy reference.

KEY TO THE MORE COMMON GENERA OF MOULDS

1. Mycelium only produced. No spores . . . . . *Mycelia sterilia* and some BASIDIOMYCETES

   Spores produced 2
2. Spores in closed receptacles 3
   Spores not in closed receptacles 7
3. Spores in sporangia. Mycelium usually coarse and non-septate . . MUCORALES (Chap. IV)
   Spores not in sporangia. Mycelium typically septate 4
4. Spores in small, thin-walled, globose or ovate, naked asci, each ascus containing 4 or 8 spores . . *Endomyces* and *Endomycopsis* (Chap. VI) *Byssochlamys* (Chap. V)

   Spores in comparatively large globose or flask-shaped receptacles 5
5. Receptacles globose, containing numerous globose or ovate asci . . *Aspergillus* (Chap. IX) *Penicillium* (Chap. X)

   Receptacles globose, stalked; asci not apparent . . . . *Monascus* (Chap. V)
   Receptacles more or less flask-shaped, pale or dark 6
6. Receptacles containing club-shaped or cylindrical asci . . . PYRENOMYCETES (Chap. V)
   Receptacles not containing asci; exuding an irregular mass of spores when squashed . . . . SPHAEROPSIDALES *Phoma* (p. 119)

7. No true mycelium produced; cells reproducing by budding or binary fission . . . . . YEASTS (Chap. VI)
   True mycelium present 8
8. Spores occurring in slimy or waxy masses, in balls on conidiophores, or in irregular clumps . . GLOIOSPORAE . 9
   Spores dry, borne singly, in chains, or in clusters . . . . XEROSPORAE . . . 17
9. Spores without septa 11
   Spores with one or more septa 10
10. Spores pale-coloured, sickle-shaped, borne in irregular masses or in sporodochia. (1-celled microconidia often present in addition) . *Fusarium* (p. 102)
    Spores dark, with 3–5 cross septa, and with a terminal crest of colourless hair-like appendages . *Pestalotia* (p. 116)

D

11. Spores formed by fragmentation of mycelium (arthrospores); colonies soft and yeast-like when young . *Geotrichum* (p. 107)

Spores borne on all parts of the mycelium (blastospores); colonies pale and slimy when young, becoming dark and leathery . . *Aureobasidium* (p. 90)

Spores borne on more or less definite conidiophores 12

12. Conidiophores much elongated phialides, arising singly from all parts of trailing hyphae; spores in balls 13

Conidiophores branched, at least at the tip, or bearing whorls of phialides 14

13. Spores colourless; colonies white or pale-coloured . . . *Cephalosporium* (p. 93)

Spores dark; colonies pale at first then dark . . . . . *Gliomastix* (p. 108)

14. Spores colourless, pale, or brightly coloured 15

Spores pale brown, borne on phialides; conidiophores dark-coloured, branched at tip . . *Phialophora* (p. 118)

Spores dark-coloured, borne on phialides with rounded tips . . *Stachybotrys* (p. 127)

15. Spores at first in chains, later forming slimy masses; conidiophores several times branched at tip . *Gliocladium* (p. 175)

Spores from the first in small balls 16

16. Branching of conidiophores irregular; colonies bright green . *Trichoderma* (p. 131)

Branching in more or less regular verticils; colonies variously coloured *Verticillium* (p. 133)

17. No true spores formed; propagation by bulbils (looking like many-celled spores or small sclerotia) formed on short lateral branches of hyphae *Papulaspora* (p. 116)

Spores, readily recognized as such, regularly produced 18

18. Spores 1-celled 19

Spores regularly 2-celled 30

Spores with more than 2 cells; dark-coloured 31

19. Spores endogenous but apparently formed by fragmentation of aerial hyphae . . . . *Sporendonema* (p. 123)

Spores distinct from mycelium 20

20. Spores (aleuriospores) borne singly and terminally on irregularly branched sporophores . *Sporotrichum* (p. 125)

Spores (radula spores) borne singly on short and very slender projections from near ends of short hyphal branches . . . *Sporothrix* (p. 123)

Spores borne in clusters or in chains 21
21. Conidiophores compacted into definite coremia 22
Conidiophores detached or, at most,
in small inconspicuous fascicles 23
22. Coremia with green fertile heads . *Penicillium* (in part)
= *Coremium* (p. 201)

Coremia white, fertile along whole
length . . . . *Isaria* (p. 111)
Coremia dark, with definite stalk and
head . . . . . *Doratomyces* (p. 100)
23. Spores (radula spores) in grape-like
clusters; colonies greyish; dark
sclerotia usually present . . *Botrytis* (p. 93)
Spores in definite chains, simple or
branched 24
Spores in terminal clusters, dark,
elongate 31
24. Spores (arthrospores) formed by
break-up of terminal branches of
tree-like conidiophores; all parts
very small . . . . . *Oidiodendron* (p. 114)
Spores multiplying by budding,
forming branched chains 25
Spores forming unbranched chains 26
25. Colonies mostly pale-coloured; spores
colourless or pale; arthrospores in
addition in old cultures . . *Monilia* (p. 114)
Colonies dark green or grey-green;
spores forming dense tree-like heads *Cladosporium* (p. 95)
26. Conidiophores arising from special-
ized foot-cells, usually non-septate, terminating in a swelling
which bears phialides . . . *Aspergillus* (p. 137)
Conidiophores not from foot-cells,
septate, bearing phialides or annellophores 27
27. Spore-bearing cells annellophores, of
indefinite length; spores with
thickened basal ring and pore . *Scopulariopsis* (p. 120)
Spore-bearing cells phialides 28
28. Phialides in definite broom-like
whorls 29
Phialides sometimes in whorls, sometimes disposed irregularly, terminating in long slender tips; colonies
pale coloured, but never green . *Paecilomyces* (p. 177)
29. Phialides thick, rounded at tip;
conidia black . . . . *Stachybotrys* (p. 127)
Phialides comparatively slender, constricted at tip . . . . *Penicillium* (p. 179)
and young stages of *Gliocladium* (p. 175)
30. Spores colourless or pale-coloured
in mass, borne in terminal clusters *Trichothecium* (p. 131)
Spores dark, forming branched

chains and eventually dense heads
(young spores usually 1-celled)   .   *Cladosporium* (p. 95)

31. Spores elongate, with several cross
septa, or not truly septate but
appearing so under low magnification, dark-coloured, produced
singly but finally forming clusters
like bunches of bananas   .   .   *Helminthosporium* (p. 109)

Spores borne similarly, with several
true cross septa, usually bent and
with middle cells thickened .   .   *Curvularia* (p. 98)

Spores with both cross and longitu-
dinal septa   32

32. Spores rounded, not beaked, in small
clusters   .   .   .   .   .   *Stemphylium* (p. 130)

Spores beaked or pointed, in short,
often branched, chains   .   .   *Alternaria* (p. 90)

Alternaria Nees ex Wallroth

(Gr. *alteres*, a kind of dumb-bell)

The taxonomy of the genus is difficult, chiefly because of the lack of stable
characteristics. The spores are not conidia but thallospores, and are liable
to vary widely in size, shape, and degree of septation, even in a single
culture. The variations found when cultures on different media are compared are even larger, the spores on rich media tending to be bigger than
on poor media, but produced less freely (Figs. 47, 48). The genus is distinguished from other genera with muriform spores by the great majority
of the spores having a definite beak, often paler than the body of the spore,
and by the spores being produced in chains. There is a taxonomic treatment of the genus by Neergaard (1945).

Most of the species grow well in culture but in nature are found chiefly
as parasites of cultivated plants. Only one species occurs regularly on
industrial materials.

A. tenuis auct. (Lat. *tenuis*, thin). When growing amongst other
moulds in Petri dishes colonies are thin velvety, almost black, with aerial
growth consisting almost entirely of spore chains. In pure cultures aerial
mycelium is always more freely produced, but, whilst some isolates continue to spore abundantly, others become sterile after only a few transfers.
Spores are produced in chains, straight or branched, of up to a dozen or
more spores, are yellow-brown to dark brown, very irregular in shape and
size, mostly with short but definite beak (Figs. 45, 46, 47). Found on
numerous kinds of organic materials in damp situations.

Aureobasidium Viala et Boyer

(Lat. *aureus*, golden yellow)

A. pullulans (De Bary) Arnaud (Lat. *pullulans*, sprouting), was originally
described by De Bary as *Dematium pullulans*. It was later made the type of

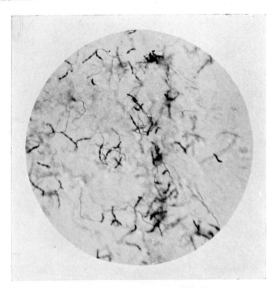

FIG. 45.—*Alternaria tenuis*—spore chains as seen in
dish culture. × 42.

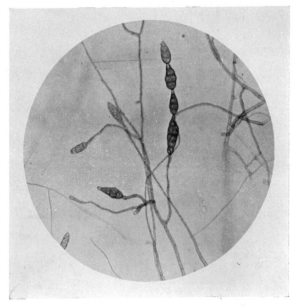

FIG. 46.—*A. tenuis*—short chain of spores (slide culture).
× 250.

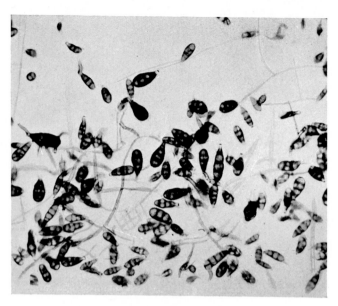

FIG. 47.—*A. tenuis*—spores from culture on poor medium.
×250.

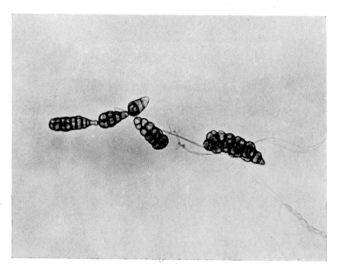

FIG. 48.—*A. tenuis*—spores from culture on rich medium.
×250.

a new genus, and became *Pullularia pullulans* (De Bary) Berkhout. It is so referred to in previous editions of this book. The generic name, *Aureobasidium*, which must be used because of priority, is a most inappropriate one, since the fungus is not golden yellow, and it does not form basidia.

A. pullulans is a fairly common mould, occurring particularly on damp cellulosic materials. Colonies at first dirty white and slimy, somewhat resembling yeast colonies, rapidly turning dirty greenish and eventually becoming black all over and leathery. Young cultures consist of fine colourless hyphae. The appearance of colour is due to the growth of thicker, dark-coloured hyphae, which are twisted, much septate, often thickened at the nodes, and bearing numerous ovate, hyaline conidia as lateral buds. Structure is far from easy to make out, since the hyphae break up and shed the conidia in fluid mounts (Fig. 49).

A good account of the occurrence and distribution of this fungus, including more particularly its role as a spoiler of painted surfaces, is given by Reynolds (1950). Ciferri, Ribaldi, and Corte (1957) give a good account of the history of this fungus, with a long list of synonyms. Cooke (1959) has also published an ecological survey, with an even longer list of synonyms.

Botrytis Persoon ex Fries

(Gr. *botrys*, a bunch of grapes)

The name is very descriptive, for the conidiophores are irregularly branched at the top, and the spores are borne, each on a little peg or sterigma, on the ends of the final branches, and are hence radula spores. Only one species is of frequent occurrence in industry.

B. cinerea Pers. ex Fr. (Lat. *cinereus*, ashy grey). Found on a wide variety of materials, and is also parasitic on many kinds of plants. Colonies floccose, pale brownish grey; conidiophores stiff, erect, much branched at the top, and bearing clusters of spores like bunches of grapes (Fig. 50); conidia ovate, pale greyish, $8-12(15) \times 6-10\mu$; sclerotia beginning to develop after a few days as small, dirty green, mycelial knots, increasing rapidly in size and turning black, eventually irregular in shape, often confluent, and several millimetres long.

Cephalosporium Corda

(Gr. *kephale*, head; hence spores in heads)

The conidiophores are really elongated phialides, arising as branches from all parts of trailing hyphae. Conidia are cut off successively and, owing to secretion of a sticky fluid, are held together in balls (Fig. 51). Spores are very numerous and the heads fall to pieces when touched. Beginners often imagine the compact spore balls seen in a living culture to be single spores. However, if the apparent diameter of the balls, as seen in the culture-tube with a $\frac{2}{3}$-inch objective, is mentally multiplied by four, and the result

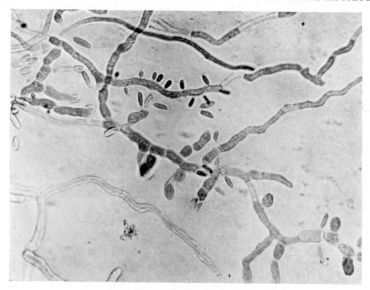

FIG. 49.—*Aureobasidium pullulans*—culture on "Cellophane".
× 500.

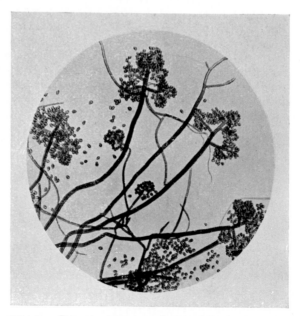

FIG. 50.—*Botrytis cinerea*—typical conidiophores. × 100.

compared with the diameter of the individual spores, as seen on a slide with a $\frac{1}{6}$-inch objective, their real nature becomes apparent.

Colonies are usually white to pinkish, at first thin and often somewhat slimy, but becoming floccose. A number of species have been described, but they are mostly ill-defined and difficult to identify. The most common infections carried by mites are species of this genus and, when introduced

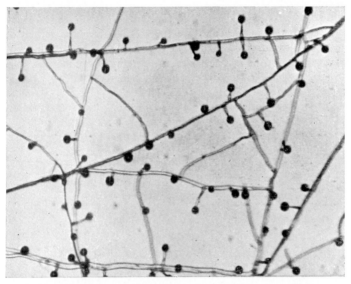

FIG. 51.—*Cephalosporium* sp.—fertile hyphae with small spore-balls, as seen in living culture. ×250.

into cultures of other moulds, are very difficult to eradicate. Some isolates tend to develop ropes of hyphae, and even coremium-like fascicles. Such have been put in a genus *Tilachlidium*, but there seems to be little reason for segregating them from the genus *Cephalosporium*.

Cladosporium Link ex Fries

(Gr. *klados*, a branch; hence branched spore chains)

The number of specific epithets which have been bestowed in this genus is very large, but the number of good species is very small. The common *C. herbarum* has, in the past, received as many different names as the plants on which it has been found. However, it is not a parasite but grows on the honeydew secreted by aphides, and hence is found on almost every plant attacked by these creatures. The most recent and best taxonomic study of the genus is by de Vries (1952).

Originally the name *Cladosporium* was supposed to be restricted to forms with septate spores only, and *Hormodendrum* was the name given to the forms with non-septate spores, but it is now recognized that almost all strains produce a high percentage of 1-celled spores in the early stages of growth, and that there are very few strains which do not produce some septate spores. This point is mentioned because the name *Hormodendrum* (often spelled incorrectly as *Hormodendron*) occurs frequently in the literature.

C. herbarum (Pers.) Link ex Fries (Lat. *herbarum*, of plants). An exceedingly common organism, being found on textiles, rubber, leather, paper, and foodstuffs of all kinds. Petri dishes exposed in the open air usually trap hundreds of spores, far more than of any other genus of moulds. It grows over a wide range of temperatures, and has been reported fairly frequently as infecting meat in cold storage.

In culture this species is easy to recognize. It grows somewhat restrictedly, thick velvety, in colour varying from deep rich green to dark grey-green, with reverse a characteristic opalescent blue-black or greenish black.

FIG. 52.—*Cladosporium herbarum*—spore heads as seen in
dish culture. × 50.

Examination of living cultures, under a low magnification, shows the spores to occur in large, tree-like clusters (Fig. 52). The sporing structures are very brittle, breaking up completely, when mounted in fluid, into spores and rod-like fragments of sporophores and mycelium. Under the micro-scope all parts of the fungus are dark-coloured, greenish brown to dark brown. Examination of young cultures, or better still slide cultures, will show that the first-formed spores increase by budding, in a manner re-miniscent of the yeasts (Fig. 53), eventually forming the tree-like masses of much-branched chains. Young spores are mostly 1-celled, older ones 2-celled or even 3-celled.

C. resinae (Lindau) de Vries (Lat. *resinae*, of resin) is best known as an inhabitant of wood impregnated with creosote or coal-tar, both of which it

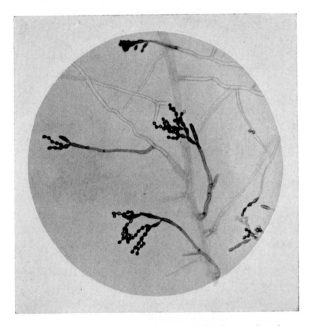

FIG. 53.—*C. herbarum*—young conidiophores showing budding of spores (slide culture). × 250.

can utilize as sources of carbon. Although it grows well on laboratory cul-ture media it is remarkable that, on creosoted timber, the growth stops short on reaching untreated wood. During recent years the fungus has become a nuisance in the petroleum industry, and in the fuel tanks of air-craft. Most storage tanks contain a small amount of water under the fuel. The fungus grows at the interface, utilizing water from below and carbon compounds from above. The tangled masses of mycelium which are formed may block supply pipes, control valves etc., and metabolites liberated may cause corrosion. See Hendy (1964).

The normal form of the mould produces thick, matted, brown colonies.

There is also a pure white form. Full descriptions are given by Christensen *et al.* (1942) and by Marsden (1954).

Fig. 54 shows a spore head produced by growing a few spores in a thin

FIG. 54.—*Cladosporium resinae* growing in paraffin. × 500.

layer of medicinal liquid paraffin under a cover-glass, where the amounts of water and air are extremely small.

C. fulvum Cooke (Lat. *fulvus*, tawny). This species occurs mainly as a leaf parasite of tomatoes in glasshouses. On laboratory media it grows very slowly, forming colonies which are matted, floccose, and usually a mixture of tawny rouge and purple.

Ciferri has put forward the view that this species does not really belong in *Cladosporium*, and has made it the type of a new genus *Fulvia*.

Curvularia Boedijn

(Name refers to shape of spores)

Both mycelium and spores are dark-coloured, the latter borne on more or less erect, septate sporophores. The tip of the sporophore develops one spore, then proceeds to grow out from just below the spore, forms a new terminal spore and pushes the first spore on one side. This process goes on until a cluster is formed. The conidiophore thus forms a typical sympodula (see Chap. II). The cluster may be tight, as in Fig. 55, or more open (Fig. 56), but always the sporophore from which spores have been shed appears

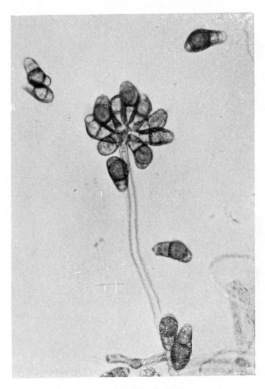

FIG. 55.—*Curvularia lunata*—sporophore and "ear" of spores. × 500.

knobby or tortuously bent along the spore-bearing part. The spores have 3 or 4 cross septa, and are mostly curved or bent about the third cell from the base, which cell is broader and darker in colour than the others. A number of species are described by Boedijn (1933), the two mentioned below being the most important, and both of frequent occurrence. All the species are difficult to maintain in good condition in culture, many isolates becoming sterile after one or two transfers.

C. lunata (Wakker) Boedijn (Lat. *lunatus*, like the crescent moon) perfect state *Cochliobolus lunatus*. Sporophores usually short, up to about

100μ long and 2–4μ diam., septate; spores 3-septate, bent 20–30 × 8–16μ, mostly 23 × 11μ.

C. geniculata (Tracy and Earle) Boedijn (Lat. *geniculatus*, knotty) perfect state *Cochliobolus geniculatus* Sporophores fairly long, 300–900μ,

FIG. 56.—*Curvularia* sp.—typical sympodula. × 500.

widening upwards and 3·5–5μ diam. at tip; spores forming dense clusters, 4-septate, with thick cell much swollen and very dark, 20–45 × 7–14μ, mostly 24 × 9μ.

Doratomyces Corda

(Gr. *doratos*, gen. of *dory*, a spear)

The genus *Stysanus* Corda, described in previous editions of this book, has been shown, by Morton and Smith (1963), to be synonymous with *Doratomyces* Corda. This was described earlier than *Stysanus*, and the name, therefore, has priority.

D. stemonitis (Pers. ex Fr.) Morton and G. Smith (*Stemonitis*, a genus of Mycetozoa with similar fructifications) is not uncommon, particularly in soil. Colonies a distinctive dark sooty grey, consisting of a mixture of simple fascicles and definite coremia (Fig. 57); coremia consisting of a stalk, which is a tight bundle of dark, septate hypae, and a head, where the conidiophores spread out and terminate in penicillate structures showing a marked resemblance to those of *Scopulariopsis* (*cf.* Figs. 58, 79 and 80). The conidia, which are borne in chains, are ovate to lemon-shaped

FIG. 57.—*Doratomyces stemonitis*—typical coremium. × 50.

almost colourless to pale brown, about 7μ in long axis. On continued cultivation in the laboratory the production of large coremia is often progressively reduced.

As in *Scopulariopsis* the spore-bearing cells are annellophores, not phialides. A number of authors have expressed the opinion that the two genera should be regarded as one, since the production of coremia has, in recent years, been regarded with little favour as a basis of classification. The situation is, however, complicated by the fact that some isolates of *Doratomyces*,[1]

[1] *Note.* The fungus pictured as *Stysanus stemonitis* in Editions 1–5, Figs. 84 and 85, has been found to be a species of the related genus *Trichurus*. This differs from *Doratomyces* in producing sterile hairs amongst the penicillate structures of the spore head.

with the typical spore form indistinguishable from *D. stemonitis*, produce
in addition a second spore form, which belongs to the genus *Echinobotryum*
Corda (Fig. 59).

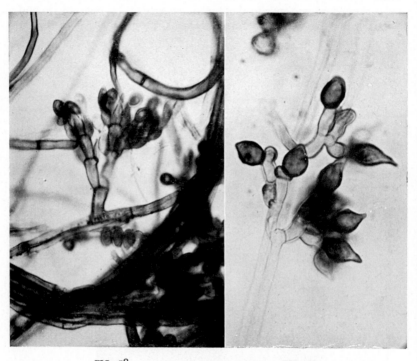

FIG. 58.   FIG. 59.

FIG. 58.—*Doratomyces stemonitis*—dissected coremium showing penicillate
sporing structures. × 1000.

FIG. 59.—*Echinobotryum atrum*, the second spore form of some strains of
Doratomyces. × 1000.

Fusarium Link ex Fries

(Lat. *fusus*, a spindle; describes shape of spores)

This is an extremely difficult genus for the taxonomist. Not only is the
number of species very large, but most species are difficult to maintain in
a state of stability. Wollenweber, who has done more than any other worker
to clear up the *Fusarium* problem, has stated that species pass through
certain well-defined stages when kept in cultivation. Under favourable
conditions they normally pass from a predominantly mycelial stage to a
freely sporing stage, and thence to a degenerate stage with spores abnormal
and sparingly produced. Another view is that the whole question of the

changes which take place can be explained by the readiness with which many species produce mutants, the relative vigour of these determining the type of growth propagated if no particular care is taken to separate the forms when sub-culturing. The latter opinion is supported by the fact, observed by many workers, that transfer of spores will usually produce sporing cultures, whilst transfer of mycelium will tend to retard sporulation and will eventually give completely sterile cultures. For culture of Fusaria so as to retain desired characteristics as far as possible Wollenweber *et al.* (1925) recommend various media made from vegetable materials. Brown (1925), however, prefers a purely synthetic medium containing: glucose, 2 g; asparagine, 2 g; K_3PO_4, 1·25 g; $MgSO_4.7H_2O$, 0·75 g; agar, 15 g; water to 1 litre. The only disadvantage of this is that, owing to its small content of solids, it dries out quickly and shrinks to a mere streak on the glass of the culture tube if cultures are kept for more than a few weeks. The best of the common media, for maintenance of stock cultures, are potato agar, carrot extract agar, and a mixture of potato and carrot extracts with agar.

Many species of *Fusarium*, when first isolated, produce mycelium alone, and it is often a matter of some difficulty to induce production of normal macroconidia for purposes of identification. Some will spore when the cultures become old and dry; others respond to a starvation diet such as plain tap-water agar; whilst some seem to need a particular medium which can only be found by trial. If such mycelial cultures are kept on media containing a fairly high percentage of sugar, such as wort or Czapek agar, they tend, after several sub-culturings, to degenerate to a slimy stage, consisting of colourless, gelatinous masses of anastomosing hyphae, and are then almost impossible to restore to the normal form.

In all systems of classification *Fusarium* is included in the family Tuberculariaceae, but many species produce, instead of sporodochia, smooth, gelatinous layers of spores (pionnotes) and are thus more nearly allied to the Melanconiales, and still others form neither sporodochia nor pionnotes, but produce conidia in scattered clusters on all parts of the aerial mycelium. The one feature common to the different types of cultures is the spore. The typical spores, or macro-conidia, of *Fusarium* are sickle-shaped, with pointed ends, usually with several cross septa, colourless or palecoloured, never dark or dematiaceous, and are quite unlike those of any other mould fungi (Figs. 60, 61, 62). The exact shape, size, and number of septa in the macro-conidia are important criteria used in identification of species. In addition to the macroconidia many species produce microconidia, which are small, ovate, elongate, pyriform, or comma-shaped, usually non-septate, more rarely 1–3-septate. A number of species form chlamydospores, which may be terminal, intercalary, or both, formed in the mycelium or in the macro-spores (Figs. 63, 64). Sclerotia are not uncommon and are often brightly coloured. As in other genera, all such features have diagnostic importance.

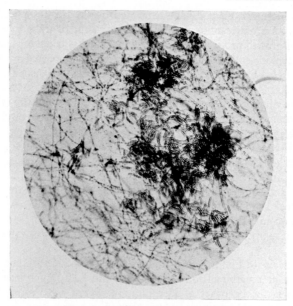

FIG. 60.—*Fusarium* sp.—clusters of conidia as seen in
dish culture. × 100.

FIG. 61.—*Fusarium* sp.—cluster of conidia,
a few containing chlamydospores. × 500.

The perfect stages of a considerable number of species of *Fusarium* are now known. They are all Ascomycetes, and belong to the genera *Hypomyces*, *Gibberella*, *Nectria*, and *Calonectria*.

Most of the taxonomic work on *Fusarium* has been done in connection with plant diseases and there is an extensive literature of such. The indentification of species occurring as parasites is now, since the publication of the monograph by Wollenweber and Reinking (1935), comparatively easy, since it is known that the number of species found on any particular plant is small, and the pathological symptoms are fairly definite for each species.

FIG. 62.—*Fusarium* sp.—typical
conidia. × 1000.

In industrial work, however, it is quite otherwise. Most of the Fusaria grow readily as saprophytes and are found in almost every situation where moulds can grow. Occasional isolates are not difficult to identify with the aid of the Wollenweber and Reinking monograph, but usually identification is far from easy and is really a job for the specialist.

However, Hansen and Snyder, in America, have, over a number of years, made an intensive study of the *Fusarium* problem. They have shown that the range of variation shown by a single species is much greater than has commonly been supposed, and, consequently, that the number of good species is small. Their final classification of the genus has not yet appeared, but a good idea of their concepts can be obtained from a series of papers already published (1941*a*, 1941*b*, 1945, 1954).

In the literature of industrial mycology there are many references to

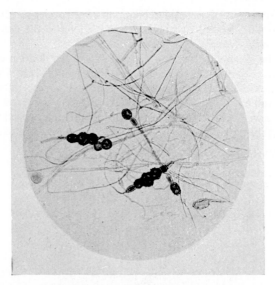

FIG. 63.—*Fusarium* sp.—chlamydospores in mycelium.
× 250.

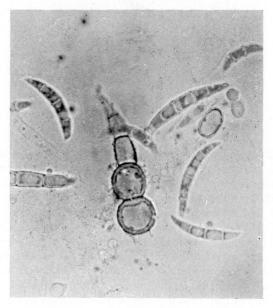

FIG. 64.—*Fusarium* sp.—chlamydospores in conidium.
× 1000.

"*Fusarium* spp.", but very little mention of named species. Undoubtedly quite a number of species are liable to cause trouble in industry, but nothing is known at present as to their relative importance. That being so, descriptions here of individual species would be out of place.

Geotrichum Link ex Pers.

(Gr. *ge*, the earth; *thrix*, hair; hence forming mycelium close to the substrate)

The main characteristic of the genus is that the mycelium breaks up almost completely into arthrospores, i.e. short lengths consisting of single cells, more or less cylindrical in shape.

G. candidum Link ex Pers. (Lat. *candidus*, white). Grows well on wort agar, less well on synthetic media; colonies thin, spreading, creamy white, soft and somewhat yeast-like in texture; in young colonies fairly long mycelial strands present, breaking up into arthrospores as colonies age; arthrospores cylindrical with rounded ends (Fig. 65). This species is common on

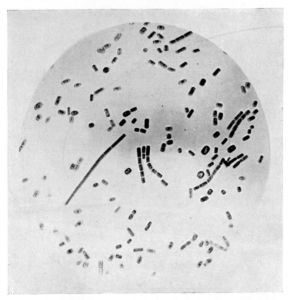

FIG. 65.—*Geotrichum candidum*—fragmenting mycelium and arthrospores. × 250.

all types of milk products. It has been known for a long time as *Oospora lactis* (Fresenius) Sacc., but it is unlike other species of the genus *Oospora* and has been identified as Link's species. It is also common, and sometimes creates quite a problem, in polluted water [Cooke (1954); Tubaki (1962)].

Windisch (1951) claims that he has found asci in many strains of the fungus, and has therefore transferred it to *Endomyces*, as *E. lactis* (Fres.) Windisch. His work, however, requires confirmation.

Carmichael (1957) has made a detailed study of the morphology and occurrence of this fungus, and provided a long synonymy.

Gliomastix Guéguen

(Gr. *gloios*, mucilage; *mastix*, a whip)

G. murorum (Corda) Hughes (Lat. *murorum*, of walls), the only common species of the genus, is frequently found in soil and on soil-contaminated products. Colonies at first white or pale pink, turning black in patches as conidia develop, and eventually black all over; conidiophores merely

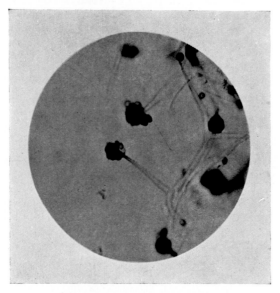

FIG. 66.—*Gliomastix murorum*—conidiospores with spore balls. × 500.

elongated phialides, arising as lateral branches from trailing hyphae or ropes of hyphae; conidia cut off successively from the tips of the phialides, forming, at least during the rapidly growing period, slimy balls, in old cultures forming whip-like twisted chains (hence the name); ripe conidia dark-coloured, ovate, $4–5 \times 3–4\mu$, with the colouring matter aggregated in small granules which tend to become detached in fluid mounts (Fig. 66).

Helminthosporium Link ex Fries

(Gr. *helmins, helminthos,* a worm; hence worm-like spores)

The genus has, until recent times, included a large number of species, the best-known of which are parasites of cereals and grasses. Many of them, however, grow well as saprophytes, and are not uncommon on plant debris. In culture some species grow normally, producing typical dark-coloured hyphae and abundant conidia. Others appear to be quite sterile, and tend to produce mostly hyaline or pale-coloured mycelium. It has been shown by Raistrick and co-workers (see Chapter XV) that certain species of the latter group, when grown on Czapek-Dox solution, synthesize red or purple pigments, which are hydroxyanthraquinones, and are closely related to some of the most valuable fast dyes.

The genus as interpreted by the older monographers, Drechsler (1923), Nisikado (1929), was not homogeneous. It included fungi with two different kinds of spores, and with two different methods of spore production. The type species of the genus, *H. velutinum* Link ex Fr., and some other

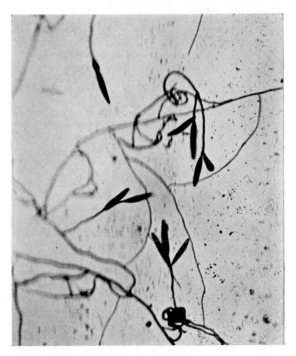

FIG. 67.—*Helminthosporium monoceras*—young conidiophores, showing "successive" formation of spores (from tube culture). × 90.

species, produce spores terminally and also in whorls from lower nodes of the stiff conidiophores. The graminicolous species, and, in addition, all the species likely to be isolated in industrial work, produce spores in quite a different manner. In all the classical keys to the Hyphomycetales *Helmintho-sporium* is described as producing spores singly on the ends of the conidio-phores, whereas examination of a sporing culture will usually show clusters of spores resembling bunches of bananas (Fig. 67). The conidiophore is actually a typical sympodula. The conidia of some of the species of this

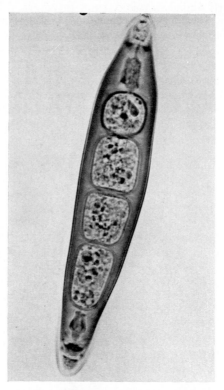

FIG. 68.—*H. monoceras*—spore, showing protoplasts. × 1000.

group are truly septate, whilst the spores of other species have no real septa, but contain a central row of structures, termed "protoplasts" by Stevens (1922), which have apparently no connection with the outer wall of the spore (Fig. 68).

Shoemaker (1959) has split the genus *Helminthosporium* into three. The first group, comprising the *H. velutinum* series, retains the name *Helmin-thosporium*. The species which form sympodula are segregated into two genera. Those which produce germ-tubes from all the cells are assigned to the genus *Drechslera* Ito, whilst those whose spores germinate only at the poles take the new generic name *Bipolaris*.

Although species of *Bipolaris* are fairly frequently isolated from mouldy

FIG. 69.—*Isaria* sp.—culture on agar slope. Natural size.

materials, and occasionally species of *Drechslera* are found, none is of great importance in industrial work.

Shoemaker (1962), in a second paper, has given descriptions of, and complete keys to, all the species of *Drechslera*.

Isaria Persoon ex Fries

(Gr. *is*, a fibre)

Species which have been referred to this genus are best known as parasites of insects, but occasionally typical isarioid colonies are isolated in the laboratory, forming long white coremia, cylindrical, occasionally branched, and fertile along the whole length (Figs. 69, 70).

The fungi which have been assigned to the genus *Isaria* include species with very different methods of spore production. Hence, in the opinion of most present-day mycologists, the genus has no real standing, but merely represents growth forms of species belonging to various better-defined genera. Here are the opinions of two mycologists who have worked with species of *Isaria*. Vuillemin (1911*b*) states: "Le genre *Isaria* n'est pas plus autonome que le genre *Coremium*. L'aggrégation des filaments sporifères

FIG. 70.—*Isaria* sp.—branched coremium. × 12·5.

en colonnes dressées, simple ou ramifiées, chargées de conidies sur toute leur surface ou seulement dans la partie supérieure la plus distante du support humide, peut s'observir dans les cultures de genres dépourvus de toute affinité entre eux."

Petch (1931), after a long study of entomogenous fungi, came to much the same conclusion. He says: "It may be noted that the genus *Isaria* is a heterogeneous mixture of species which have little in common, except that they produce erect fascicles of conidiophores. As regards the entomo-

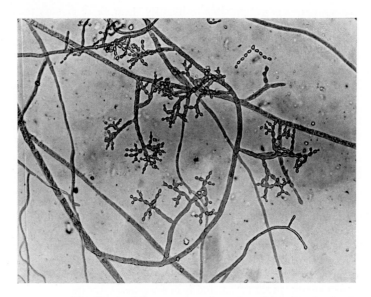

FIG. 71.—*Monilia sitophila*—young culture (on slide) showing budding of spores. × 200.

FIG. 72.—*M. sitophila*—spore masses as seen in culture tube. × 100.

genous species, it is not possible to identify them, or to classify them in
homogeneous groups, from their descriptions, because the actual conidiophore, i.e. the ultimate branch of the fascicle, has, in general, not been
described. It is highly desirable that these species should be classified on
the character of the conidiophore, i.e. not as *Isaria*, but as aggregate forms
of *Oospora, Sporotrichum, Spicaria*, etc."

Monilia Persoon ex Fries

(Lat. *monile*, a necklace of beads)

This is rather a heterogeneous genus, characterized by the production of
long branched chains of bead-like spores, produced in basifugal succession,
i.e. with the youngest spore at the outermost end of the chain. *M. fructigena*
and *M. cinerea*, imperfect states of *Sclerotinia*, are well-known parasites of
tree fruits, forming greyish pustules consisting chiefly of masses of spores,
and producing soft rot.

  M. sitophila (Montagne) Saccardo (Gr. *sitos*, corn; *phileo*, to love).
Best known as the red bread mould. From time to time outbreaks of the
infection occur in bakeries but, more often, the fungus attacks wrapped
and, particularly, sliced bread, and is noticed only after the bread has left
the bakery. In an amazingly short time the bread becomes covered with a
characteristic pink, loose-textured growth. According to Shear and Dodge
(1927) it also attacks sugar-cane bagasse in store, forming long pink festoons
from the bales. It is sometimes found as an aerial contaminant and can then
be a great nuisance in a culture room, since it spreads with extreme rapidity
on suitable media and often sheds spores outside a Petri dish in which it is
growing.

  It grows best on organic culture media, forming loose floccose masses
of a pale pink to salmon-pink colour. Ovate conidia are formed at the ends
of aerial hyphae (Fig. 71); these increase by budding, and eventually form
enormous, irregular masses (Fig. 72). Later the mycelium breaks up to
some extent at the septa, forming arthrospores. Shear and Dodge (1927)
have shown that *M. sitophila* is the imperfect form of an Ascomycete,
Neurospora sitophila, but, since the latter is heterothallic, it is seldom that
perithecia are found. They have also shown that there are several species
of *Neurospora* with very similar conidial states. Of these, *N. tetrasperma*
has asci containing only 4 spores and is homothallic, but single strains of
the others cannot be recognized with certainty unless they are grown in a
series of cultures along with authentic + and − strains of the various
species.

Oidiodendron Robak

(*oidio-*, like *Oidium*; Gr. *dendron*, a tree)

Superficially this genus has the appearance of a miniature *Cladosporium*,
but differs in the method of spore formation. The erect conidiophores

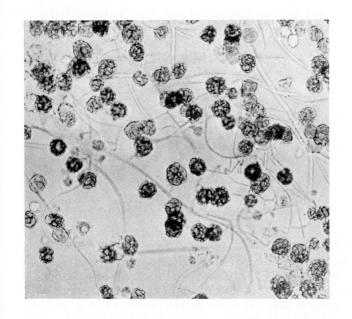

FIG. 74.—*Papulospora coprophila*—bulbils. × 150.

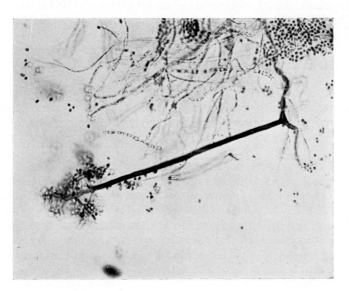

FIG. 73.—*Oidiodendron fuscum*—conidiophore. × 500.

branch freely in a tree-like manner and then the branches gradually split up, from the tips backwards, into spores. All the species are very slow-growing, and hence are easily missed when mouldy materials are plated out. Robak (1932) has described 4 species, all from Norwegian wood-pulp, and one of them, *O. fuscum*, has been isolated from a number of sources in this country. It is probably not uncommon.

O. fuscum Robak (Lat. *fuscus*, dull dark brown). Colonies very slow-growing on all media, brownish grey, powdery, with reverse gradually turning dark brown; all parts of the fungus very small; conidiophores brown, septate, branched at apex, with branches eventually becoming chains of spores; spores irregularly ovate, $1\cdot6-3\cdot6 \times 1\cdot5-2\cdot2\mu$ (Fig. 73).

Papulospora Preuss

(Lat. *papula*, pimple, pustule)

No true spores are formed but reproduction is by structures known as "bulbils". These are multicellular structures, resembling in the young stage the spores of such genera as *Stemphylium*, except that they are mostly paler in colour. The bulbils increase in size until they are somewhat like miniature sclerotia (Fig. 74). Some species have been shown to be the imperfect stages of Ascomycetes or Basidiomycetes, whilst in other cases no such connections are known. Species belonging to this genus are found on dung, in soil, on rotting plant debris, and on paper. One species, *P. byssina* Hotson, is of importance as the brown plaster mould of commercial mushroom beds. A large number of species are known, for further details of which, and for keys, the reader is referred to a number of papers by Hotson (1912, 1917, 1929, 1942).

Pestalotia de Notaris

(F. Pestalozzi, Italian mycologist)

The name of the genus is often spelt *Pestalozzia*, but this is not the original spelling, and it is not in accordance with the usual method of latinizing names.

Some species of the genus are weak parasites of plants, whilst several have been found as causes of damage to materials made of cellulose.

The genus is easy to identify since the spores are quite characteristic (Fig. 75). They have 3–5, mostly 4, cross septa, with the terminal cells colourless or nearly so and the median cells dark-coloured, often with the lowest median cell paler than the others. The lower end of the spore bears a hair-like stalk or pedicel, and the upper end is furnished with a crest of 2–5 long colourless hairs. The appendages are not easily seen in lacto-phenol mounts, but show up well in water.

In culture the mycelium is usually submerged or superficial, seldom aerial, becoming thickly dotted with small, black, cushion-shaped pustules consisting chiefly of masses of spores. Some species retain this appearance in successive sub-cultures, but others rapidly degenerate and become sterile, producing only matted growths of whitish or yellowish mycelium.

Determination of species is not easy. The criteria are: shape and size of

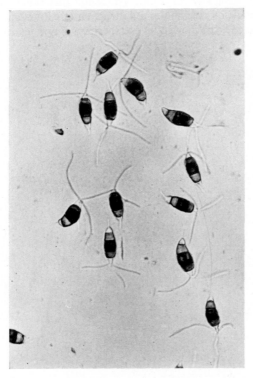

FIG. 75.—*Pestalotia gracilis*—spores. × 500.

the pustules; shape, size, and septation of the spores; range of colour in the median cells; and number and dimensions of the terminal hairs. For a long time the standard taxonomic treatment of the genus was that of Guba (1929, 1932). More recently Steyaert (1949, 1953) has maintained that the genus *Pestalotia* includes one species only, and has transferred all the remainder, including all the common saprophytes, to two new genera, *Truncatella* with 4-celled conidia, and *Pestalotiopsis* with 5-celled conidia. However, Servazzi (1954) gives a number of reasons for retaining the name *Pestalotia* to include all the three genera of Steyaert, and his views have been supported by Guba (1955). Steyaert has replied to both mycologists (1955, 1956) and, at present, the matter remains *sub judice*.

In the meantime Guba (1961) has published a second monographic treatment of the genus. In the opinion of most mycologists who have occasion to study the genus, Guba's treatment is somewhat disappointing. He has erected far too many new species, many of them separated on the most trivial grounds.

During the second world war it was found that some species of the genus are capable of causing rapid destruction of cotton and jute fabrics, but, unfortunately, the identity of the species was not determined.

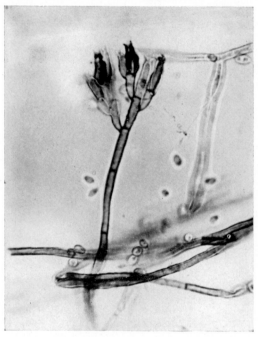

FIG. 76.—*Phialophora melinii*—showing typical phialides. × 1000.

Phialophora Medlar
(Gr. bearing phialides)

This is not a very well defined genus, but fungi which obviously belong here are encountered fairly often, particularly on paper, wood-pulp, and other cellulosic materials.

Colonies are dark coloured, usually matted floccose, but sometimes forming a blackish stroma with little aerial mycelium. The fructifications are normally irregularly branched, but are sometimes truly verticillate, sometimes reduced to solitary phialides, with all parts dark coloured. The phialides are typically bottle-shaped, and some of them at least terminate in a small flared cup, out of which the spores immerge. The spores are small, ellipsoidal, paler than the hyphae, and form irregular balls (Fig. 76).

The best taxonomic treatment of the genus is by van Beyma (1943), who attributes it to Thaxter, not Medlar.

Phoma Sensu Sacc.

(Probably a corruption of Gr. *phyma*, a wart or pustule)

All the published keys to the Sphaeropsidales distinguish three related genera, *Phoma*, with spores less than 15μ long, occurring on stems of plants, *Macrophoma*, with spores over 15μ long, and *Phyllosticta*, distinguished from *Phoma* only by being found on leaves. The segregation of

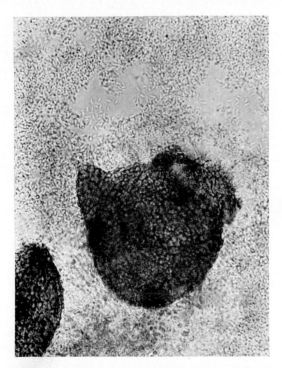

FIG. 77.—*Phoma terrestris*—crushed pycnidium. × 250.

Macrophoma from *Phoma* is purely artificial, and it has been shown that the distinction between *Phoma* and *Phyllosticta* is invalid, because there are many species which grow equally well on either leaves or stems. The

E

number of so-called species which have been described is enormous (Saccardo lists 1700 of *Phoma* and 1500 of *Phyllosticta*), but many of the descriptions are inadequate and many of the names are undoubtedly synonyms.

In culture, most species produce a fair amount of floccose aerial mycelium, white at first then darkening, with pycnidia formed slowly, and best in the light, mostly scattered and close to the surface of the medium. Pycnidia are more or less flask-shaped, sometimes irregular, with short necks, dark brown, fairly thin-walled and easily crushed, emitting a mass of hyaline, ovate to elongate, 1-celled spores (Fig. 77).

P. violacea (Bertel) Eveleigh (Lat. *violaceus*, violet-coloured), commonly referred to as *P. pigmentivora* Massee, the latter epithet referring to its habit of growing on painted surfaces, on which it produces unsightly pink to purple spots, often several centimetres in diameter. Pycnidia detached or clustered, almost globose with scarcely any neck, 125–150μ diam.; spores ellipsoid, 4–6 × 2–2·5μ.

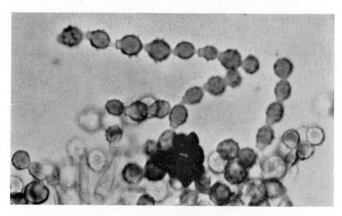

FIG. 78.—*Scopulariopsis brevicaulis*—spores, some showing the typical thickened ring. × 1000.

Scopulariopsis Bainier

(Gr. *opsis*, like; hence like *Scopularia*, another and older genus of fungi. The latter name is derived from Lat. *scopula*, a little broom)

Colony colour varies from white through various shades of brownish yellow to deep brown, greyish brown, or practically black; it is never green, this point being one of the reasons given by Thom (1930) and Raper and Thom (1949) for separating the genus from *Penicillium*. Actually the genus is not related to *Penicillium* at all, the occasional production of more or less penicillately branched conidiophores being the only point of

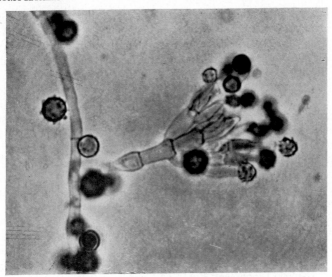

FIG. 79.—*S. brevicaulis*—penicillus.   × 1000.

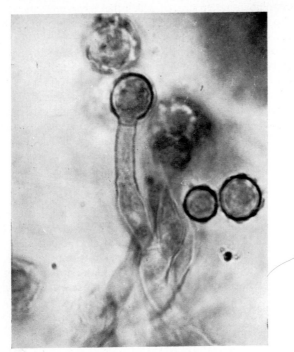

FIG. 80.—*Scopulariopsis brevicaulis*—annellophore.   × 2000.

resemblance. The spore-producing cell is not a phialide (or sterigma), but, as shown by Hughes (1953), an annellophore (see p. 81 and Fig. 80).

Fruiting structures in this genus are very irregular, varying from definitely penicillate structures (Fig. 79) to single annellophores sessile on aerial hyphae. Conidiophores, when present, are always very short. The most distinctive feature of the genus is the conidium, which varies in form from almost spherical to lemon-shaped, but which always has, at the point of attachment, a thickened ring with central pore (Fig. 78). The conidia may be rough or smooth.

The perfect state, known for several species, is *Microascus* Zukal. The perithecia are black and carbonaceous, usually almost globose, but always with a definite opening (the *ostiole*) through which the spores are forcibly expelled when ripe, often forming long, twisted, more or less columnar masses, usually termed cirrhi.

S. brevicaulis (Sacc.) Bainier (Lat. *brevis*, short; *caulis*, a stalk). Synonym, *Penicillium brevicaule* Saccardo. Colonies usually thin and smooth velvety at first, furrowed, greyish white then yellowish brown, becoming overgrown with loosely floccose to funiculose hyphae; conidia lemon-shaped, coarsely roughened, 6–7μ in long axis.

Raper and Thom divided the cultures of *Scopulariopsis* which they were able to examine into 9 groups, of which the first represents *S. brevicaulis*. They were not prepared, at the time, to assign specific epithets to the remainder.

More recently Morton and Smith (1963) have made a detailed study of the genus, of its perfect state *Microascus*, and of the related genus *Doratomyces* (= *Stysanus*).

S. brevicaulis is by far the commonest type encountered. This species is found growing on all kinds of decomposing organic matter and, unlike many moulds, flourishes on substrates containing a high percentage of protein, such as meat and ripening cheese. It is also found as a human parasite, causing a serious infection of the nails.

A special point of interest is that most strains of *Scopulariopsis* (and of *Paecilomyces* too) can liberate arsenic in the form of very poisonous gaseous compounds from any substrate containing even only a trace of this element. When grown on ordinary gelatine, for example, the garlic-like odour common to arsine and its alkyl derivatives is distinctly noticeable. In the past there have been one or two serious cases of arsenic poisoning due to the growth of *S. brevicaulis* on wallpapers coloured with paris green, and it has been proposed to use this species for detecting minute traces of arsenic in suspected materials, instead of employing the usual chemical methods.

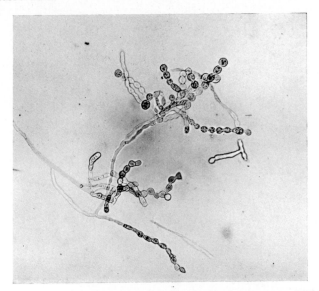

FIG. 81.—*Sporendonema casei*—chains of endogenous conidia
(from slide culture). ×250.

Sporendonema Desmazières ex Fries

(Gr. *endon*, within; *nema*, a filament)

The name refers to the fact that the conidia are endospores, i.e. formed
within the conidiophore, and either extruded or liberated finally by the
decay of the conidiophore walls.

S. casei Desm. (Lat. *casei*, of cheese). Found principally on cheese-rind,
where it forms orange-coloured colonies; in culture grows best at temperatures not exceeding 18° C, spreading slowly at first then more rapidly,
usually orange to reddish, with aerial chains of rounded spores which
appear to be formed by fragmentation of the hyphae (Fig. 81). Examination
of young cultures reveals the true method of spore production. This species
was formerly known as *Oospora crustacea* [Bull. ex] sacc.

S. sebi Fries (Lat. *sebi*, of tallow). A species with many synonyms,
found chiefly in highly sugared materials. It forms characteristic "buttons"
in sweetened condensed milk. In cultures it grows very slowly, forming
velvety, chocolate-coloured colonies, with long rows of endospores, which
round up and become free at maturity, and are mostly 2–2·5μ, but occasionally up to 4μ of even 5μ in diameter (Fig. 82).

Sporothrix Hektoen et Perkins

(Gr. *thrix*, hair; i.e. hyphae covered with spores)

Spores are borne, often in small clusters, on short branches from trailing

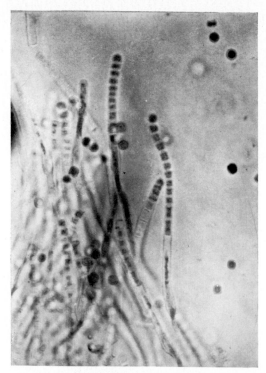

FIG. 82.—*Sporendonema sebi*—endospore.   × 1000.

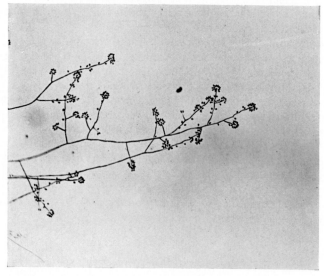

FIG. 83.—*Sporothrix schenckii*—conidia (slide culture).   × 250.

hyphae, each spore being produced on a small peg or sterigma. The best-known species is *S. schenckii* Hektoen et Perkins, which causes the human disease known as sporotrichosis. The fungus is also known as a saprophyte, occurring chiefly on wood, and it is probable that most human infections originate from handling infected timber. Fig. 83 shows spore production in *S. schenckii*.

The name *Sporotrichum*, commonly used for this species by medical mycologists, is incorrect in this context, since the fungus is quite different from Link's genus.

Sporotrichum Link ex Fries

(derivation of name as *Sporothrix*)

There has been much confusion in the nomenclature of organisms which are included in this genus. In the fifth edition of this book the name *Aleurisma* Link was used. Hughes (1953) has stated that *Aleurisma* Link = *Trichoderma* Persoon (1801). Carmichael (1962) maintains that the fungi known as *Aleurisma sensu* Vuillemin (1911*a*), are to be placed in *Chrysosporium* Corda (1833) and has published a long paper on the genus *Chrysosporium*. However, there is no doubt that the correct name is *Sporotrichum* Link ex Fries. As noted above, the genus known as *Sporotrichum* by

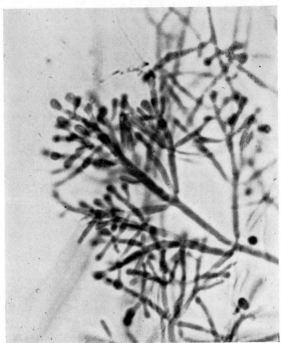

FIG. 84.—*Sporotrichum* sp. × 1500.

medical men is quite distinct from Link's genus, and its correct name is *Sporothrix* Hektoen et Perkins (1900).

The spore-bearing hyphae are often branched irregularly at the tip and the spores appear, under low magnification, to form a cluster, but each final branch bears one spore, attached by the full width of the branch, so that the spores resemble very small chlamydospores. All the species are pale or brightly coloured, usually white, yellow, or greyish, sometimes floccose but more often forming a close mat only a few millimetres high, and becoming mealy in age as the spores are liberated by decay of the hyphae which bear them.

S. carnis Brooks and Hansford (Lat. *carnis*, of flesh). In culture grows slowly but eventually spreads widely, white with pale yellow edge, turning

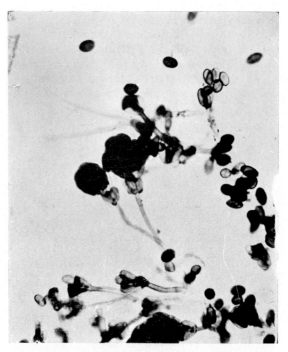

FIG. 85.—*Stachybotrys atra*—conidiophores, showing phialides and dense masses of dark spores. × 500.

greyish, almost velvety, wrinkled; reverse yellow to orange-yellow; spores colourless, ovate to pear-shaped, 2–5 × 2–4μ. This species has been reported from a variety of sources. It grows well at low temperatures and infects meat in cold storage at temperatures as low as −6° C (Fig. 84).

Stachybotrys Corda

(Gr. *stachys*, progeny; *botrys*, a bunch of grapes)

S. atra Corda (Lat. *ater*, black) is the only common species of the genus (see Bisby, 1943). Vegetative hyphae hyaline; conidiophores at first hyaline then dark, bearing a terminal whorl of phialides; phialides thick, rounded at the top, hyaline at first then brownish, producing slimy spores which collect in a dense, irregular mass; spores ellipsoidal to sub-globose, often slightly curved, dark brown to black, smooth when young,

FIG. 86.—*Stachybotrys echinata.* × 500.

but mostly becoming rough in old cultures, 8–12 × 5–7μ (Fig. 85). Common on paper and other cellulosic materials; has been found fairly frequently on rotting canvas and is able to decompose cellulose rapidly in presence of small amounts of more available nutrients.

S. echinata (Rivolta) G. Smith (Lat. *echinatus*, spiny). Syn. *Penicillium echinatum* Rivolta; *Memnoniella echinata* (Rivolta) Galloway (1933).

In the fifth edition of this book this species was described under *Memnoniella*, and the opinion was expressed that the difference between this genus and *Stachybotrys* is very slight, and not of generic importance. It was transferred to *Stachybotrys* in 1962.

Colonies at first consisting of fine mycelium growing close to the substrate, more or less wet in appearance, white or pale dirty pink; conidial development proceeding gradually from the centre outwards, the entire colony eventually becoming black; spore production most abundant on cellulosic materials; conidial heads superficially like those of a monoverticillate *Penicillium*, but dark-coloured; conidiophores arising as short branches from creeping hyphae, septate, roughened when ripe with dark granules, 70–100 × 3·2–4μ, slightly swollen at the tip; phialides forming a dense terminal whorl, rounded at the tip (like *Stachybotrys*), 8–10 × 3–4μ; conidia in parallel chains, dark brown to black, rough, globose to ovate, 4–6 × 4μ (Fig. 86).

Described by Galloway (1933) as occurring on cotton yarn. Stated by Bisby (1943) to be a "suspiciously rare fungus", but was found frequently,

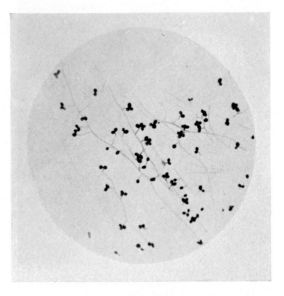

FIG. 87.—*Stemphylium lanuginosum*—as seen in dish culture. × 65.

during the second world war, in tropical and sub-tropical countries, on sand-bags, fibre board, tentage, etc., and admitted by Bisby (1945) to be "not uncommon". It rots cellulosic materials fairly rapidly.

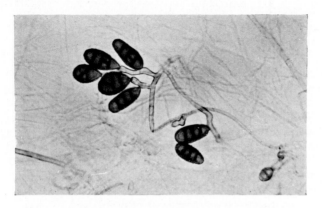

FIG. 88.—*S. lanuginosum*—short sporophore with cluster of spores. × 500.

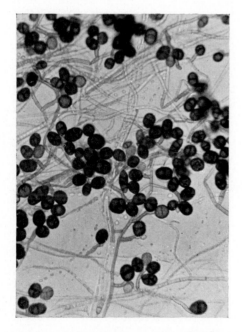

FIG. 89.—*S. lanuginosum*—spores showing a very usual type of septation. × 250.

Stemphylium Wallroth

(Gr. *stemphylon*, a mass of pressed grapes; from the
many-celled spores)

The genus is characterized by muriform spores (dictyospores) which are
not beaked and not borne in chains (cf. *Alternaria*). The type species of
the genus, *S. botryosum* Wallr., is the imperfect stage of *Pleospora herbarum*
(p. 52). Wiltshire (1938) has shown that this differs, both in shape of spores
and in method of spore production, from the common saprophytic species,
but considers that both types should, for the present, be retained in the
genus. The latest taxonomic treatment of the genus is by Neergaard (1945).

S. lanuginosum Harz (Lat. *lanuginosus*, downy) is the only species
commonly found on industrial products. It attacks cellulose and is found
frequently on coarse fabrics, paper, strawboard, and the like. Colonies
spreading rapidly, dark olive when young, becoming almost black as spores
develop; sporophores arising as short branches from aerial hyphae, bearing
spores in small clusters; spores ellipsoid to almost globose, very dark-
coloured, and often with comparatively few septa (Figs. 87, 88, 89).

FIG. 90.—*Trichoderma viride*—conidiophore with
balls of spores. × 500.

Trichoderma Persoon ex Fries

(Gr. *thrix*, hair; *derma*, skin)

Conidia are produced in slimy balls, as in *Cephalosporium* and *Verticillium*, but the conidiophores are irregularly branched, the final branches being true phialides, abstricting the conidia successively. Three names are frequently found in the literature, *T. lignorum* (Tode) Harz, *T. koningii* Oudemans, and *T. viride* Pers. ex Fr., distinguished, according to various authors, by colony colour and size of spores. Bisby has shown (1939) that, when a large number of strains are examined, there are no sharp differences to be observed, and that the range of variation does not exceed that which may be expected in a single widespread species. He considers the genus to be monotypic, and the correct name of the species to be *T. viride* Pers. ex Fr.

This species is commonly found on fallen timber and is also of widespread occurrence in the soil, especially very damp soils, where it can parasitize other fungi (see Chapter XII). It appears fairly frequently in the laboratory on a variety of materials, and grows well on most common media. Colonies spread rapidly, forming a somewhat thin mycelial layer with irregularly shaped patches of verdigris green, this being the colour of the ripe spores in mass. The spores are produced most freely at the shallow end of slopes, and the spore masses in some strains remain white for some time, becoming green only tardily. On some media the reverse and medium are bright yellow to brownish, and some, but not all, strains have a distinct smell of coco-nut. The fragile, spherical spore-heads each contain 10 to 20 conidia, globose or slightly ovate, $2 \cdot 5 - 3 \mu$ diam. (Fig. 90).

Trichothecium Link ex Fries

(Gr. *thrix*, hair; *theka*, a case or receptacle)

T. roseum (Pers.) Link ex Fries (Lat. *roseus*, pink); synonym *Cephalothecium roseum* Corda. This, the only important species of the genus, causes a rot of apples in storage and is found, somewhat less frequently, on other substrata. Colonies somewhat thin, floccose, wide-spreading, white at first, then slowly clear pink; sporophores erect, bearing terminal clusters of spores directly attached to the tip (Fig. 91); spores 2-celled, roughly ovate with a nipple-like projection at the point of attachment, $18-20 \times 8-10 \mu$ (Fig. 92).

The actual method of spore formation is distinctive, and has been well described and illustrated by Ingold (1956). The spores form a chain, with each spore disposed so that the long axis is almost horizontal, alternate spores being turned in opposite directions. The spores are attached to one another through slight thickenings on each side of the base.

FIG. 91.—*Trichothecium roseum*—conidiophore as seen in dish culture. × 200.

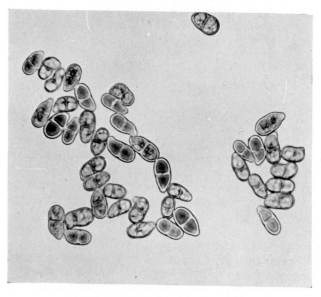

FIG. 92.—*T. roseum*—spores. × 500.

Verticillium Nees ex Wallroth

(Lat. *verticillus*, a whorl of branches)

Conidia are produced in slimy balls, as in *Trichoderma*, but the conidiophores are branched, sometimes several times, in definite whorls (Fig. 93).

V. lateritium (Ehrenberg) Rabenhorst (Lat. *lateritius*, brick-coloured). Syn. *Acrostalagmus cinnabarinus* Corda; *V. cinnabarinum* (Corda) Reinke et Berthold. This is not uncommon, particularly in soil. Colonies thin, spreading, dull brick red; conidiophores long and stiff, mostly several times branched; conidia ellipsoidal, about $3 \times 1 \cdot 5\mu$ (Fig. 94).

It is the conidial state of an Ascomycete, *Nectria inventa* Pethybridge, and is sometimes referred to as "conidial *Nectria inventa*".

V. alboatrum Reinke and Berth. (Lat. *albus*, white; *ater*, black) is a serious parasite of many plants, causing wilt. Colonies at first white, then turning black in patches (hence the name); conidiophores usually much simpler than in the last species, often being reminiscent of *Cephalosporium*. Cultures are difficult to maintain in good condition, usually becoming persistently white.

REFERENCES

Bisby, G. R. (1939). *Trichoderma viride* Pers. ex Fries, and notes on *Hypocrea*. *Trans. Brit. mycol. Soc.*, **23**, 149–68.

Bisby, G. R. (1943). *Stachybotrys*. *Ibid.*, **26**, 133–43.

Bisby, G. R. (1945). *Stachybotrys* and *Memnoniella*. *Ibid.*, **28**, 11–12.

Boedijn, K. (1933). Über einige phragmosporen Dematiazeen. *Bull. Jard. bot. Buitenzorg, Ser. 3*, **13**, 120–34.

Brown, W. (1925). Studies in the genus *Fusarium*. II. An analysis of factors which determine the growth-forms of certain strains. *Ann. Bot., Lond.*, **39**, 373–408.

Carmichael, J. W. (1957). *Geotrichum candidum*. *Mycologia*, **49**, 820–29.

Carmichael, J. W. (1962). *Chrysosporium* and some other aleuriosporic Hyphomycetes. *Canad. J. Bot.*, **40**, 1137–73.

Christensen, C. M., Kaufert, F. H., Schmitz, H., and Allison, J. L. (1942). *Hormodendrum resinae* (Lindau), an inhabitant of wood impregnated with creosote and coal tar. *Amer. J. Bot.*, **29**, 552–8.

Ciferri, R., Ribaldi, M., and Corte, A. (1957). Revision of 23 strains of *Aureobasidium pullulans* (De By) Arn. (=*Pullularia pullulans*). *Atti Ist. bot. Univ. Pavia.* Ser. 5, **14**, 78–90.

Cooke, W. B. (1954). *Sewage ind. Wastes*, **26**, 139–49, 661–74, 790–4.

Cooke, W. B. (1957). *Ibid.*, **29**, 1243–51.

Cooke, W. B. (1959). An ecological life history of *Aureobasidium pullulans* (De Bary) Arnaud. *Mycopathologia*, **12**, 1–45.

de Vries, G. A. (1952). *Contribution to the knowledge of the genus Cladosporium Link ex Fr.* Baarn: Uitgeverij & Drukkerij Hollandia.

FIG. 93.—*Verticillium theobromae*—conidiophores as seen in living
culture. × 100.

FIG. 94.—*Verticillium lateritium*—conidiophores.
× 250.

Drechsler, C. (1923). Some graminicolous species of *Helminthosporium* I. *J. agric. Res.*, **24**, 641–740, with 33 plates.

Galloway, L. D. (1933). Notes on an unusual mould fungus. *Trans. Brit. mycol. Soc.*, **18**, 163–6.

Guba, E. F. (1929). Monograph of the genus *Pestalotia* de Notaris. Part I. *Phytopathology*, **19**, 191–232.

Guba, E. F. (1932). *Ibid.* Part II. *Mycologia*, **24**, 355–97.

Guba, E. F. (1955). *Monochaetia* and *Pestalotia*. *Mycologia*, **47**, 920–1.

Guba, E. F. (1961). *Monograph of Monochaetia and Pestalotia.* Camb. Mass.: Harvard Univ. Press.

Hendy, N. I. (1964). Some observations on *Cladosporium resinae* as a fuel contaminant and its possible role in the corrosion of aluminium alloy fuel tanks. *Trans. Brit. mycol. Soc.* **47**, 467–75.

Hotson, J. W. (1912). Cultural studies of fungi producing bulbils and similar propagative bodies. *Proc. Amer. Acad. Arts Sci.*, **48**, 227–306.

Hotson, J. W. (1917). Notes on bulbiferous fungi with a key to described species. *Bot. Gaz.*, **44**, 265–84.

Hotson, J. W. (1929). *Papulospora atra* n. sp. *Amer. J. Bot.*, **16**, 219–20.

Hotson, J. W. (1942). Some species of *Papulospora* associated with rots of *Gladiolus* bulbs. *Mycologia*, **34**, 391–9.

Hughes, S. J. (1953). Conidiophores, conidia and classification. *Canad. J. Bot.*, **31**, 577–659.

Ingold, C. T. (1956). The conidial apparatus of *Trichothecium roseum*. *Trans. Brit. mycol. Soc.*, **39**, 460–4.

Marsden, D. H. (1954). Studies of the creosote fungus, *Hormodendrum resinae*. *Mycologia*, **46**, 161–83.

Morton, F. J., and Smith, G. (1963). The genera *Scopulariopsis* Bainier, *Microascus* Zukal, and *Doratomyces* Corda. *Mycol. Pap.*, No. 86.

Neergaard, P. (1945). *Danish species of Alternaria and Stemphylium.* London: Oxford Univ. Press.

Nisikado, Y. (1929). Studies on the *Helminthosporium* diseases of Gramineae in Japan. *Ber. Ohara Inst.*, **4**, 111–26.

Petch, T. (1931). Notes on entomogenous fungi. *Trans. Brit. mycol. Soc.*, **16**, p. 24.

Reynolds, E. S. (1950). *Pullularia* as a cause of deterioration of paint and plastic surfaces in South Florida. *Mycologia*, **42**, 432–48.

Robak, H. (1932). Investigations regarding fungi on Norwegian ground woodpulp and fungal infections at woodpulp mills. *Nyt. Mag. Naturv.*, **71**, 185–330.

Servazzi, O. (1954). *Pestalotia* o *Pestalotiopsis* ? *Nuovo G. bot. ital.*, N.S. **60**, (1953), 943–7.

Shear, C. L., and Dodge, B. O. (1927). Life histories and heterothallism of the red bread-mold fungi of the *Monilia sitophila* group. *J. agric. Res.*, **34**, 1019–42.

Shoemaker, R. A. (1959). Nomenclature of *Drechslera* and *Bipolaris*, grass parasites segregated from *Helminthosporium*. *Canad. J. Bot.*, **37**, 879–87.

Shoemaker, R. A. (1962). *Drechslera* Ito. *Canad. J. Bot.*, **40**, 809–36.

Smith, G. (1963). Some new and interesting species of micro-fungi. III. *Trans. Brit. mycol. Soc.*, **45**, 387–94.

Snyder, W. C., and Hansen, H. N. (1941a). The effect of light on taxonomic characters in *Fusarium*. *Mycologia*, **33**, 580–91.

Snyder, W. C., and Hansen, H. N. (1941b). The species concept in *Fusarium* with reference to section Martiella. *Amer. J. Bot.*, **28**, 738–42.

Snyder, W. C., and Hansen, H. N. (1945). The species concept in *Fusarium* with reference to Discolor and other sections. *Amer. J. Bot.*, **32**, 657–66.

Snyder, W. C., and Hansen, H. N. (1954). Variation and speciation in the genus *Fusarium*. *Ann. N.Y. Acad. Sci.*, **60**, 16–23.

Stevens, F. L. (1922). The *Helminthosporium* foot-rot of wheat, with observations on the morphology of *Helminthosporium* and on the occurrence of saltations in the genus. *Bull. Ill. nat. Hist. Surv.*, **14**, Art. 5.

Steyaert, R. L. (1949). Contribution a l'étude monographique de *Pestalotia* de Not. et *Monochaetia* Sacc. (*Truncatella* gen. nov. et *Pestalotiopsis* gen. nov). *Bull. Jard. bot. Brux.*, **19**, 285–354.

Steyaert, R. L. (1953). New and old species of *Pestalotiopsis*. *Trans. Brit. mycol. Soc.*, **36**, 81–9.

Steyaert, R. L. (1955). *Pestalotia, Pestalotiopsis* et *Truncatella*. *Bull. Jard. bot. Brux.*, **25**, 191–9.

Steyaert, R. L. (1956). A reply and an appeal to Professor Guba. *Mycologia*, **48**, 767–8.

Tubaki, K. (1962). Studies on a slime-forming fungus in polluted water. *Trans. mycol. Soc. Japan*, **3**, 29–35.

Valenta, V. (1948). Notes on *Verticillium cinnabarinum*. *Studia bot. čechoslov.*, **9**, 160–73.

van Beyma, F. H. (1943). Beschreibung der in Centraalbureau voor Schimmelcultures vorhandenen Arten der Gattung *Phialophora* Thaxter und *Margarinomyces* Laxa, nebst Schlüssel zu ihrer Bestimmung. *Leeuwenhoek ned. Tijdschr.*, **9**, 51–76.

Vuillemin, P. (1911a). Les Aleuriosporés. *Bull. Soc. sci. Nancy.*, **12**, 151–75.

Vuillemin, P. (1911b). Les *Isaria* de la famille des Verticillacées (*Spicaria* et *Gibellula*). *Bull. Soc. mycol. Fr.*, **27**, p. 75.

Wiltshire, S. P. (1938). The original and modern conceptions of *Stemphylium*. *Trans. Brit. mycol. Soc.*, **21**, 211–38.

Windisch, S. (1951). Zur Biologie und Systematik des Milchschimmels und einiger ähnlicher Formen. I. Der Milchschimmel (*Endomyces lactis*) und *Endomyces magnussii*. *Beitr. biol. Pft.*, **28**, 124.

Wollenweber, H. W. *et al.* (1925). Fundamentals for taxonomic studies of *Fusarium*. *J. agric. Res.*, **30**, 833–43.

Wollenweber, H. W., and Reinking, O. A. (1935). *Die Fusarien*. Berlin: Paul Parey.

Chapter IX

Aspergillus

Species of the great group *Aspergillus* . . . together with the
Penicillia and the Mucors, furnish the "weeds" of the
culture room.

Thom and Church. *The Aspergilli*, 1926.

Species of the genus *Aspergillus* Micheli ex Fries form a very large proportion of all the moulds encountered in industrial work. They are to be found
almost everywhere on every conceivable type of substratum, and a good
working knowledge of the genus is essential to anyone undertaking serious
work on problems of mouldy deterioration. The activities of the Aspergilli,
however, are not entirely destructive in nature, for several species have had
their fermentative powers harnessed for commercial purposes. In the East,
strains of *A. oryzae* have long been used for the saccharification of rice
starch in the production of saké and similar potable liquors. Strains belonging to the same group are used in the manufacture of soy sauce and of
the enzymic mixtures sold under the names "Takadiastase" and "Polyzime". The black Aspergilli, long known to be capable of producing considerable amounts of oxalic acid from sugar, are now used for the successful
commercial production of citric acid, following Currie's discovery that, by
suitable adjustment of conditions, the formation of oxalic acid can be
inhibited and the production of the more valuable citric acid encouraged.
The same group of Aspergilli are also used for the production of gluconic
acid, and of gallic acid from tannin.

The following is a generic diagnosis sufficiently broad to cover all the
species of any importance.

Aspergillus (Lat. *aspergillum*, a mop for distributing holy water). Mycelium colourless or bright or pale coloured, or bearing surface concretions of
colouring matter, never dematiaceous (black or smoky brown), septate,
partly submerged and partly aerial; fertile branches (stalks) arising from,
and more or less perpendicular to, specialized thick-walled enlarged mycelial cells (the foot-cells), mostly non-septate, smooth, roughened or pitted,
frequently enlarging towards the apex and terminating in a swelling (the
vesicle) which may be variously globose, sub-globose, club-shaped, hemispherical, or a mere thickening of the stalk; vesicle bearing, from whole
surface or from upper part only, either phialides or more or less cylindrical
intermediate cells (metulae) which themselves bear clusters of phialides;
phialides or metulae produced simultaneously from surface of vesicle;
conidia successively cut off from the continuously elongating tips of the

phialides, thus forming unbranched chains, variously shaped and coloured; ripe spore heads of various shapes—globose, radiate, club-shaped or columnar—varying much in size in different species; sclerotia formed by several species but of minor diagnostic importance; perithecia of the perfect state produced by only a few species, thin-walled, breaking up to liberate the ascospores.

Some of the smaller and more delicate species bear a strong resemblance to certain of the monoverticillate Penicillia, but there are two criteria by which such may be separated. Thom and Church (1926) and Thom and Raper (1945) regard the presence of foot-cells as the important distinguishing feature and, in spite of the fact that Smith (1933) has described two species of *Penicillium* with foot-cells to the conidiophores, the criterion is quite valid for the borderline species of the two genera. Fig. 95 shows a typical foot-cell with the stalk and head arising from it, and this may be compared with several photographs of Penicillia in the next chapter. The French school of mycologists consider the method of production of phialides, simultaneous in *Aspergillus* (see Fig. 96), successive in *Penicillium*, to be of primary generic significance. In the opinion of the Author, both distinctions are equally to be trusted, but it is sometimes very difficult to demonstrate the presence of foot-cells in the borderline species of *Aspergillus*, owing to their compact habit of growth and the persistence of tangled masses of mycelium in microscopic mounts. On the other hand, it is usually easy to find, especially near the edge of a growing colony, partially developed heads showing various stages in the growth of the phialides.

Sterigmatocystis

Various authorities in the past have considered *Aspergillus* to include only the species with phialides borne directly on the vesicle, and have separated as a separate genus, *Sterigmatocystis*, all the species with metulae. More recently Biourge (1933) has upheld the separation. However, Wehmer and, later, Thom and Church (1926) have rejected *Sterigmatocystis* on the ground that, in several species, the production of metulae is not constant. In particular, strains of *A. flavus* and *A. tamarii* regularly produce two kinds of heads, with and without metulae respectively, and it is not unusual to find heads with both phialides and metulae borne directly on the vesicle.

Eurotium

The connection of the *A. glaucus* series with the ascomycetous genus *Eurotium* has been mentioned in the chapter on the Ascomycetes. From the systematic point of view there is much to be said for retaining the name of the perfect form but, in accordance with the consensus of modern opinion, the group of species is considered here as a section of *Aspergillus*. It should be noted that many references in the literature to this important group will be found under the generic name *Eurotium*.

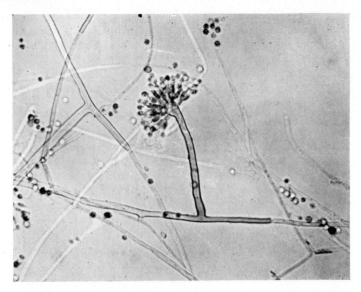

FIG. 95.—*Aspergillus ustus*—conidiophore with foot-cell. × 500.

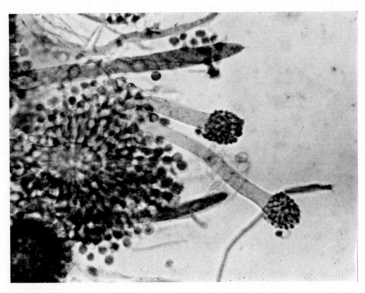

FIG. 96.—*A. repens*—showing "simultaneous" production of phialides.
× 500.

Benjamin (1955) has suggested reviving the use of the name *Eurotium* for the *A. glaucus* series, and also the names *Emericella* Berk et Br. for the perfect state of the *A. nidulans* series, and *Sartorya* Vuill. for the perfect state of the *A. fumigatus* series. Raper (1957) has strongly criticized this view of the situation, and advanced reasons for retaining the name *Aspergillus* for all species, perithecial and purely conidial, and Raper and Fennell (1965) still follow this practice.

The situation has become more complicated since the discovery of a perfect state in members of other series. *A. citrisporus* von Höhnel emended, and a new species *A. ornatus*, are described by Raper *et al.* (1953), and placed originally in the *A. tamarii* series. Fennell and Warcup (1959) found that the sclerotia of *A. alliaceus*, in the *A. wentii* series, develop into multi-locular perithecia on long storage. If all the perithecial species of *Aspergillus* were to have names based on the perfect state, it would be necessary to erect two new genera for the above-mentioned species, since, in both cases, the perithecia are quite different from those found in other series of the genus *Aspergillus*. This would bring the total of generic names to six. Here it is considered to be most convenient to treat them all under the name *Aspergillus*.[1]

Determination of species. The taxonomy of the genus is, up to a point, not difficult owing to the excellent scheme of classification given in the monograph by Thom and Church. The later Manual, by Thom and Raper (1945) followed the same classificatory scheme with little modification. The most recent revision of the genus, by Raper and Fennell (1965), makes a number of improvements in the classification, and takes into account the large number of new names introduced since 1945. One factor which facilitates identifications is the wide range of colony colour exhibited, the genus being probably unique amongst the common Hyphomycetales in this respect. The colours, moreover, are comparatively stable, and are correlated with morphological and biochemical characteristics.

One of the most important factors in the clarification, by Thom and Church, of the taxonomy of this important genus, is their conception of what may be termed the "group species". Some species are fairly definite and of stable morphology. A single strain may be described in definite terms and other strains encountered will fit the description with considerable accuracy, justifying the conclusion that here we have a true species. In other cases, as in the series represented by *A. versicolor*, *A. candidus*, *A. flavus* and others, different strains, whilst clearly related, differ somewhat widely in minor details. Anyone who collects only a few strains belonging in such a series may observe differences which seem to justify the bestowal of several specific epithets but, when scores or hundreds of strains

---

[1] As it is contrary to the International Code of Botanical Nomenclature to use *Aspergillus* as the name for perfect states, perfect state names have been inserted in square brackets wherever appropriate in the descriptions which follow.

are examined, the sharp lines of demarcation disappear and, instead of a few definite species, there is a perfectly graduated series such that no one could describe any particular strain in terms which would permit of its certain identification by another who had not the same series of strains before him. Delimitation of species in such a group is also complicated by the fact that a single strain may cover a considerable range in the series in response to variations in cultural conditions, or even in successive cultures grown under the same conditions. The specialist who is interested in one particular group will collect a number of strains and will know them sufficiently well to be able to separate them on grounds dictated by his own requirements. For most purposes, however, it is sufficient to be able to identify any particular culture as a member of one of the group species. After all, this so-called "lumping" of strains is only the equivalent of the recognition of variation amongst individuals in any one species of flowering plant or animal.

KEY TO GROUPS

1. No metulae present; phialides borne
   directly on vesicle 2
   Metulae present in some heads 7
   Metulae always present 11
2. Vesicles elongated club-shaped . *A. clavatus* group (p. 152)
   Vesicles otherwise 3
3. Heads radiate or scarcely columnar 4
   Heads definitely columnar 5
4. Heads radiate; spore masses some
   shade of green; yellow perithecia
   usually abundant . . . *A. glaucus* group (p. 142)
   Heads radiate to loosely columnar,
   large, mostly dull green to greyish;
   perithecia white, purplish or olive *A. ornatus* group
   Heads radiate, small, in pinkish fawn
   shades; no perithecia . . . *A. cervinus* group
5. Colonies slow-growing, dull green to
   dull brownish green; young conidia
   cylindrical, becoming ovate, rough *A. restrictus* group (p. 148)
   Colonies spreading, bluish green to
   smoky green . . . . *A. fumigatus* group . 6
6. No perithecia produced . . . *A. fumigatus* series (p. 150)
   Whitish perithecia present . . *A. fischeri* series (p. 151)
7. Conidial heads globose to radiate,
   splitting in age; stalks not constricted below vesicle 8
   Conidial heads globose to radiate;
   stalks constricted below vesicle 10
   Conidial heads radiate, but often becoming more or less columnar 9
8. Heads dark brown to black; stalks
   smooth, splitting like cane . . *A. niger* group (p. 161)
   Heads white to cream; stalks smooth *A. candidus* group (p. 161)
   Heads yellow to ochraceous . . *A. ochraceus* group (p. 165)

9. Stalks rough; heads yellow green to
   deep olive brown   .   .   .   *A. flavus* group (p. 167)
   Stalks smooth or only slightly rough;
   heads yellow brown to dull buff   .   *A. wentii* group (p. 163)
10. Heads of one type, green to buff; peri-
    thecia in two species   .   .   .   *A. cremeus* group
    Conidial structures of two types;
    large, grey-green to buff, with
    brown stalks; fragmentary struc-
    tures near or under agar surface   .   *A. sparsus* group
11. Heads some shade of green   12
    Heads some other colour   14
12. Heads radiate; stalks mostly un-
    coloured   .   .   .   .   .   *A. versicolor* group   .   13
    Heads columnar; stalks brown; purple
    perithecia with coloured ascospores;
    hülle cells usually present   .   .   *A. nidulans* group (p. 154)
13. Heads all of one colour, some shade
    of green   .   .   .   .   .   *A. versicolor* series (p. 155)
    Heads of two types, white and blue
    green   .   .   .   .   .   *A. janus* series
14. Heads radiate to broadly columnar,
    in drab olive to dull brown shades,
    stalks brown-walled   .   .   .   *A. ustus* group (p. 159)
    Heads broadly to irregularly colum-
    nar, white to buff or vinaceus, stalks
    yellow to brownish yellow or un-
    coloured   .   .   .   .   .   *A. flavipes* group (p. 160)
    Heads compactly columnar, cinnamon
    to orange-brown or pale buff, stalks
    colourless   .   .   .   .   *A. terreus* group (p. 157)

Four of the groups will not be considered further, since all the species are uncommon, and many of them are represented by single isolates. These groups are: *ornatus* (6 species), *cervinus* (4 species), *cremeus* (5 species), and *sparsus* (4 species).

A. glaucus group

(Lat. *glaucus*, grey-green)

Several species belonging to this group are amongst the most commonly occurring and most destructive of all moulds. They require less moisture for germination of spores and subsequent growth than practically all other moulds, or, what often amounts to the same thing, they will grow on sub-strata of high osmotic concentration. Thus they are commonly found, in pure culture, on jams which contain only a little less than the safe per-centage of sugar, they grow on textiles which contain very little more moisture than the amount usually considered to be normal for any parti-cular material, they attack tobacco and, in general, seem to be capable of infecting any organic material which is in equilibrium with an atmosphere of relative humidity greater than 70 per cent.

The various species can usually be recognized as belonging to this group

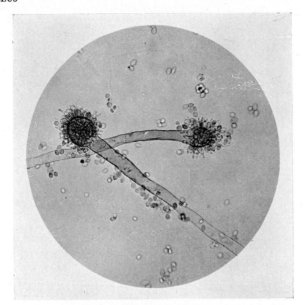

FIG. 97.—*A. glaucus* series (*A. ruber*)—heads with small vesicles. × 250.

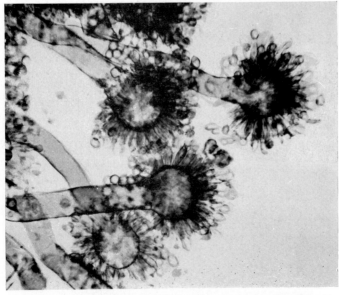

FIG. 98.—*A. glaucus* series (*A. chevalieri* var. *intermedius*)—heads with globose vesicles. × 500.

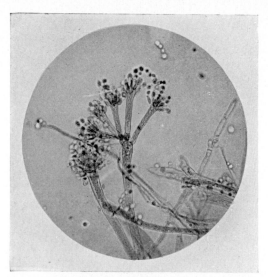

FIG. 99.—*A. glaucus* sp.—proliferation of phialides
to form secondary heads. ×250.

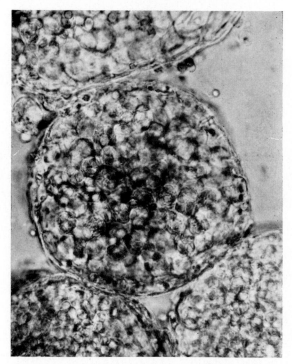

FIG. 100.—*A. amstelodami*—perithecia. ×600.

—complete identification is more difficult—even in their natural habitats, and without isolation in pure culture, by the presence of both bluish green or greyish green conidial heads and bright yellow perithecia, the latter being easily visible to the naked eye.

Group diagnosis. Conidial areas bright bluish green in young colonies, becoming dull green to brownish green, with the colour often partially masked by red or brown pigments encrusting the mycelium; medium often coloured red, brown or purplish; stalks usually septate, smooth, thin-walled often collapsing in mounts; vesicles varying from mere thickenings of the stalk (Fig. 97) to more or less globose (Fig. 98); phialides borne directly on the vesicle, covering the whole of it or only the upper part, usually broad, often proliferating to form slender short stalks bearing secondary heads (Fig. 99); conidia usually spiny, rarely smooth, mostly greater than 5μ in long axis, but varying much in size and attaining, particularly at low temperatures, 15μ or even 20μ, usually ovate or pear-shaped; heads loosely radiate, having a ragged appearance when old; in the perfect state perithecia (Fig. 100) yellow to orange, globose or nearly so, varying in different species from about 80μ to 250μ in diameter, containing 8-spored asci in irregular arrangement; asci usually breaking up as spores mature; ascospores (Fig. 101) colourless, biconvex, usually with a furrow marking the line along which the wall splits on germination with the sides of the furrow rounded or raised into crests or frills.

The group was for long a difficult one for taxonomists. Simplification has come only after it has been recognized that satisfactory delimitation of

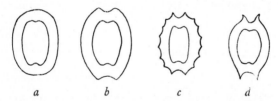

a   *b*   *c*   *d*

FIG. 101.—Ascospores of perfect states of the *A. glaucus* series.
a. A. repens; b. A. ruber; c. A. amstelodami; d. A. chevalieri.

species must be based on adequate study of a very large number of strains, and that a too narrow conception of what constitutes a species results only in confusion.

Mangin (1909) was the first to make a comparative study of a number of strains from various sources. He followed earlier workers in regarding size and markings of ascospores as of primary significance in separating species, but the number of strains studied (23) was not sufficient for a really adequate treatment of the group. Thom and Church (1926) adopted Mangin's scheme of classification and fitted into it a number of species described by Bainier and Sartory (1911, 1912), but this section of their monograph has not proved to be entirely satisfactory in use. More recently Thom and Raper (1941) have published the results of a careful re-study of the group and have later (1945) reproduced their new classification in their Manual of the Aspergilli. The following key to the more common species is based on the keys given in the Manual. It includes the small-spored species, i.e. those with ascospores 6μ or less in long axis, and one of the large-spored species. Experience has shown that most species of the latter group are of rare occurrence. In this connection, however, Thom and Raper (1941, p. 42) point out that the large-spored species have much lower temperature optima than the small-spored forms, and suggest that they might be isolated more frequently if the incubation temperature were 15–20°, instead of the more usual 24–25°.

1. Ascospores 8–10μ in long axis   *A. echinulatus* (Delacroix) Thom and Church [*Eurotium echinulatum* Delacroix]

   Ascospores 6μ or less in long axis   2

2. Ascospores with curved surfaces smooth or nearly so 3

   Ascospores with curved surfaces rough . . . .   *A. amstelodami* (Mangin) T. and C. [*Eurotium amstelodami* Mangin]

3. Equatorial ridges (crests) lacking: furrow nil or very shallow and inconstant . . . .   *A. repens* (Corda) Sacc. [*Eurotium repens* Corda]

   Ridges low and rounded; furrow broad and shallow . .   *A. ruber* (Konig, Spieckermann and Bremer) T. and C. [*Eurotium rubrum* Konig, Spieckermann and Bremer]

   Ridges thin, crest-like; spore like a pulley . . . .   *A. chevalieri* (Mangin) T. and C. [*Eurotium chevalieri* Mangin]

All the species were originally included in *Eurotium* and have been transferred to *Aspergillus* as indicated.

A. repens [*Eurotium repens*] (Lat. *repens*, creeping; refers to the stolon-like hyphae at the edges of colonies). Probably the commonest and most cosmopolitan member of the group, forming characteristic colonies, the surface being an intimate mixture of dirty green and yellow, with reverse and medium gradually becoming dirty brown on wort agar and yellow to brownish red on Czapek agar containing 20 per cent of sucrose (the medium favoured by Thom and Raper); conidial heads large, with conidia mostly 5–6·5μ in diameter, rough; perithecia of the perfect state mostly 75–100μ in diameter; ascospores 4·8–5·6 × 3·8–4·4μ, smooth, with usually a trace of furrow but no crests.

A. ruber [*Eurotium rubrum*] (Lat. *ruber*, red). Also found very frequently on all kinds of substrata; colonies rusty red to deep red, the colour being due to incrustations of pigment on the mycelium, with reverse deep red to intense red-brown; conidial heads pale blue-green turning greyish; conidia elliptical, rough, 5–6·5μ in long axis; perithecia of the perfect state abundant, 80–120μ in diameter; ascospores 5·2–6·0 × 4·4–4·8μ, with broad shallow furrow.

A. chevalieri [*Eurotium chevalieri*] (F. Chevalier, French mycologist). Colonies show patches of pale blue-green conidial heads and areas of orange hyphae in which perithecia are embedded, with reverse orange-red; conidial heads radiate, fairly large; conidia almost globose, spiny, 4·5–5·5μ in diameter; perithecia of the perfect state 100–140μ in diameter; ascospores smooth-walled, with thin prominent crests, 4·6–5·0 × 3·4–3·8μ.

A. amstelodami [*Eurotium amstelodami*] (Lat., of Amsterdam). Colonies usually deep rich green, speckled with sulphur yellow, with reverse almost colourless or at most pale yellow; conidial heads radiate or more or less columnar, large; conidia sub-globose, delicately roughened, mostly about 4μ in diameter; perithecia of the perfect state up to 140μ or even 160μ in diameter; ascospores rough all over, with prominent furrow and rounded crests (Fig. 102). Occasional cultures vary from type in being almost entirely perithecial, with conidial heads produced tardily and sparingly.

A. echinulatus [*Eurotium echinulatum*] (Lat. *echinulatus*, with little spines). Colonies slow-growing, mottled with a mixture of blue-green conidial heads and orange-red perithecial masses, with reverse pale to deep reddish brown; conidiophores long, up to about 1 mm., and mostly broadening upwards; conidia elliptical, pear-shaped or sub-globose, rough, usually 8–10μ in long axis; perithecia of the perfect state yellow, embedded in red hyphae growing close to surface of medium, mostly 150–160μ diam.; ascospores in some strains 9–10 × 6·5–7·5μ, in others 8–9 × 6–7μ, with broad shallow furrow and thin frills with irregular edges.

Compared with other species of the large-spored series, *A. echinulatus* is not uncommon.

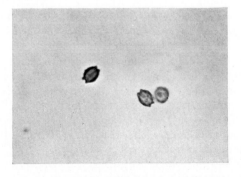

FIG. 102.—*A. amstelodami*—ascospores as
actually seen. × 1000.

A. restrictus group

Species belonging in this group are of common occurrence but are easily missed in plates from infected material owing to their very slow rate of growth, even on rich culture media. Some of them have been described as species of *Penicillium* and certainly bear a striking resemblance to certain of the monoverticillate members of that genus. In the author's experience they are amongst the most resistant of all moulds to the inhibiting effect of antiseptics. Like the *A. glaucus* group, they are able to grow on comparatively dry materials and are of common occurrence on textiles which contain little more than the normal amount of moisture. They are easy to recognize by their restricted growth, their dark green or grey-green colour, and their very long slender columnar heads, composed of chains of fairly large, ovate to barrel-shaped, rough conidia. They resemble *A. fumigatus* in disposition of phialides and shape of heads, and the *A. glaucus* group in size and markings of spores.

The following five species are regarded as distinct by Raper and Fennell.

1. Heads columnar; vesicles small, fertile
   on upper surface only 2
   Heads radiate when young, tardily
   loosely columnar; vesicles fertile over
   upper half to two-thirds . . *A. penicilloides* Spegazzini
2. Colonies growing moderately well on
   standard Czapek agar . . . *A. caesiellus* Saito
   Colonies growing very slowly on
   Czapek agar 3
3. Colonies dark olive green; spore columns long, often twisted . . *A. restrictus* G. Smith
   Colonies light grey green; columns
   shorter and more delicate 4
4. Conidia elliptical at first, often remaining so, 4·0–4·5 × 3·0–3·5 μ . . *A. conicus* Blochwitz
   Conidia subglobose to pyriform, mostly
   3·0–3·5 μ diam. . . . . *A. gracilis* Bainier

A. caesiellus (Lat. diminutive of *caesius*, bluish grey). This species has been isolated only very infrequently. It is less osmophilic than other species of this series.

A. conicus (Lat. *conicus*, conical; presumably referring to the vesicle). Found occasionally in soil. Colonies rather pale grey green at first, often becoming wet and slimy, with reverse in dark green shades; conidia mostly 4·0–4·5 × 3·0–3·5 μ but occasionally larger, rough.

A. gracilis (Lat. *gracilis*, slender). Colonies growing slowly on all media, pale grey green, with reverse greenish; spore chains in very slender columns; spores delicately roughened, small.

A. penicilloides (Gr. *eidos*, like; hence like *Penicillium*). Differs from other members of the series in that the spore heads are radiate rather than columnar. Grows scarcely at all on Czapek agar; for normal growth re-

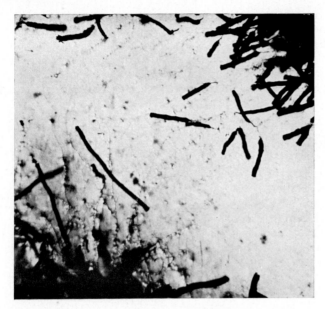

FIG. 103.—*A. restrictus*—columnar heads as seen in Petri dish.
× 50.

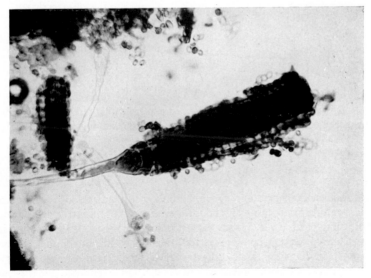

FIG. 104.—*A. restrictus*—typical conidiophore. × 500.

quires a medium containing a high percentage of sugar. Conidia at first elliptical, becoming barrel-shaped to globose, very rough, 3·0–4·0μ or even 5·0μ diam.

A. restrictus (Lat. *restrictus*, close, tight; from the habit of growth). Figs. 103, 104, 105. Colonies dull green to brownish green; vesicles more or less conical; heads forming very long compact columns; conidia almost

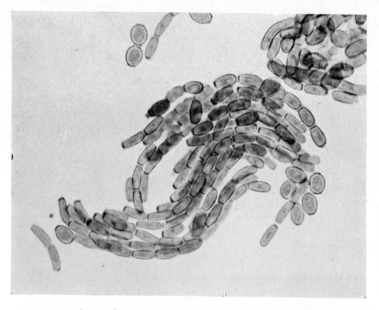

FIG. 105.—*A. restrictus*—spores, mostly young but with a few older ones showing roughening. × 1000.

colourless and cylindrical when young, becoming pear-shaped and dark brownish green, 4·5–6·0 × 3–4μ, in occasional strains larger, up to 10μ long. This is the commonest of the four species and appears to be cosmopolitan.

A. *fumigatus* Fresenius

(Lat. *fumigatus*, smoky)

Colonies usually dark smoky green, becoming still darker in age, and more or less velvety; in occasional strains definitely floccose, sporing tardily and with young heads a very bluish green; conidial heads columnar, varying in length and about 40μ broad (Fig. 106); stalks smooth, short, often greenish, 2–8μ in diameter; vesicles flask-shaped, fertile on upper half or two-thirds; phialides borne directly on vesicle, close packed, 6–8 × 2–3μ; conidia globose, rough, mostly 2·5–3μ in diameter. Heads mounted in

actophenol have a very characteristic appearance (Fig. 107). This species
s readily distinguished from the last group by its smaller, globose conidia,

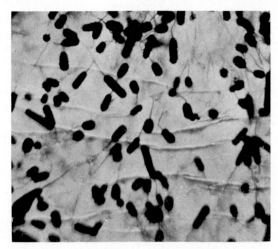

FIG. 106.—*A. fumigatus*—columnar heads as seen in
Petri dish. × 50.

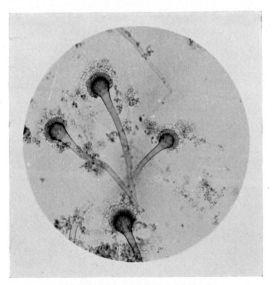

FIG. 107.—*A. fumigatus*—typical conidiophores. × 250.

nd by its much more rapid growth. Three species with a perfect state and
vith the conidial morphology of *A. fumigatus* are known, *A. fischeri*

F

Wehmer [*Sartorya fumigata* Vuillemin], *A. malignus* Lindt, and *A. quadri-cinctus* Yuill. Of these the first two are not of common occurrence, and the third is known only as the type culture.

A. fumigatus is the causal agent of a disease of birds, known as Aspergillosis. The disease also occurs in lambs, and can be very serious (Austwick *et al.*, 1960). Cases in which human beings have contracted the disease, which has many of the symptoms of tuberculosis, are rare. Nevertheless, care should be taken, when handling cultures, not to distribute spores in the air.

This species grows well over a wide range of temperature. It flourishes at 45° C and often at higher temperatures, and is commonly found in decomposing compost.

A. clavatus Desmazières

(Lat. *clavatus*, club-shaped)

Colonies grow rapidly, almost velvety or forming a fairly dense felt, bluish grey-green in colour. The most typical feature of the species is the elongated, club-shaped vesicle, bearing short, densely packed phialides over its whole surface (Fig. 108). It is sometimes possible to find, in ordinary

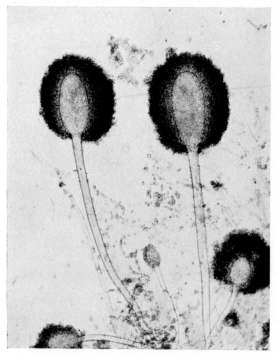

FIG. 108.—*A. clavatus*—conidiophores. × 100.

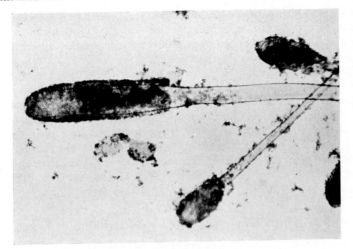

FIG. 111.—*A. giganteus*—head with elongate vesicle. × 100.

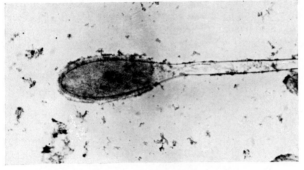

FIG. 110.—*A. giganteus*—head with clavate vesicle. × 100.

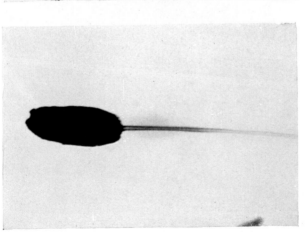

FIG. 109.—*A. giganteus*—single head of characteristic shape as seen in living culture. × 25.

cultures, mature heads which are definitely clavate, but mostly the chains
of conidia form dense masses of irregular shape. Large club-shaped heads,
on coarse stalks, can be obtained by growing on special media. Conidia are
elliptical, $3 \cdot 5$–$4 \cdot 5 \times 2 \cdot 5$–$3\mu$.

A closely related species, *A. giganteus* Wehmer (Figs. 109, 110, 111),
differs little from *A. clavatus* when grown on ordinary media but produces
enormous heads, on stalks several centimetres long, under special conditions of culture. Thom and Church specify organic nitrogen in the medium,
whilst Biourge (1933) states that typical gigantic heads are formed only
when the medium contains zinc and the cultures are exposed to the light.
Some authors have stated that *A. giganteus* should be regarded merely as
a growth form of *A. clavatus*, but Raper and Fennell accept it as a good
species.

A. nidulans (Eidam) Winter [*Emericella nidulans* (Eidam) Vuillemin]

(Lat. *nidulus*, a little nest; refers to the way the perithecia are
embedded)

In most isolations colonies are at first smooth velvety and of a beautiful
clear green colour, developing dirty white spots from the centre outwards
as perithecia are formed; reverse deep red to purple. Occasional strains are
purely conidial or may remain so through a series of cultures and then
suddenly begin to produce perithecia of the perfect state. Stalks sinuous,

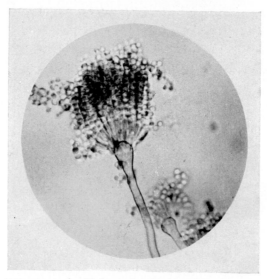

FIG. 112.—*A. nidulans*—head with columnar mass
of spores (the spores have remained *in situ* much
better than in most mounted specimens). × 250.

cinnamon brown; heads short columnar; vesicles more or less hemispherical; phialides borne on metulae, $5-6 \times 2-2 \cdot 5\mu$; conidia globose, rough,
$3-3 \cdot 5\mu$ in diameter; perithecia of the perfect state roughly globose, dark
purplish red, usually about 150μ in diameter; asci 8-spored, disintegrating
as the perithecia ripen; ascospores red to purple, with crests, about $5 \times 4\mu$;
perithecia surrounded by masses of thick-walled cells, known as hülle cells
(Fig. 114), up to 25μ in diameter. Fig. 112 shows a somewhat unusually
large head and Fig. 113 a more usual appearance of a mounted specimen.

Thom and Raper (1939) described five species and two varieties which
all have the general characteristics of *A. nidulans*, but which differ in size
and markings of ascospores. Raper and Fennell (1965) describe 18 species
and 5 varieties. The majority of strains encountered, however, all belong
to *A. nidulans* proper.

FIG. 113.—*A. nidulans*—a more usual type of
appearance. × 250.

A. versicolor (Vuillemin) Tiraboschi

(Lat. *versicolor*, multi-coloured or changing colour)

This common organism is of world-wide distribution and is found in much
the same situations as species of the *A. glaucus* group, but requires slightly

more moisture than the latter for growth. As the name implies, cultures show a considerable range of colour. Different strains may be variously pale green, greyish green, buff, or even pink in small areas, whilst a single culture may, on occasion, show patches of yellow, pink, white and green on the surface, and a deep red or plum colour in reverse. The texture of the colony may be velvety, floccose, or both in patches. Stalks smooth, colourless; heads loosely radiate; vesicles globose or ovate, fertile over the upper half or three-quarters; phialides borne on metulae, $5-10 \times 2-2\cdot5\mu$; conidia globose, delicately roughened, usually $2\cdot5-3\mu$ but occasionally somewhat larger. Whatever the colour of the colony, the spores turn a beautiful emerald green when mounted in lacto-phenol; see Fig. 115. Raper and Fennell regard the *A. versicolor* series as containing 17 species and one variety, but a number of the species are uncommon or are represented by single isolates.

A. *sydowii* (Bainier and Sartory) Thom and Church
(P. Sydow, German mycologist)

This species is closely allied morphologically to *A. versicolor*, but forms velvety colonies of a deep bluish green or greenish blue colour with reverse usually very deep red. It is easily mistaken for a *Penicillium* at first sight and until examined under the microscope. Dwarfed heads and even isolated metulae with their clusters of phialides are found in nearly all strains, and in some isolates are produced almost exclusively.

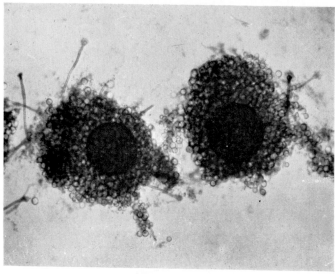

FIG. 114.—*A. nidulans*—perithecia of perfect state with masses of hülle cells. × 100.

A. *terreus* group

Thom and Raper recognized 3 species and 3 varieties in this group. Raper and Fennell have drastically reduced the number to one species and two varieties. Common characteristics are columnar heads, smooth colourless stalks, hemispherical vesicles fertile over the upper half or two-thirds, phialides borne on metulae, and small smooth conidia.

1. Colonies velvety, cinnamon to orange-
      brown; conidiophores short   2
   Colonies floccose, golden yellow; heads
      small; conidiophores up to 500μ
      long or even longer   .   .   .   *A. terreus* var. *aureus*
                                                  Thom & Raper

2. Sclerotium-like masses produced on
      malt agar   .   .   .   .   .   *A. terreus* var. *africanus*
                                                  Fennell & Raper
   Sclerotium-like masses absent   .   .   *A. terreus* Thom

The majority of isolates belonging to this group are *A. terreus* Thom.

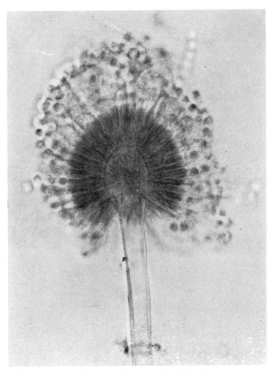

FIG. 115.—*A. versicolor*—typical head. × 1000.
A. sydowii is very similar.

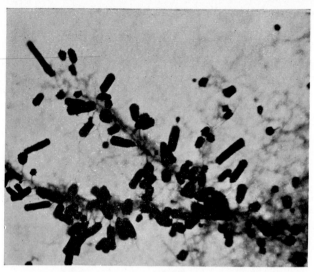

FIG. 116.—*A. terreus*—columnar heads as seen in Petri dish. × 50.

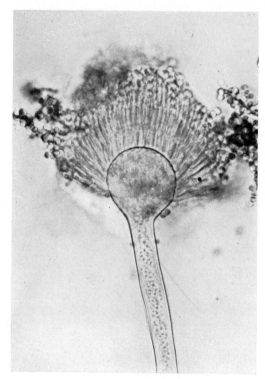

FIG. 117.—*A. terreus*—typical head. × 800.

Nevertheless, variants from type occur sufficiently frequently to justify their mention here.

A. terreus (Lat. *terreus*, of the earth). Colonies cinnamon to sand-brown, velvety; reverse and medium yellow to deep dirty brown, with brown colour more marked in the presence of zinc; heads columnar, up to 500μ long and 30–50μ in diameter; vesicles dome-shaped, bearing closely packed metulae; phialides 5·5–7·5 × 1·5–2·0μ; conidia globose or nearly so, about 2μ in diameter. Fig. 116 shows the heads as seen in a living culture and Fig. 117, taken from a mounted specimen, the shape of the vesicle and disposition of phialides.

A. terreus is a common soil organism and is found frequently on all kinds of vegetable materials. The author has found it on scores of samples of Egyptian cotton, and less frequently on American cottons.

A. ustus (Bainier) Thom and Church

(Lat. *ustus*, burnt; from the dark colour)

Colonies floccose, at first brownish yellow, becoming brownish or purplish grey to almost neutral grey, with reverse yellow, dull reddish or purplish; stalks short, smooth, pale brown, 5–10μ in diameter, arising from prominent foot-cells (Fig. 95); vesicles globose, fertile over upper two-thirds; metulae 5–6μ and phialides 6–9μ long; conidia globose, rough, 3–3·5μ or occasionally up to 5μ in diameter; perithecia unknown, but hülle cells are produced in abundance by some strains, rarely globose, mostly ovate or sausage-shaped, and frequently bent or twisted (Fig. 118). Raper

FIG. 118.—*A. ustus*—hülle cells. × 250.

and Fennell recognize 5 species in the *A. ustus* series. The other four are not of sufficient importance to justify inclusion here.

A. flavipes (Bainier and Sartory) Thom and Church

(Lat. *flavus*, yellow; *pes*, foot)

The usual type of colony is white or silvery white, sometimes with patches of pale pink or mauve mycelium, with the white colour persisting or

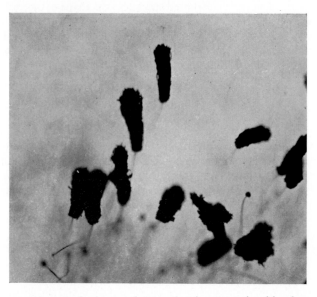

FIG. 119.—*A. flavipes*—columnar heads as seen in old culture.
× 50.

changing to pale buff or greyish buff. Some strains produce mycelial knots suggestive of sclerotia. The buff forms, in old culture, often approach the appearance of *A. terreus*, but the two are readily distinguished by examination of young vigorously growing colonies. The persistently white strains are not likely to be confused with *A. candidus* owing to the difference in shape of the heads. Stalks yellow or brownish yellow (hence the specific epithet); heads, at least in old cultures, definitely columnar (Fig. 119); phialides borne on metulae, $5–8 \times 1\cdot5–2\mu$; conidia smooth, globose, $2–3\mu$ in diameter. Raper and Fennell recognize three species belonging to the *A. flavipes* series, two of them transferred from the *A. terreus* series as described by Thom and Raper. These are *A. niveus* Blochwitz, easily recognized by its persistently white colonies and loosely columnar heads; and *A. carneus* (van Tieghem) Blochwitz, with heads at first white, then vinaceous fawn, and stalks colourless or nearly so.

A. *candidus* Link ex Fr.

(Lat. *candidus*, shining white)

The name covers a number of somewhat diverse forms with white or creamy white globose heads. Most strains grow rather slowly on ordinary media, more rapidly on media containing a high percentage of sugar. Some isolates produce only scanty mycelium, with numerous fertile stalks arising from the substrate. Others produce more or less floccose colonies, with sporing structures arising from both submerged and aerial mycelium. Stalks erect, colourless, or faintly yellowish near the vesicle, smooth, in some strains short, in others up to 1 mm. long; heads mostly globose, but often splitting in age as long chains of conidia are formed, varying from less than 100μ to 250μ in diameter; vesicles globose, mostly producing metulae over the whole surface, but occasionally with only the upper surface fertile, and then producing heads more or less columnar; metulae varying in different heads from about 5μ long to $15-20\mu$, or even 30μ, thick, wedge-shaped; phialides $5-8 \times 2-3\mu$; conidia globose, or somewhat barrel-shaped, smooth, $2 \cdot 5 - 3 \cdot 5\mu$ in diameter. Some strains produce numerous sclerotia, purple to black, but such are not common. In mounted specimens heads are very similar to those of *A. niger* (Fig. 121) except for colour.

This species resembles the *A. glaucus* group in its ability to grow on materials of very low water content, and in its preference for culture media containing high concentrations of sugar.

A. *niger* group

(Lat. *niger*, black)

Black Aspergilli are of very common occurrence in all parts of the world and, as is usual in any group of widespread distribution and omnivorous habits, there are a large number of races, strains, or species. The following diagnosis will make clear the type of variation to be expected.

Colonies spreading rapidly, with mycelium white at first, frequently developing areas of bright yellow; stalks arising from the substratum, varying from 200μ to several millimetres in length and from about 10μ to 20μ in diameter, mostly colourless, smooth, splitting when crushed like pieces of cane; heads globose, often splitting at the periphery (Fig. 120); vesicles globose, colourless or yellowish brown, fertile over the whole surface, up to about 50μ in diameter; phialides borne directly on the vesicle in a few strains but usually metulae present; phialides fairly uniform in size, about $7 \times 2-3\mu$, usually brown; metulae varying in length from $10-15\mu$ in *A. niger* to 120μ in *A. carbonarius*, usually deep brown; conidia globose, brown to black or purplish brown, with the colour aggregated into bars and nodules between the inner and outer walls, appearing spiny except at very high magnification, varying from $2 \cdot 5\mu$ to 10μ in diameter. Fig. 121 shows a typical head after decolorization. Sclerotia are sometimes found in freshly isolated strains, these being round, white, and about 1 mm. in diameter, but it is often found that such strains become purely conidial after a few transfers. On the other hand, some strains may remain purely conidial for years and then suddenly begin to produce sclerotia.

The dimensions of metulae and of conidia are made the basis of separation of the group into a number of fairly definite species. Thom and Church (1926) list 7 species, Thom and Raper (1945) several more. Raper and Fennell recognize 10 species and 2 varieties with metulae in the heads, and 2 species without metulae. The majority of isolates agree reasonably well with the diagnosis of *A. niger*, and, whilst some of the other species are found occasionally, it is felt that a detailed list would be out of place here.

A. niger van Tieghem is distinguished by comparatively short metulae, 20–30μ long, and small conidia, 2·5–4μ, occasionally 5μ, in diameter. Heads vary in colour, different strains showing dark brown, purplish brown, or black.

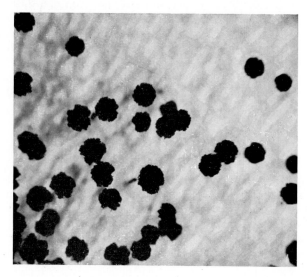

FIG. 120.—*A. niger*—globose heads as seen in Petri dish.
× 20.

It is hardly surprising that, with a species of such widespread and common occurrence, mutations have from time to time been obtained. Thom and Church mention several stable mutants obtained experimentally by growing *A. niger* on media containing small amounts of poisonous substances, the most distinctive of these being *A. niger* mut. *cinnamomeus* (Schiemann) T. and R., almost colourless, and *A. niger* mut. *schiemannii* (Thom) T. and R., with brownish colonies. More recently Yuill and Yuill (1938) have described as *Cladosarum olivaceum* a mutant of a very different type. Development is normal up to the production of phialides. Then, instead of unbranched chains of black spores being produced, branched outgrowths are formed from the phialides, these sometimes producing single terminal conidia. Mature colonies are grey to greenish black.

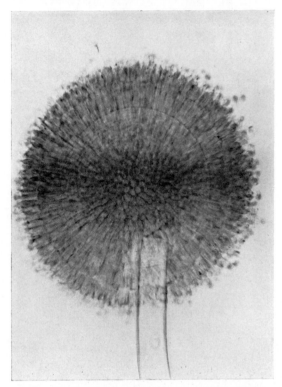

FIG. 121.—*A. niger*—young head not yet fully pigmented.
× 500.

As indicated earlier, species belonging to this group have interesting biochemical characteristics, and workers who are concerned with this side of their activities will need to separate strains on biochemical rather than on morphological grounds.

A. wentii Wehmer
(F. A. F. C. Went, Dutch mycologist)

This species is well marked and easy to recognize. The aerial mycelium forms dense floccose masses, white or tinged with pink to rose when cultures are exposed to light, completely filling culture tubes when the food supply is abundant. On some natural substrata it is found as a dense, furry,

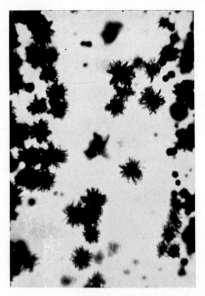

FIG. 122.—*A. wentii*—mature heads
as seen in Petri dish. ×20.

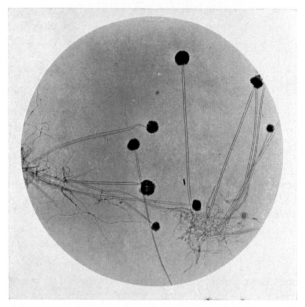

FIG. 123.—*A. wentii*—portion of young culture (mounted).
×25.

polychromatic growth. The conspicuous spore heads are coffee-coloured, up to 500μ or even 800μ in diameter, very ragged-looking when mature, and borne on long slender stalks. Stalks smooth and colourless, up to several millimetres in length and 10–25μ in diameter; vesicles globose, fertile over the whole surface; metulae varying much in length, but normally 10–20μ long; phialides mostly 6–8 × 3μ; conidia globose or slightly

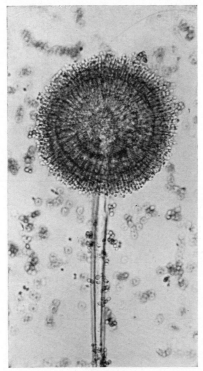

FIG. 124.—*A. wentii*—typical head. ×250.

elliptical, usually rough but occasionally almost smooth, about 5μ in diameter. Fig. 122 shows mature heads in a living colony, Fig. 123 two small portions picked off from a young colony and showing the long smooth stalks, Fig. 124 a specimen mounted in lacto-phenol.

Raper and Fennell describe a number of related species, but these are not sufficiently common to merit description here.

A. *ochraceus* Wilhelm

(Lat. *ochraceus*, ochre-coloured)

The name covers a series of forms which are of common occurrence and wide distribution. These are separated by Thom and Raper into a number

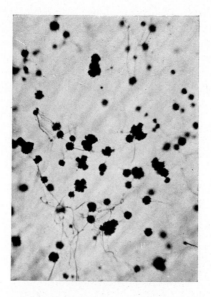

FIG. 125.—*A. ochraceus*—heads as seen
in Petri dish. ×20.

FIG. 126.—*A. ochraceus*—single large head in old
culture, showing characteristic splitting. ×25.

f definite species but, except for the specialist, are conveniently lumped
as *A. ochraceus*. Colonies ochre-coloured or buff, with submerged mycelium
colourless or tinted to various degrees yellow or purplish; spore heads
large and conspicuous, at first globose, then splitting into two or more
columnar masses, frequently of a shape and colour recalling sheaves of
corn (Figs. 125 and 126); stalks long, 10μ or more in diameter, yellow,
pitted so as to appear definitely rough (Fig. 127); vesicles globose, fertile

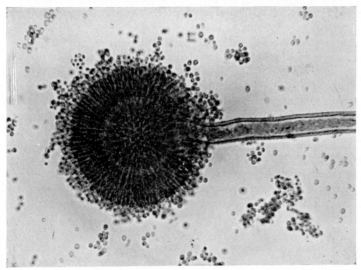

FIG. 127.—*A. ochraceus*—typical head. × 500.

all over; metulae varying much in length but normally $15–30\mu$; phialides
$7–10 \times 1\cdot5–2\cdot5\mu$; conidia globose or nearly so, usually delicately roughened, $3–5\mu$ in diameter. Some strains form abundant heads and no
sclerotia; others produce masses of orange to brownish sclerotia and
comparatively few heads.

A. sulphureus (Fresenius) Thom and Church
(Lat. *sulphureus*, sulphur-coloured)

A number of strains, which are conveniently grouped as *A. sulphureus*, resemble *A. ochraceus* in general aspect and dimensions but differ in that the
conidial colour is clear yellow instead of buff. As in *A. ochraceus*, some
strains produce abundant sclerotia.

A. flavus group

Thom and Raper recognize five species in this group, of which three are of
sufficient importance to mention here. In all of them stalks are rough,

vesicles bear metulae, phialides, or both side by side, and the conidia are rough.

A. flavus Link ex Fr. (Lat. *flavus*, yellow; a misnomer since the colony colour is green). Colonies spreading rapidly, yellowish at first, then yellowish green, darkening with age and becoming finally brownish green; stalks arising from the substrate, up to 1 mm. long and up to 15μ in diameter with walls coarsely roughened; heads varying much in size, loosely radiate or becoming somewhat columnar; vesicles dome-shaped to flask-shaped bearing metulae, phialides, or both; phialides usually 7–10 × 2·5–3μ; conidia globose or somewhat pear-shaped, rough, 3·5–5μ in diameter. Some strains produce brownish sclerotia. This is the commonest of the species in the group, being found on many kinds of organic matter in all parts of the world.

A. oryzae (Ahlburg) Cohn (Lat. *oryzae*, of rice). Colonies seldom show true green colour but are yellowish at first, becoming pale yellowish brown stalks up to 2 mm. long, rather thin-walled, rough; conidia varying much in size, from about 4μ to 9–10μ in diameter.

The original strain was isolated from starters for the manufacture of fermented drinks in the Far East. Since then it has been found that strains of *A. flavus*, or strains intermediate between the two species, are just as frequently used. Strains belonging to this group which are associated with the preparation of native beverages in Africa have all proved to be typical *A. flavus*.

A. oryzae (Ahlb.) Cohn var. *effusus* (Tiraboschi) Ohara (Lat. *effusus* loose in texture). This species, not uncommon, differs considerably in appearance from *A. flavus*. Colonies deeply floccose, white at first, then pale dirty greenish yellow, and finally dull tan, with reverse yellow to rose. The stalks arise from the aerial mycelium, and it is not unusual to find fertile hyphae consisting of almost uninterrupted series of foot-cells. Heads otherwise are very similar to those of *A. flavus* (Fig. 128).

A. parasiticus Speare (Lat. *parasiticus*, parasitic). This forms rich green colonies, deeper in shade than the normal *A. flavus*, with comparatively short stalks, mostly less than 400μ, no metulae, and conidia which are globose to somewhat pyriform, mostly 4–5μ diam., very rough.

In 1960 100,000 young turkeys died of a mysterious disease, which was eventually found to be caused by the infection of groundnuts feed with strains of the *A. flavus* series. Later tests showed that different strains vary much in toxicity, the ones producing most toxin corresponding well with *A. parasiticus*. The toxic substance has been called "Aflatoxin". A report of the Working Party on Groundnut Toxicity (1962) gives a good account of the early work on the subject. Allcroft and Carnaghan (1963), and Sargeant, Carnaghan and Allcroft (1963) give useful reviews of both biological and chemical aspects. Feuell (1966) discusses a number of toxic factors of mould origin. Coveney, Peck and Townsend (1965) summarize recent work, chiefly carried out at the Tropical Products Institute, on what are

termed "mycotoxicoses". They describe toxic substances produced by a
number of moulds other than those belonging to the *A. flavus* series.

   A. tamarii Kita (Lat. *tamarii*, of Tamari, a kind of soy sauce). Colony
colour in mature cultures is brown to deep brown or reddish brown, but
young cultures frequently show a distinct greenish tone and have the
peculiarly lush appearance of a typical *A. flavus*, indicating a very near
relationship to the latter species. Some strains, when first isolated, rapidly

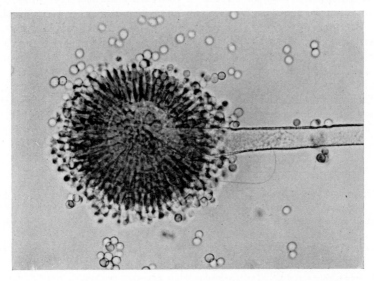

FIG. 128.—*A. oryzae* var. *effusus*—the type of head commonly seen in
slides of any strain of the *A. flavus-oryzae* series. × 500.

become deep brown, with only an evanescent greenish tone in very young
cultures, but are more persistently greenish after several transfers and
eventually develop the typical brown shade only very tardily. Heads large,
at first globose, then becoming loosely radiate, with single chains of spores
visible under low magnification; vesicles thin-walled, easily crushed when
mounting; metulae normally present but absent in some heads, not unusually mixed with phialides; metulae usually $7–10\mu$ long but occasionally
much longer, up to about 30μ; phialides $7–10 \times 3\mu$; conidia globose to
more or less pear-shaped, coarsely roughened from nodules of brown
colouring matter between the two walls, $5–7{\cdot}5\mu$ in diameter. *A. tamarii* resembles the *A. flavus-oryzae* group in producing kojic acid. This substance
was first obtained by Saito in 1907 from the powdered mycelium of *A.
oryzae*. It was later investigated by Yabuta (1924), who showed that it could
be obtained by growing the mould on a synthetic medium with glucose as
the sole source of carbon, and eventually proved it to be 5-hydroxy-2-

hydroxy-methyl-γ-pyrone. Raistrick and co-workers independently discovered kojic acid in 1923, and their work is of great interest in showing that a number of Aspergilli, when grown on Czapek's solution, produce kojic acid, and that all of these belong in the series represented by *A. flavus*, *A. oryzae*, and *A. tamarii* (Birkinshaw *et al.*, 1931). Kojic acid, in water solution, gives with ferric chloride a very intense and characteristic blood-red colour, and this reaction serves as a useful diagnostic test for strains in this group. The simplest method of conducting the test is to grow the mould on the liquid medium for any length of time between two and four weeks, filte rthe solution and add a few drops of a 10 per cent solution of ferric chloride. Raistrick and co-workers grew all their strains on Czapek's solution, but found that the intensity of the reaction varied, and that one strain of *A. flavus* gave no reaction. In the author's experience the test is more certain, and the reaction more intense and uniform, if the Raulin-Thom medium is used instead. *All* strains so far examined give a strongly positive reaction under these conditions. Reference is made in Chapter XV to the industrial uses of *A. flavus-oryzae*.

REFERENCES

Allcroft, R., and Carnaghan, R. B. A. (1963). Toxic products in groundnuts. Biological effects. *Chem & Ind. (Rev.)*, 1963, No. 2, 50–3.

Austwick, P. K. C., Gitter, M., and Watkins, C. V. (1960). Pulmonary aspergillosis in lambs. *Vet. Rec.*, **72**, 19–21.

Bainier, G., and Sartory, A. (1911). Étude biologique et morphologique de certains *Aspergillus*. *Bull. Soc. mycol. Fr.*, **27**, 453–68.

Bainier, G., and Sartory, A. (1912). *Ibid.*, **28**, 257–69.

Benjamin, C. R. (1955). Ascocarps of *Aspergillus* and *Penicillium*. *Mycologia*, **47**, 669–87.

Biourge, P. (1933). Sur les champignons dits "moisissures". A quoi bon leur étude et comment la faire. *Rev. Quest. Sci.*, Jan. 1933.

Birkinshaw, J. H., Charles, J. H. V., Lilly, G. H. and Raistrick, H. (1931). Kojic acid (5-hydroxy-2-hydroxymethyl-γ-pyrone). *Phil. Trans.*, Ser. B, **220**, 127–38.

Coveney, R. D., Peck, H. M., and Townsend, R. J. (1965). Recent advances in mycotoxicoses. *Soc. chem. Ind. Symposium Microbiol. Deterioration in Tropics, 1965.*

Fennell, D. I., and Warcup, J. H. (1959). The ascocarps of *Aspergillus alliaceus*. *Mycologia*, **51**, 409–15.

Feuell, A. J. (1966). Toxic factors of mould origin. *Canad. med. Assoc. J.*, **94**, 574–81.

Report (1962). Interdepartmental Working Party on Groundnut Toxicity Research, *Toxicity associated with certain batches of groundnuts.*

Mangin, L. (1909). Qu'est ce que l'*Aspergillus glaucus*? *Ann. Sci. nat. (Bot.)*, 9 Ser., **10**, 303–71.

Raper, K. B. (1957). Nomenclature in *Aspergillus* and *Penicillium*. *Mycologia*, **49**, 644–62.

Raper, K. B., Fennell, D. I., and Tresner, H. D. (1953). The ascosporic stage of *Aspergillus citrisporus* and related forms. *Mycologia*, **45**, 671–92.

Raper, K. B., and Fennell, D. I. (1965). *The genus Aspergillus*. Baltimore: The Williams & Wilkins Co.

Sargeant, K., and Carnaghan, R. B. A. (1963). Toxic products in groundnuts. Chemistry and origin. *Chem. & Ind. (Rev.)*, 1963, No. 2, 53–5.

Smith, G. (1933). Some new species of *Penicillium*. *Trans. Brit. mycol. Soc.*, **18**, 88–91.

Thom, C., and Raper, K. B. (1939). The *Aspergillus nidulans* group. *Mycologia*, **31**, 653–69.

Thom, C., and Raper, K. B. (1941). The *Aspergillus glaucus* group. *Misc. Publ. U.S. Dep. Agric.*, **426**, 1–46.

Thom, C., and Raper, K. B. (1945). *Manual of the Aspergilli*. London: Baillière, Tindall & Cox.

Wehmer, C. (1899–1901). Die Pilzgattung *Aspergillus* in morphologischer, physiologischer und systematischer Beziehung. *Mém. Soc. Phys. Genève*, **33**, (2), 1–157.

Yabuta, T. (1924). The constitution of kojic acid, a γ-pyrone derivative formed by *Aspergillus oryzae* from carbohydrates. *J. chem. Soc.*, **125**, 575–87.

Yuill, E., and Yuill, J. L. (1938). *Cladosarum olivaceum*. A new Hyphomycete. *Trans. Brit. mycol. Soc.*, **22**, 194–200.

Chapter X

Penicillium and Related Genera

Alive and actively growing, they have individuality as pronounced as their capacities for evil, but the elements of that individuality, color, odor, and habit of growth, are as evanescent as frost designs on a window pane in winter.

C. Thom. *The Penicillia*, 1930

The Penicillia are closely related to the Aspergilli and are just as widespread and omnivorous. Like the Aspergilli, many of them are common and serious agents of destruction, but, on the other hand, there are several species which are best known as the means of conducting industrial fermentations. Probably the most important of these are the cheese moulds, the *P. roqueforti* and *P. camemberti* series. Thom's studies of the occurrence and mode of action of these moulds has made it possible for production of cheeses of the Gorgonzola, Stilton, Roquefort, and Camembert types to be carried on anywhere with success, instead of being confined in special regions. The first attempts to produce citric acid commercially by mould fermentation used Penicillia, of the group formerly known as *Citromyces*, although modern citric acid manufacture is carried on with the aid of *Aspergillus niger*, owing to the better yields obtained. Other successful large-scale fermentations are the manufacture of gluconic acid, using a member of the *P. purpurogenum* series, and of penicillin, using *P. chrysogenum* or *P. notatum*. The studies of Raistrick and his school have shown that the Penicillia exhibit the most varied metabolic activity, different species being capable of synthesizing from glucose a bewildering variety of substances of complex chemical constitution, and it is conceivable that such work will lay the foundations of future commercial processes employing moulds. Amongst many interesting results already obtained may be mentioned the production of ergosterol, the parent substance of Vitamin D, and of a number of substances closely related to ascorbic acid (Vitamin C).

Unfortunately the taxonomy of the genus presents much greater difficulty than that of the related genus *Aspergillus*. In the latter genus there is a wide range of colour to serve as a convenient basis of classification, whereas the great majority of the species of *Penicillium* are some shade of green. Also, whereas the conidial colours of the Aspergilli are comparatively stable throughout the growing period, the greens of the Penicillia change as colonies age and, in addition, vary with changes in cultural conditions. Further, the number of species of *Penicillium* is appreciably greater than the number in *Aspergillus*. The number of specific epithets which have been bestowed runs to many hundreds and, even after many of these have been relegated to synonymy, and others abandoned because of hopelessly inadequate descriptions, there remain, according to the latest taxonomic study of the genus, 137 species and a few varieties.

In view of the ubiquity and importance of the Penicillia it is not surprising that many taxonomists have attempted to devise a satisfactory scheme of classification. Most of the papers and monographs which have been published on the subject during the last half-century are now mainly of historical interest and there is no point in citing them here. The latest publication, *Manual of the Penicillia* by Raper and Thom (1949), is the only one which is of real use to those wishing to make a serious study of the genus. The following is a diagnosis of the genus, in accordance with the ideas of Raper and Thom, and introducing a few terms which are commonly used in descriptions of species.

Penicillium Link ex Fries

(Lat. *penicillus*, an artist's brush)

Vegetative mycelium colourless or pale or brightly coloured, never demati-aceous, septate, either predominantly submerged or partly submerged and partly aerial, with aerial portion closely matted, loosely floccose, or partially as ropes of hyphae; fertile branches (conidiophores) arising from, and more or less perpendicular to, submerged or aerial hyphae, either detached from one another or to some degree aggregated into fascicles or compacted into definite coremia, septate, smooth or rough, terminating in a broom-like whorl of branches (the *penicillus*), the latter consisting of a single whorl of spore-bearing organs (phialides), or twice to several times verticillately branched, with the branching system symmetrical or asymmetrical, the final branches being the phialides; conidia produced (as in *Aspergillus*) by abscission, forming unbranched chains, globose, ovoid, elliptical, or pyriform, smooth or rough, in most cases green during the growing period but sometimes colourless or in other pale colours; perithecia of the perfect state produced by some species, either sclerotium-like, ripening tardily from the centre outwards, or soft, ripening quickly; sclerotia produced by several species.

In penicilli with more than one stage of branching, the branches bearing the phialides are, as in *Aspergillus*, called **metulae**. The branches supporting the metulae, if comparatively short and obviously part of the penicillus, are known as **rami** (Fig. 129).

Langeron (1922) proposed the generic name *Carpenteles* for a species, described by Brefeld in 1874 as *Penicillium glaucum*, producing perithecia which originated as sclerotium-like bodies and gradually formed asci from

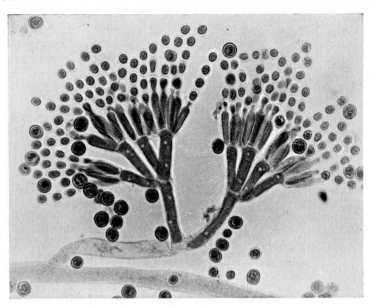

FIG. 129.—*Penicillium*—a typical "penicillus" with several stages of branching and showing the mode of formation of the conidia. × 1000.

FIG. 130.—*Gliocladium roseum*—conidiophores. × 250.

the centre outwards, and Shear (1934) has revived the name to cover a number of species of similar morphology. Raper and Thom do not recognize the genus but have adopted the name *Carpenteles* for a series of species which produce ascospores. More recently Benjamin (1955) has proposed reviving the use of the name *Carpenteles*. In addition, he has erected the new genus *Talaromyces*, to include the species of the *P. luteum* series, which in the perfect state form soft, mostly yellow, perithecia. Raper (1957) has expressed strong disapproval of these suggestions, and argues that the needs of taxonomy are best met by retaining the name *Penicillium* for all the species, whether perithecial or entirely conidial.

Three other genera, closely related to *Penicillium*, and including a few common and important species, were recognized by Thom in his monograph of 1930 and are accepted and dealt with somewhat briefly by Raper and Thom in the Manual. Of these, *Paecilomyces* and *Gliocladium* are, for the present, retained in the *Penicillium* group, since they have many features in common with *Penicillium sensu stricto*, and there are transitional forms linking both of the genera to *Penicillium*. The case of the third genus, *Scopulariopsis*, is quite different. The spore-producing cells in this genus are not phialides but, as Hughes (1953) has shown, annellophores. In addition, the perfect state of *Scopulariopsis*, where known, is *Microascus*, which has perithecia very different from the perfect states of *Penicillium*. *Scopulariopsis* has therefore been described in Chapter VIII.

Gliocladium Corda

(Gr. *gloios*, mucilage; *klados*, a branch)

This genus differs from *Penicillium* in that a mucilaginous substance is produced by the fruiting organs, and the conidia, instead of standing in detached chains, adhere together, or, in the most typical species, lose entirely the catenary formation and form solid, slimy balls. The line of separation from *Penicillium* is not a sharp one, but in the more common species the distinctive character is well marked and unmistakable.

G. roseum (Link) Bainier Lat. *roseus*, pink (Fig. 130). Colonies are loosely floccose in texture, pale pink to salmon when first isolated, often becoming almost white on continued cultivation, with conidia 5–7 × 3–5μ, forming slime balls, in some strains quickly, in others tardily. Barnett and Lilly (1962) have shown that this species is a destructive parasite of a number of other fungi.

The perfect state of *G. roseum* was discovered and described by Smalley and Hansen (1957). It is a species of *Nectria*, to which they gave the name *N. gliocladioides*. It is interesting that the three genera, originally separated from *Penicillium* entirely on account of differences in conidial morphology, should prove to have perfect states differing from the known perfect states of *Penicillium*, and differing widely from each other. The perfect state of *Gliocladium* is *Nectria*, with bright coloured perithecia and two-celled ascospores; that of *Paecilomyces* is *Byssochlamys*, which produces irregular clusters of asci, without any trace of peridium (containing wall); whilst that of *Scopulariopsis* is *Microascus*, with dark, carbonaceous perithecia and one-celled ascospores.

G. *deliquescens* Sopp (Lat. *deliquescens*, melting, dissolving) (Fig. 131). Colonies thin, with little aerial mycelium, soon forming dark green fruiting

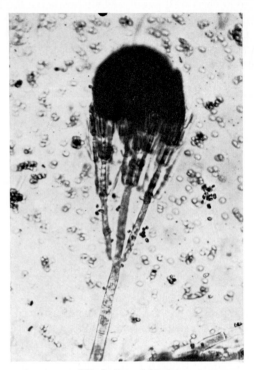

FIG. 131.—*Gliocladium deliquescens.* × 400.

areas and eventually becoming completely slimy; conidia green, 3–3·5 × 2–2·5μ.

G. *catenulatum* Gilman and Abbott (Lat. *catenula*, a little chain). Colonies floccose to funiculose, white at first, becoming clear dark green in scattered patches; conidia chains in long slimy columns; conidia greenish, 4–7·5 × 3–4μ.

In all the species the penicilli are mostly three to four times verticillate, with the branches becoming more slender in successive whorls (Figs. 130 and 131).

The three species described are all fairly common in soil and have been isolated many times from mouldy fabrics and various items of military equipment.

Paecilomyces Bainier

(Gr. *poikilo-*, varied)

Colony colour white, pale pink, lilac, yellow brown, or dirty greyish brown, never green; texture closely matted, loosely floccose, or funiculose. Fruiting organs vary considerably in complexity (much as in *Scopulariopsis*), and the conidia are usually ovate and borne in very long tangled chains. The

FIG. 132.—*Paecilomyces variotii*—ropes of hyphae and various types of spore-bearing structures. ×250.

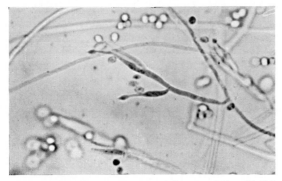

FIG. 133.—*Paecilomyces carneus*—characteristic phialides. ×800.

phialides are of very characteristic shape, narrow flask-shaped at the base and terminating in long slender spore-bearing tips, which are usually bent away from the main axis (Figs. 132, 133). Most species also produce so-called "macrospores", which are comparatively large, globose or ovate, aleuriospores, borne singly or in small clusters, and usually found on the mycelium close to the substrate or even in the submerged mycelium (Fig. 134).

Brown and Smith (1957) have published the results of a taxonomic study of the genus. They have shown that the name *Spicaria*, as used by Gilman

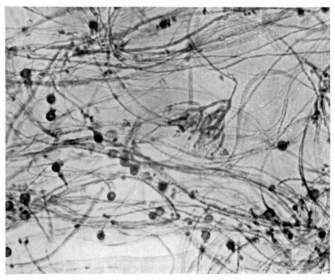

FIG. 134.—*Paecilomyces puntonii*—macrospores. × 500.

in a well-known book, and by many other authors, is a *nomen ambiguum*, and have transferred a number of species, originally described as *Spicaria*, to *Paecilomyces*. Four species are of common occurrence.

P. variotii Bainier (Dr. Variot, French physician). Colonies grow well on almost any kind of substrate, and are pale dull brown or yellowish brown, loosely floccose and mostly ropy; phialides 15–20μ long; conidia elliptical, 5–7 × 2·5–3μ, formed in very long chains. Some isolates produce abundant macrospores, others few or none. Also the complexity of the sporing structures varies considerably.

Byssochlamys fulva (see Chapter V) has a very similar conidial state and it is possible that strains of *P. variotii* are merely strains of *B. fulva* which have lost the ability to produce asci.

P. carneus (Duché and Heim) Brown and Smith (Lat. *carneus*, flesh coloured). Colonies grow slowly, compact, matted, showing very little

ndency to form funicles of hyphae, pure white at first, becoming pale
ear pink, especially if exposed to light, with reverse slowly turning deep
een; sporing structures simple or fairly complex; phialides mostly 12–16μ
ng; conidia elliptical, definitely rough, 3–4 × 2–2·5μ. Found on a wide
riety of substrates.

P. puntonii (Vuillemin) Nannizzi (V. Puntoni, Italian mycologist).
olonies growing rapidly, spreading, funiculose, dirty white often tinged
ith yellow, with reverse usually pale greyish buff; sporing structures
nsisting of solitary phialides or simple verticils of phialides, which are
–16μ long; conidia almost cylindrical, 3–5 × 1·5–3μ; macrospores abun-
nt, 4–8 × 4–6μ. Not very common, but widely distributed.

P. fumosoroseus (Wize) Brown and Smith (Lat. *fumosoroseus*, smoky
nk). Colonies spreading, deeply floccose, white at first then gradually
ll pink, with reverse slowly turning pale yellow; sporing structures con-
sting of solitary phialides or simple verticils; phialides 7–18μ long;
nidia almost cylindrical, 3–4 × 1–2μ. Best known as a parasite of insects,
t it has been found on a number of other substrata.

P. marquandii (Massee) Hughes. See p. 189.

The Penicillia proper

he large number of species fall naturally into three main sections, recog-
zed as a primary basis of separation by all modern authorities. The
onoverticillata, equivalent to the section Aspergilloides of some authors
d to Wehmer's genus *Citromyces*, have penicilli consisting of a single
horl of phialides (Fig. 135). The Biverticillata-Symmetrica have a com-
ct whorl of metulae, each bearing phialides, the whole penicillus being
proximately symmetrical about the axis. In addition, the spores are
most always ovate or pear-shaped and are borne on slender, tapering
cuminate) phialides (Fig. 136). The third and largest section, the Asym-
etrica, includes all species in which the penicillus is branched more than
ice and is asymmetrical about the axis, or if approximately symmetrical,
s not the compacted structure and the tapering phialides of the Biverti-
lata-Symmetrica.

Each section is subdivided in various ways—presence or absence of
rithecia or sclerotia, texture of colonies, and so on. Both of the general
ys given by Raper and Thom and the key which follows below do not
d to individual species but to "series", a series including one to several
sely related species. For many purposes it is sufficient to place a parti-
lar isolate in its correct series, rather than attempt a closer identification.
Perhaps the most difficult terms for the beginner to understand are those
ferring to texture, such as velvety, lanose, funiculose, and fasciculate.
he velvety colony has little aerial mycelium, the visible growth consisting
most entirely of conidiophores arising from submerged hyphae, and
ving exactly the appearance of a short pile velvet. In lanose (woolly)

species there is a tangled mass of aerial mycelium and the conidiophor
arise mainly from these aerial hyphae, the points of origin being qui
away from the substratum. In the Funiculosa subsections part or most
the aerial hyphae are aggregated into trailing ropes, a character which
usually not difficult to recognize under a low power of the microscope,
even with a good hand-lens. In fasciculate colonies the conidiophore

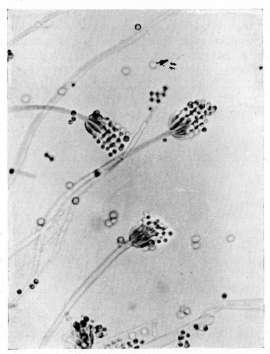

FIG. 135.—*Penicillium spinulosum*—typical monoverticillate
penicilli. × 500.

mostly arising from the substratum, are not evenly distributed, as in t
velvety type of colony, but are more or less in clusters, giving a granular
mealy appearance to the colony, or may even be compacted into defini
coremia.

One of the great difficulties in attempting to identify Penicillia is th
most of the cosmopolitan species appear to be to some extent unstabl
Some isolates tend to vary slightly in successive cultures made, in t
laboratory, under apparently identical conditions, but these, with few e:
ceptions, are approximately stable over periods long enough to permit
identification, and the instability in question is something more fund.
mental. When trying to identify a *Penicillium* it is frequently impossible

find an exact fit for the strain in any published description, although it is readily placed as "near to" two, or perhaps three, species previously described. Biourge (1923), with a comparatively small number of cultures in his hands, gave specific epithets to all the members of such related series examined by him, but Thom (1930) showed that, when hundreds of strains are examined, differences become less perceptible, and the only alternatives are the bestowal of new specific epithets for almost every strain isolated, and the more practical one of lumping together nearly related strains as a group species. Most living things show, under natural conditions, some tendency to mutate, and it is reasonable to expect that a mould of common and widespread occurrence should be found as a series of slightly different

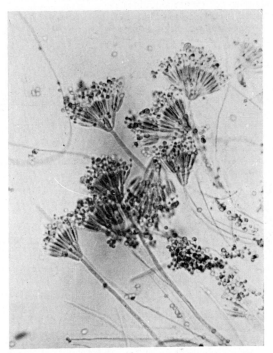

FIG. 136.—*P. variabile*—symmetrically biverticillate penicilli. × 500.

strains, representing the tendency to mutate within certain limits, limits which are narrow for some species, wider for others. It is emphasized again and again by Raper and Thom that the accepted species in any series represent little more than convenient centres around which to group the numerous different strains, and it is a great merit of their book that the amount and type of variation to expect is discussed for the great majority of species, and certainly for all the common species.

KEY TO THE SERIES

1. Penicilli consisting of phialides only. Conidiophores unbranched or branched irregularly, with each branch bearing a whorl of phialides only . . . . . MONOVERTICILLATA (p. 184) 2

  Penicilli with phialides and metulae; usually asymmetric. If symmetrical then with abruptly tapering phialides . . . . ASYMMETRICA (p. 188) . 10

  Penicilli normally biverticillate, funnel-shaped with slender metulae and slender acuminate phialides. (Fractional penicilli common in some species but phialides always typical) . . . . BIVERTICILLATA-SYMMETRICA (p. 209) . 27

2. Perithecia or sclerotia formed 3
  No perithecia or sclerotia 4
3. Perithecia, ripening slowly . . *P. javanicum* series (p. 184)
  Pink sclerotia formed . . . *P. thomii* series (p. 184)
4. Conidiophores unbranched (occasional branches in some species) 5
  Cphs. irregularly branched, each branch bearing a typical monovert. penicillus . . . . RAMIGENA series (p. 188)
5. Colonies velvety (or appearing so) 6
  Colonies floccose or funiculose 9
6. Cphs. mostly from substrate 7
  Cphs. as short branches from aerial hyphae . . . . *P. decumbens* series (p. 185)
7. Colonies spreading widely 8
  Colonies restricted in growth . *P. implicatum* series (p. 188)
8. Conidia globose or nearly so . . *P. frequentans* series (p. 185)
  Conidia definitely elliptical . . *P. lividum* series (p. 186)
9. Colonies floccose, with little tendency to form hyphal ropes . . *P. restrictum* series (p. 188)
  Colonies definitely funiculose . *P. adametzi* series (p. 188)
10. Branching of penicillus irregular, with rami and metulae divergent DIVARICATA (p. 188) . 11
  Branching fairly compact 15
11. Perithecia (slow-ripening) formed . *Carpenteles* series (p. 188)
  Sclerotia, or masses of thick-walled cells produced . . . *P. raistrickii* series (p. 188)
  Neither perithecia nor sclerotia 12
12. Colonies showing no green colour; lilac or fawn . . . *P. lilacinum* series (p. 189)
  Colonies some shade of green 13
13. Colonies blue-green to grey-green 14
  Colonies grey to olive-grey; reverse dull yellow to brown . *P. nigricans* series (p. 191)
14. Chains of conidia divergent; phialides with comparatively long tips *P. janthinellum* series (p. 189)

Conidial chains tending to form
columns; phialides with short tips   *P. canescens* series (p. 191)
15. Colonies velvety .   .   .   .   VELUTINA (p. 191) .   .   16
     Colonies floccose .   .   .   .   LANATA (p. 200)   .   21
     Colonies funiculose (ropy)   .   .   FUNICULOSA (p. 200)   .   22
     Colonies mealy, granular or coremi-
     form .   .   .   .   .   FASCICULATA (p. 201)   .   23
16. Penicilli not usually branched below
     metulae, hence biverticillate, but
     with the appearance of a bunch of
     monoverticillate heads   *P. citrinum* series (p. 191)
     Penicilli usually more complex   17
17. Penicilli somewhat loose in structure   18
     Penicilli short and compact; conidial
     chains becoming much tangled   .   *P. brevi-compactum* series (p. 196)
18. Colony outline regular; conidio-
     phores usually smooth   19
     Colony outline cobwebby; cphs.
     rough .   .   .   .   .   *P. roqueforti* series (p. 197)
 9. Reverse of colonies yellow with pig-
     ment often diffusing .   .   .   *P. chrysogenum* series (p. 193)
     No yellow pigment   20
20. Colonies bluish green; found in soil   *P. oxalicum* series (p. 198)
     Pale olive-green; on citrus fruits   .   *P. digitatum* (p. 199)
21. Colonies persistently white or tardily
     pale grey-green   .   .   .   *P. camemberti* series (p. 200)
     Colonies quickly some shade of green   *P. commune* series (p. 200)
22. Colonies some shade of green; peni-
     cilli large; conidia not markedly
     elliptical .   .   .   .   *P. terrestre* series (p. 201)
     Colonies variously coloured but not
     green; conidia elliptical to cylin-
     drical .   .   .   .   *P. pallidum* series (p. 201)
23. Sclerotia produced   .   .   .   *P. gladioli* (p. 203)
     Sclerotia not produced   24
24. Colonies showing some fasciculation
     but simple cphs. predominating   25
     Colonies with most of cphs. in
     fascicles or small coremia   .   *P. granulatum* series (p. 207)
     Definite coremia predominant   .   *P. claviforme* series (p. 209)
25. Colonies not showing true green   .   *P. ochraceum* series (p. 203)
     Colonies in bright yellowish green
     shades; cphs. rough   .   .   *P. viridicatum* series (p. 203)
     Colonies in blue-green shades; cphs.
     usually rough   .   .   .   *P. cyclopium* series (p. 204)
     Colonies in grey-green shades; cphs.
     rough or smooth   .   .   *P. expansum* series (p. 204)
     Colonies pale grey-green to greenish
     grey; cphs. smooth   26
26. Phialides 8μ or more in length;
     young conidia almost cylindrical .   *P. italicum* (p. 207)
     Phialides 6μ or less in length   .   *P. patulum* (p. 207)
27. Perithecia produced, soft, white or
     yellow .   .   .   .   *P. luteum* series (p. 209)
     Sclerotia produced   .   .   *P. novae-zeelandiae*

G

Neither perithecia nor sclerotia 28
28. Colonies predominantly coremial . *P. duclauxi* (p. 210)
    Colonies more or less funiculose . *P. funiculosum* series (p. 210)
    Colonies neither coremial nor funi-
    culose 29
29. Reverse and medium red or pur-
    plish; colonies velvety . . *P. purpurogenum* series (p. 212)
    Reverse slowly orange; colonies
    restricted, velvety or more or less
    floccose . . . . . *P. rugulosum* series (p. 214)

The above key is substantially that of Raper and Thom and includes all the series described by them in the Manual. Where a series contains only one species the word "series" has been omitted from the key.

In attempting to identify any mould it is advisable to get as much information as possible from a study of living colonies before making any microscopic slides. This procedure is absolutely necessary in the case of the Penicillia. Such data as gross characteristics of colonies, origin of conidiophores, and disposition of spore chains are not only essential parts of the diagnoses of species but are used freely in constructing the key.

The diagnoses given in the following pages are mostly in standard form. They are brief but should be adequate to enable the student to recognize all the commoner species. Details of rami and metulae have been omitted in order to save space, because these are less important than the characteristics of the whole penicillus and of the phialides and conidia. Unless otherwise stated the descriptions all refer to colonies on Czapek agar.

It has not been possible to give the etymology of all the specific epithets in this genus. The omissions are all personal genitives, which, unless derived from names of well-known mycologists, are of little interest. Most of them are names of species described by Zaleski, who, in his monograph (1927), used personal genitives almost exclusively for the names of his new species, but without dedications.

The Monoverticillata

P. javanicum series (Lat. *javanicum*, javanese) includes 5 species, none of which is common or of economic importance.

P. thomii series, characterized by the production of sclerotia, includes 5 species, of which 2 are fairly common.
    P. thomii Maire (named after Charles Thom). Colonies pale blue-green, almost velvety at first, with reverse yellow to brownish, soon developing masses of pinkish sclerotia. Fresh isolates usually produce abundant sclerotia but, when kept in cultivation, soon become entirely conidial. This species has been reported as parasitic on orchid leaves (Moore, 1941).
    P. lapidosum Raper and Fennell (Lat. *lapidosus*, stony, full of stones. So

named because of the stone-like sclerotia). Easy to recognize because of the striking colours developed in culture media. On Czapek agar spreads rapidly, pale greyish yellow then dull orange, consisting almost entirely of hard sclerotia, with scattered conidiophores not obvious but visible under the microscope; reverse deep yellow, with the colour diffusing through the agar, occasionally becoming reddish in age. On malt agar the surface is similar but the reverse, at first yellow, soon turns a deep mahogany red. Penicilli are described by Raper and Thom as mostly monoverticillate, but with occasionally 2 metulae. Fresh isolates usually produce penicilli with clusters of metulae, similar to those of the *P. citrinum* series. On long cultivation the penicilli become simpler in structure, and finally are almost exclusively monoverticillate. The conidia are smooth, $2 \cdot 5 - 3 \times 2 - 2 \cdot 5 \mu$. Not uncommon, particularly in soil.

P. decumbens series (Lat. *decumbens*, lying down) is characterized by velvety to slightly floccose colonies in which the conidiophores are borne on aerial hyphae close to the substrate or on a tough felt of mycelium. There are 5 species, of which 2 have been found fairly frequently in industrial work.

P. citreo-viride Biourge (Lat. *citreus*, lemon yellow; *viridis*, green). Colonies are slow-growing and of a pale yellowish-grey colour, with only a suggestion of green, velvety to very slightly floccose, with reverse yellow and sometimes spotted, especially near the edge of the colony, with deep brown; conidial chains loosely parallel or divergent; phialides $9-12 \mu$ by about $2 \cdot 5 \mu$; conidia thin-walled, smooth, $2 \cdot 2 - 2 \cdot 8 \mu$ diam.

P. fellutanum Biourge (Fellut, old name of Feluy, Belgian village). Colonies are slow-growing, blue-green to sage green or fairly dark green in some strains, often with centre of colony almost white, usually with a submerged outer zone; conidial chains in somewhat ragged columns; phialides closely packed, $6-8 \times 1 \cdot 5 - 2 \mu$; conidia fairly thick-walled, subglobose to somewhat elliptical, $2 \cdot 5 - 3 \times 2 - 2 \cdot 2 \mu$.

P. frequentans series (Lat. *frequentans*, frequent, common). There are 3 species, of which 2 are of widespread distribution and exceedingly common occurrence on all kinds of substrata. Colonies vary from strictly velvety to very slightly floccose and the conidial chains are in columns.

P. frequentans Westling grows rapidly on all culture media, with colonies normally a rich green, becoming darker and browner in age, and reverse at first yellow then deep brown; conidiophores usually smooth, appreciably swollen at the tip; phialides mostly $9-10 \times 3-3 \cdot 5 \mu$; conidia globose or almost so, smooth or slightly rough, about 3μ in diameter, packed into long solid columns (Fig. 137). See p. 215 for biochemical characteristics.

Some strains of this common organism differ somewhat from the usual type, the conidial colour being a rather pale blue-green and reverse colourless or nearly so. Before the appearance of the Manual these would

have been identified as *P. flavi-dorsum* Biourge. However, the latter is
now regarded as a synonym of *P. frequentans*.

P. spinulosum Thom (Lat. *spinulosus*, with little spines). Colonies are
broadly spreading, and consist of a somewhat loose felt, blue-green or
grey-green, becoming definitely greyish in old cultures, with broad white
edge during the growing period; reverse white to cream or faintly pink;
conidiophores arising from either submerged or aerial hyphae, smooth
or slightly rough, with apices definitely swollen; phialides not very

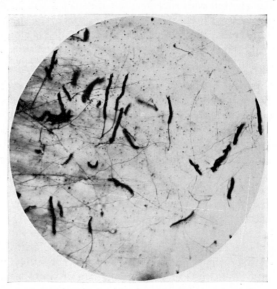

FIG. 137.—*P. frequentans*—columns of spore chains
as seen in Petri dish. × 50.

numerous, 6–9 × 2·5–3μ; conidia globose or nearly so, spinulose, 3–3·5μ
diam., borne in loose columns. Certain strains differ from the normal in
producing an intense purple colour in certain culture media. On solid
media the appearance is very distinctive, there being a central patch of
green, then a ring of bright yellow, and an outermost zone of purple.
Birkinshaw and Raistrick (1931) have isolated the purple pigment and
shown it to be a methoxydihydroxytoluquinone. A solution of the pigment is turned yellow on acidification and the yellow colour which
develops in culture media is due to the gradual accumulation of citric
acid. Such strains, when kept in culture, invariably lose, after a short
time, the power of producing the pigment.

P. lividum series (Lat. *lividus*, dark blue). The 3 species of this series all
produce very bluish green colonies, with elliptical rough spores. None is
of sufficiently common occurrence to merit further description here.

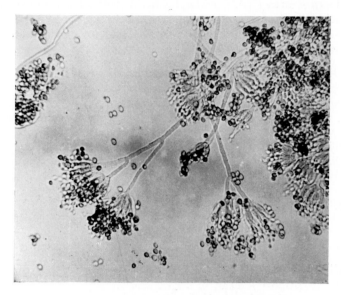

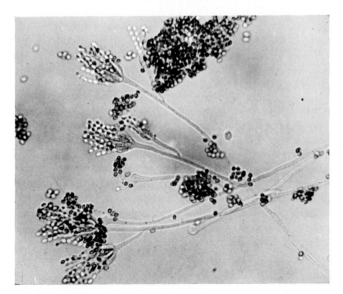

FIGS. 138, 139.—*P. charlesii*—variously branched conidiophores characteristic of the Monoverticillata-Ramigena. × 500.

P. restrictum series (Lat. *restrictus*, restricted, close). The 2 species are not closely related to each other, but both are characterized by floccose colonies, coarsely roughened globose conidia, and conidiophores often suggestive of *Aspergillus sydowii*.

P. implicatum series (Lat. *implicatus*, tangled). The type species, *P. implicatum*, is the only one of the 3 species which is at all common.

> *P. implicatum* Biourge grows somewhat restrictedly, rich blue-green, often with the green areas on a background of yellow to orange mycelium, velvety or nearly so: reverse bright orange-red; conidiophores mostly smooth, with enlarged apices; phialides 8–10μ by about 2μ; conidia globose to sub-globose, 2–2·5μ diam., finely roughened, in loosely parallel chains, not true columns.

P. adametzi series. The 4 species are all fairly common in soil. The characteristics of the series are funiculose colonies with reverse orange to brown or purple, and rough globose conidia. The species may be distinguished as follows:

> *P. adametzii* Zaleski. Colonies very ropy; reverse orange-yellow; conidia slightly rough, 2–2·5μ diam., in tangled chains.
>
> *P. terlikowskii* Zaleski. Colonies less definitely ropy; reverse orange-red to brown; conidia about 3μ, rough, in columns.
>
> *P. vinaceum* Gilman and Abbott (Lat. *vinaceus*, of wine, hence winecoloured). Colonies restricted, floccose funiculose, with large vinaceous drops; reverse red to purple; conidia 2–2·5μ, slightly rough.
>
> *P. phoeniceum* van Beyma (Lat. *phoeniceus*, of Phoenicia). Colonies scarcely funiculose, almost velvety, without respiration drops; reverse intense purple; conidia 2·5–3μ, smooth or nearly so.

RAMIGENA series. There are 5 recognized species but none is at all common. Nevertheless they occur occasionally on mouldy material, especially material of tropical origin. The fruiting structure consists of an irregular and loose branching system, each branch terminating in a monoverticillate penicillus (see Figs. 138, 139). The student may find some difficulty in distinguishing between this series and the Divaricata, since the branching systems are similar in the two groups. A useful point of distinction is that in the Ramigena the tips of all the branches are usually definitely swollen, as are the tips of the conidiophores of the true Monoverticillata, and are not swollen in the Divaricata.

The Asymmetrica

Subsection DIVARICATA

The *Carpenteles* series with 3 species, and the *P. raistrickii* series with 4, do not include any common or important species.

P. lilacinum series includes 3 species, all characterized by complete lack of green colour.

P. lilacinum Thom (Lat. *lilacinus*, lilac-coloured) is a common soil organism and is also found on a variety of organic substrates. Colonies are floccose, at first white then gradually purplish or pale greyish violet; reverse in some strains fairly dark purple, in others only slightly coloured; penicilli varying from complex structures to simple clusters of phialides on short branches from trailing hyphae; phialides 7–8μ long; conidia elliptical, smooth, 2·5–3 × 2μ.

Spicaria violacea Abbott. The organism described under this name in the Manual is a *Paecilomyces*, and its correct name is *Paec. marquandii* (Massee) Hughes. It is readily distinguished from *P. lilacinum* by forming colonies of bluer surface colour, and with reverse bright yellow instead of purple.

P. janthinellum series. There are 7 species in this series, but only 2 are of common occurrence.

P. janthinellum Biourge (Lat. *janthinellus*, somewhat violet) is an extremely common organism. It is widely distributed in soil and, like most common soil Penicillia, is found on a wide variety of substrates. For the beginner it is a puzzling species because different isolates may show very considerable differences in appearance of colonies. Colonies usually form a tough felt, with surface slightly floccose or ropy, radially furrowed, pale grey-green or greenish grey; reverse in many freshly isolated strains reddish to purple, in others, and in most old strains, colourless to pale dirty yellow; respiration drops pale brownish or lacking; penicilli irregular, sometimes monoverticillate, sometimes forming fairly compact heads with metulae and phialides, but with the metulae often of different lengths, sometimes with loose irregular branching suggestive of the Ramigena series (Fig. 140); conidiophores smooth or slightly rough, 3–3·5μ diam.; phialides divergent, 8–10 × 2–2·2μ, tapering abruptly, but with comparatively long slender tips (this species forms one of the transitions between *Penicillium* and *Paecilomyces*); conidia mostly elliptical but sometimes nearly globose, delicately roughened, 3–3·5μ in long axis, borne in divergent or tangled chains.

P. simplicissimum (Oudemans) Thom (Lat. *simplicissimus*, very simple). Colonies pale blue-green, thick velvety with a tough basal felt; reverse colourless to pale dull yellow; penicilli usually consisting of a group of 2–4 metulae and appearing as a bunch of monoverticillate heads, or as several short branches from a trailing hypha, each bearing a cluster of phialides; conidiophores rough, about 2·5μ diam.; phialides tapering abruptly at tip, 8–10 × 2–2·5μ; conidia elliptical to subglobose, 2·5–3μ in long axis, delicately roughened, in divergent to tangled chains. Differs from *P. janthinellum* in its brighter conidial colour, lack of red colour in reverse, simpler penicilli, and somewhat smaller spores.

FIG. 140.—*P. janthinellum*—divaricate penicillus as seen in living culture. × 100.

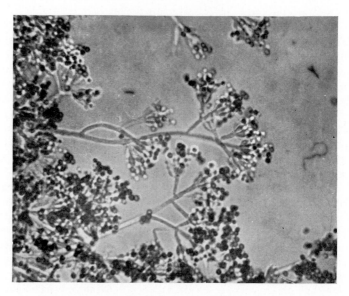

FIG. 141.—*P. nigricans*—penicillus. × 500.

P. canescens series (Lat. *canescens*, becoming whitish). None of the 3 species is common. They are distinguished from the last series by their production of fairly definite columns of spore chains.

P. nigricans series. There are 5 species, very closely related and difficult to distinguish.

P. nigricans Bainier (Lat. *nigricans*, becoming black). A soil organism of wide distribution. Colonies thick velvety, in fresh isolates almost pure grey becoming very dark in age, in old strains usually a definite greenish shade at first, becoming slowly dark grey; reverse deep orange-red to brownish orange; penicilli very divaricate, consisting either of a fairly definite cluster of metulae, or metulae scattered along the terminal portion of the conidiophore, or of two or more small clusters of metulae; conidiophores smooth, 2·5–3μ diam.; phialides 7–8 × 2μ approximately; conidia globose, very rough, spiny, 3–3·5μ diam., in chains at first roughly parallel then tangled (Fig. 141).

Subsection VELUTINA

The 6 series in this subsection are all of importance. Most isolates are readily placed here because of the smooth velvety appearance of colonies. However, occasional strains, particularly of the *P. chrysogenum* series, tend to produce somewhat floccose colonies, and these can be puzzling to the student until he has a clear mental picture of the sum-total of the characteristics of the various series.

P. citrinum series. The typical penicillus consists of a cluster of metulae, often of somewhat different lengths, each bearing a compact group of phialides, the whole appearing as a verticil of monoverticillate heads (Fig. 142). There are 3 species, of which *P. citrinum* is abundant and the other 2 fairly common.

P. corylophilum Dierckx (Gr. *korylos*, hazel tree; *phileo*, to love). Colonies on Czapek agar somewhat restricted, blue-green to grey-green, and finally olive-brown, with reverse pale dirty brown; on malt agar remaining green, with reverse very dark, almost black; penicilli with 2–3 unequal metulae; conidiophores smooth, 2–2·5μ diam.; phialides 8–12 × 2–2·5μ; conidia subglobose to elliptical, 2·5–3μ in long axis, smooth, in roughly parallel or tangled chains, not in true columns.

P. citrinum Thom (Lat. *citrinus*, lemon yellow). Colonies restricted, dull grey-green with very narrow white edge, velvety; drops pale dull yellow; reverse yellow to dull orange with agar coloured similarly or pinkish; penicilli typical, with several metulae; conidiophores smooth, 2·5–3μ diam.; phialides crowded, 8–11 × 2–2·5μ; conidia globose to subglobose, 2·5–3μ diam., packed in solid divergent columns, one to each metula (Fig. 143). Abundant everywhere on a wide variety of substrata. On Czapek's solution and some other liquid media a golden yellow colour is

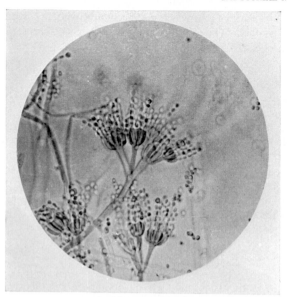

FIG. 142.—*P. citrinum*—divaricate penicillus. × 500.

FIG. 143.—*P. raistrickii*—divergent columns of spore chains as seen in Petri dish culture. × 100. *P. citrinum* and *P. steckii* are similar in appearance.

produced. The pigment, citrinin, has been isolated by Hetherington and Raistrick (1931), but it is only comparatively recently that its chemical structure has been completely elucidated. It is readily obtained, as a yellow crystalline precipitate, by acidification of the solution on which the mould has been grown. Raistrick and Smith (1941) have shown that the substance has marked anti-bacterial properties.

P. steckii Zaleski. Resembles *P. citrinum* in producing divergent columns of spore chains. Differs in not producing citrinin. Colonies dull yellow-green, becoming almost pure grey; reverse pale dull yellow to buff; conidiophores smooth, $2·8–3·2\mu$ diam.; phialides $8–10 \times 2\mu$; conidia globose, smooth or slightly rough, $2–2·5\mu$ diam.

P. chrysogenum series, formerly known as the Radiata, characterized by forming spreading colonies with a clean circular outline. The typical penicillus consists of a compact verticil of metulae and phialides, with one or more branches from lower nodes of the conidiophore, these bearing phialides alone or metulae and phialides, each metula producing a fairly definite column of spore chains, but occasionally with the chains of the terminal cluster of metulae united into one broad columnar mass. The 4 species recognized in the Manual are little more than landmarks in an extremely variable series. However, the majority of isolates fit fairly well with *P. chrysogenum* or *P. notatum*.

P. chrysogenum Thom (Lat. *geno*, to cause; Lat. ex Gr. *chrysos*, golden yellow). Colonies broadly spreading, blue-green to bright green, with broad white margin during the growing period, smooth velvety, usually becoming greyish or purplish brown in age with overgrowth of white or rosy hyphae; reverse yellow, with colour diffusing somewhat; drops usually numerous, colourless to bright yellow; penicilli fairly complex (Fig. 144) with all parts smooth; conidiophores $3–3·5\mu$ diam.; phialides $8–10 \times 2–2·5\mu$; conidia elliptical to subglobose, $3–4\mu$ in long axis, smooth; milk digested to give a golden yellow fluid; some penicillin produced by most strains. Some strains are stable in culture, especially if kept on organic media, but others tend to produce more and more floccose mycelium in successive cultures and eventually become quite atypical.

P. notatum Westling (Lat. *notatus*, marked, noted). Differs from *P. chrysogenum* in producing simpler penicilli, seldom with rami, and in having globose spores about 3μ diam. or slightly larger, borne in ill-defined columns or even in tangled chains (Fig. 145).

P. herquei Bainier and Sartory (epithet from Col. Herqué, who sent the specimen to Bainier). Colonies thick velvety, bright to dull yellow-green, with reverse intense yellow-green to almost black. On malt agar masses of orange-red pigment are deposited in the medium. Conidiophores $3·5–4·5\mu$ diam., smooth or with the upper portion coarsely roughened; metulae $10–15 \times 4·0–4·5\mu$, often uneven in length in the same verticil; phialides $9–12 \times 3·0–4·0\mu$, abruptly tapered at the tip; conidia elliptical,

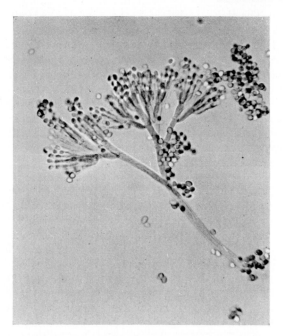

FIG. 144.—*P. chrysogenum*—penicillus. ×500.

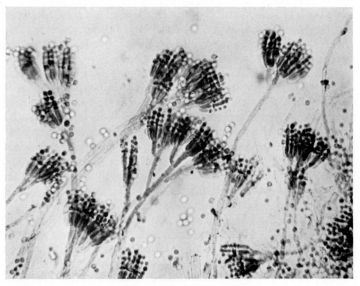

FIG. 145.—*P. notatum* (Fleming's strain). ×500.

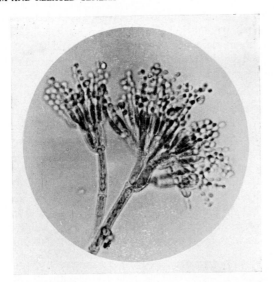

FIG. 146.—*P. brevi compactum*—normal penicillus.
× 600.

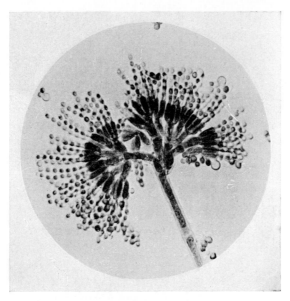

FIG. 147.—*P. brevi compactum*—very large penicillus.
× 600. Such are not uncommon, particularly in
freshly isolated cultures.

smooth or very faintly roughened, 3·5–4·0μ in long axis, borne in loosely parallel or tangled chains.

Raper and Thom place this species in the Biverticillata-Symmetrica. It does not produce the acuminate phialides, which are so characteristic of this series, and is much better classified with the Asymmetrica-Velutina.

P. brevicompactum series. There are 3 species, with many synonyms, recognized at present. The name very aptly describes the penicilli characteristic of these species. There are frequently three or even four stages of branching, and yet the total length of the penicillus is only 35μ to 45μ at the most (Figs. 146, 147). Invariably the metulae and rami are closely appressed to the main axis, giving a very distinctive appearance, even to penicilli which show comparatively little branching. Colonies usually appear velvety, but the conidiophores do not arise directly from submerged hyphae but from a thin surface mycelial felt. In young cultures of all three species the penicilli may be suggestive of *P. citrinum*, but the metulae are

FIG. 148.—*P. brevi compactum*—heads in living culture, showing the very characteristic tangled spore chains. × 100.

always more compacted together and are shorter in proportion to the phialides. Older cultures are unmistakable, since the spore chains are no longer parallel but are twisted and tangled in a very characteristic way (Fig. 148). In all three species, particularly if growing under very moist conditions, marginal extension of colonies is often by stolon-like hyphae. See p. 215 for biochemical characteristics.

P. brevi compactum Dierckx (Lat. *brevis*, short; *compactus*, compact). Colonies restricted in growth, grey-green with narrow edge shading through pale blue-green to white, becoming greyer and often brownish in age; drops dull yellow to brownish; reverse dull yellow or greenish brown; penicilli mostly complex, particularly in freshly isolated cultures; conidiophores coarse, 4–5μ diam., smooth or slightly rough, often with apex enlarged; rami and metulae enlarging upwards, more or less wedge-shaped, short, mostly 4–5μ diam. but often inflated and reaching 6–7μ; phialides 7–10 × 3–3·5μ, occasionally broader; conidia globose or sub-globose, slightly rough, 3·5–4μ diam. Occasionally a secondary growth appears on top of an established culture, pale blue-green, with long stiff conidiophores and enormous conidial heads, appearing almost like an *Aspergillus*. This is the commonest of the three species. In some seasons it occurs abundantly as a parasite of agarics, it is found growing on textiles and a variety of other manufactured products, and is a common infection of stored maize, the consumption of infected grains being supposed at one time to be the cause of Pellagra.

P. stoloniferum Thom (Lat. *fero*, to bear; hence, bearing stolons or runners). Differs from the above in colonies being thinner, more yellowish green, and more wrinkled; conidiophores thinner and sinuous; and elements of the penicillus not inflated.

P. paxilli Bainier (of *Paxillus*, a genus of agarics). Differs from the preceding species in producing penicilli which are rarely branched below the metulae, but which nevertheless have the typical compact structure and form tangled chains of conidia.

P. roqueforti series, formerly the Stellata. There are 2 species both associated with cheese ripening.

P. roqueforti Thom (of Roquefort, a variety of blue-veined cheese). Colonies usually broadly spreading, blue-green then duller dark green, smooth velvety with irregular margin consisting of radiating lines of conidiophores, usually termed "arachnoid" from its similarity to a spider's web; reverse sometimes almost colourless but more usually very dark green, almost black; penicilli distinctly asymmetric, commonly with three stages of branching (Fig. 149); conidiophores rough, 4–6μ diam.; metulae often rough; phialides 8–12 × 3–3·5μ; conidia globose or nearly so, mostly 4–5μ diam. but occasionally larger, smooth, borne in loose columns or tangled chains. Most isolates grow normally on almost any culture medium but occasional strains, especially those from

French Roquefort, grow very poorly on Czapek agar. All strains grow
well on malt agar and many of them produce smooth conidiophores.
This species is used for ripening cheeses of the Roquefort, Stilton and
Gorgonzola types. Many local varieties of blue-veined cheese also owe
their characteristic appearance and flavour to *P. roqueforti* although it
cannot be said that the mould is consciously used. During recent years a
tan-coloured mutant of the mould has appeared, and has caused a good
deal of trouble in the industry, because it detracts from the appearance
of the cheese. The remedy is to pasteurise the milk, since this appears to
be the main source of infection, clean the air of the factory, and work
with pure cultures of the green mould. It is also common in silage and
compost heaps.

P. casei Staub (Lat. *casei*, of cheese). Differs from the above in growing
more slowly, lacking the arachnoid margin, and in the reverse of colonies
being yellow to brown instead of green. It is associated with certain
Swiss cheeses.

P. oxalicum series. There are 2 species in the series, but only one is of
any importance.

P. oxalicum Currie and Thom (name refers to production of oxalic acid).
Colonies broadly spreading, dull green with margin shading through

FIG. 149.—*P. roqueforti*—penicillus. × 500.

pale blue-green to white, velvety, with dense masses of spores which tend to break off in crusts when the culture is jarred; reverse colourless, pale yellow, or pinkish; penicilli irregularly biverticillate, often with rami and metulae arising from the same point; conidiophores smooth, $3\cdot5-4\cdot5\mu$ diam.; phialides usually about 10μ, but sometimes up to 15μ by $3-3\cdot5\mu$; conidia elliptical, smooth, mostly $5-5\cdot5 \times 3-3\cdot5\mu$, with chains massed into columns. Easily recognized by its heavily sporing, dark green colonies and large elliptical conidia. When kept in cultivation it tends to become floccose, with conidium production much reduced. Common in soil, and occurs occasionally as a parasite of corn seedlings.

P. digitatum series. Includes only one species and one variety. *P. digitatum* Saccardo (Lat. *digitus*, a finger). This occurs chiefly on citrus fruits of all kinds and is the cause of serious financial loss in the industry. It can readily be obtained from mouldy oranges, lemons, or grapefruit on which the infected areas are olive to dull yellowish green, not blue-green. It grows very poorly on Czapek agar, but spreads very rapidly on malt agar or potato agar. Colonies smooth velvety, dull yellowish green; penicilli comparatively simple but with all parts large;

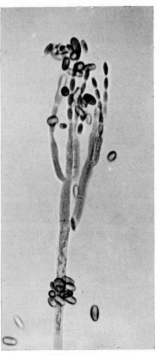

FIG. 150.—*P. digitatum*—conidio-phores. × 250.

FIG. 151.—*P. digitatum*—coni-diophore. × 500.

conidiophores short and mostly 4–5μ diam.; metulae and phialides few in the verticil, 15–30μ long; conidia ovate, smooth, mostly 6–8 × 4–7μ but sometimes up to 12μ or even more in long axis. The penicilli are very fragile and it is exceedingly difficult to prepare good slides. When unbroken penicilli are found the aptness of the specific epithet is apparent, for they have a distinct resemblance to skeleton hands (Figs. 150, 151). *P. digitatum* Sacc. var. *californicum* Thom differs from the parent species in producing pure white colonies.

Subsection LANATA

The colony texture is definitely floccose or woolly, at least in young cultures. Colonies usually spread at first as a white felted mass of hyphae, which slowly becomes coloured as conidial heads ripen, the coloration starting at the centre and spreading outwards. Older colonies may be more or less velvety at the edges, and in very old cultures the floccose central areas tend to die down to a compact felted layer. Spore chains tend to form tangled masses rather than compact columns. There are 2 series, the moulds of Camembert and similar cheeses, and the *P. commune* series.

P. camemberti series. There are 2 species, both associated with the ripening of Camembert, Brie, and similar cheeses.

    P. camemberti Thom (of Camembert cheese). Colonies white, densely floccose, slowly turning pale greyish green from the centre outwards; reverse uncoloured or cream; penicilli irregular, with rami and metulae at the same level and with few branches at each stage; conidiophores slightly rough, 2·5–3·5μ diam.; phialides 9–14μ by about 2·5μ; conidia elliptical at first then subglobose, up to 5 × 4·5μ, smooth, borne in tangled chains.

    P. caseicola Bainier (Lat. *caseus*, cheese; *colo*, to inhabit). Differs from the above in producing colonies which are persistently white. Both species are used in the cheese industry and produce cheeses of somewhat different flavours. *P. caseicola* appears to be the more common.

P. commune series. Of the 7 species in this series 2 are known only as type strains and, of the remainder, only one is frequently isolated.

    P. lanosum Westling (Lat. *lanosus*, woolly). Colonies fairly deeply floccose, grey-green to almost pure grey, becoming brownish in age; reverse colourless or pale dirty yellow; penicilli large, irregularly branched; conidiophores slightly rough, 2·5–3μ diam.; phialides 7–8 × 2–2·5μ; conidia globose, slightly rough, 2·5–3μ diam., in tangled chains.

Subsection FUNICULOSA

The 2 series placed here are not in any way closely related, but are conveniently considered together because the production of ropes of aerial hyphae is common to both.

P. terrestre series. Of the 4 species only one is of frequent occurrence.
P. terrestre Jensen (Lat. *terrestris*, of the earth). Colonies broadly spreading, fairly deeply floccose, with ropiness evident under low magnification, rather yellowish grey-green, becoming brownish in age; reverse
uncoloured or pale tan; penicilli compact, mostly with three stages of
branching; conidiophores rough, about 3μ diam.; phialides crowded,
10–12 × 2·5–3μ; conidia at first elliptical, becoming subglobose, mostly
3·5–4μ in long axis, smooth, in loose columns at first then in tangled
chains; odour usually of moist garden soil, occasionally somewhat sour.
As its name implies, common in soil.

P. pallidum series. All the 4 species are uncommon, and unlikely to be
of interest except to the specialist.

Subsection FASCICULATA

Classified in this subsection are some of the most interesting species of the
genus, species which are of widespread occurrence and of considerable
economic importance. The distinguishing feature of the group is the production of erect bundles (fascicles) of conidiophores, either small and
simple bundles or true coremia consisting of tightly compacted bundles.
Species which regularly produce coremia are readily recognized as belonging here, and the same may be said of certain others whose colonies present
a rough, granular appearance. There are, however, a number of species

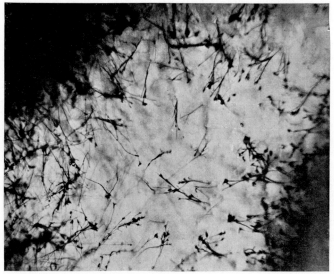

FIG. 152.—*P. aurantiovirens*—edge of colony in Petri dish, showing
fasciculation. × 25.

with a more or less velvety appearance when grown on ordinary laboratory media and which show obvious fasciculation only at the edges of colonies, or only when grown under special conditions or on special media. Most such species have a suggestion of mealiness on the surface of colonies, and examination of the edges of growing colonies will usually reveal small tufts of conidiophores (Fig. 152). Another way of demonstrating fasciculation in

FIG. 153.—*P. cyclopium*—mounted specimen, showing fascicles. × 250.

cases of doubt is to pick off a small portion of a colony, from the edge or just inside it, immerse in a drop of alcohol on a slide, then mount in lacto-phenol without teasing out. Small bundles of conidiophores are thus readily observed (Fig. 153).

Although the number of species is not large (18), differences are suffi-ciently marked to make it necessary to accommodate them in 9 series. Some of the commonest species are separated mainly on the basis of colony colour, and it is here that the beginner, and sometimes the experienced worker, may have trouble, because the differences in colour are not clear cut, the various series shading imperceptibly into one another. There are, however, other characteristics which are taken into account and the most significant of these are emphasized below.

P. gladioli series. There is only one species, which is outstanding in this subsection because of its production of abundant sclerotia.

P. gladioli Machacek (of *Gladiolus*; from its parasitic habit). When grown at temperatures above 20° C colonies consist almost entirely of whitish, buff, or pinkish sclerotia, produced in concentric rings if the cultures are made in Petri dishes. At 15° C or lower growth is predominantly conidial at first, with sclerotia produced tardily. Conidial areas blue-green, fasciculate; penicilli with usually three stages of branching and few elements at each stage; conidiophores slightly rough, 3–3·5μ diam.; phialides 9–12 × 2–2·5μ; conidia elliptical to subglobose, smooth, 2·8–3·5μ in long axis, borne in tangled chains.

P. ochraceum series includes 2 species which produce colonies lacking true green colour. Neither is of common occurrence.

P. viridicatum series. Characterized by bright yellow-green to dark yellow-green colonies, with fasciculation usually evident, particularly in marginal areas. There are 3 recognized species but only one is of importance.

P. viridicatum Westling (Lat. *viridicatus*, made green). Colonies bright yellow-green, sometimes with a narrow somewhat bluish green zone just inside the white margin, fairly thick, usually distinctly granular, becoming dull brown in age; reverse pale dull yellow to dull brown; penicilli with normally three stages of branching, often with rami

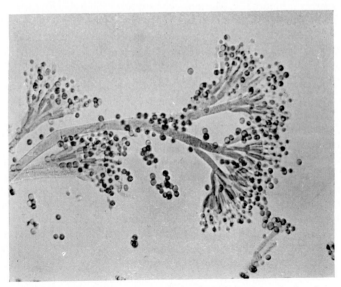

FIG. 154.—*P. viridicatum*—penicillus. × 500. *P. cyclopium* is very similar in appearance.

and metulae borne at the same level (Fig. 154); conidiophores mostly
3·5–4·5μ diam., but sometimes up to 6μ, rough to very rough; phialides
7–10 × 2·5–3μ; conidia elliptical up to 4·5 × 3·3μ, or subglobose about
3·5μ diam., slightly rough, borne in tangled chains or ill-defined loose
columns. Not uncommon and fairly widely distributed.

P. cyclopium series. Colonies typically blue-green, usually distinctly
granular, at least in marginal areas. There are 4 species and one variety, of
which 2 species are of very common occurrence.

P. cyclopium Westling (Gr. *kyklops*, round-eye; probably refers to the
clean circular outline of colonies). Colonies rather dull blue-green, with
brighter zone inside the white margin, almost velvety but showing distinct fasciculation in the younger areas; reverse usually pale peach but occasionally fairly bright yellow or purplish brown; penicilli compact, with
three stages of branching; conidiophores slightly rough (more definitely
rough on malt agar), 3–3·5μ diam.; phialides 7–10 by about 2·5μ; conidia globose to subglobose, smooth or faintly roughened, 3–4μ diam.,
borne in tangled chains. This is probably the commonest of all the
Penicillia, at least in this country. It is found on almost every conceivable
type of organic substrate; it occurs, along with members of the *P. brevi-
compactum* series, on various species of agarics; and it is a parasite of
bulbous plants. Tulips, in particular, are liable to be attacked at the
growing point, producing, as a result, distorted growth with blooms
mis-shapen or lacking.

P. martensii Biourge (P. Martens, who supplied Biourge with the material
from which the mould was isolated). Distinguished from *P. cyclopium* by
its bluer colour, and by the reverse of colonies becoming gradually deep
reddish brown, often with a purplish tone, and with the pigment diffusing into the agar.

P. expansum series. The colony colour is normally grey-green, but may
tend towards yellow-green or blue-green. Zonation is usually evident and
may be marked. There are 2 species.

P. expansum Link ex F. S. Gray (Lat. *expansus*, spread out). Different
strains vary considerably in the freedom with which they produce fascicles or coremia, and in the degree of zonation of colonies when grown
in Petri dishes. Colonies may be granular from the first, or may be
almost velvety and show fasciculation only in age. Fig. 155 shows a large
coremium produced in an old culture, on an agar slope, which for the
most part was definitely velvety. The same strain, grown on a thin layer
of agar in a Petri dish, and producing concentric rings of small coremia,
is shown in Fig. 156.

Colonies spreading rapidly, dull green with white margin, becoming
eventually brownish; reverse in some strains colourless, in others becoming deep brown, with the colour often patchy; penicilli long and
compact (Fig. 157); conidiophores normally smooth but occasionally

slightly rough, 3–3·5μ diam.; phialides 8–12, sometimes up to 15μ by 3μ; conidia elliptical at first, remaining so or becoming subglobose, smooth, 3–3·5μ in long axis, borne in loose columns or in tangled chains; conidial chains often persistent in fluid mounts. Occurs on a variety of substrates, including textiles and paper, but is best known as the cause of a very distinctive brown rot of apples in storage, and also, to a lesser

FIG. 155.—*P. expansum*—single large coremium on agar slope. ×25.

extent, of a number of soft fruits. In advanced stages of the disease conidial areas appear on the brown sodden patches in the form of blue-green coremia.

Some strains of this common organism are with difficulty distinguished from *P. cyclopium*. The main points of difference are: the conidiophores of *P. cyclopium* are usually rough and of *P. expansum* smooth; the penicilli of *P. expansum* have a long drawn-out appearance (Fig. 157) whereas those of *P. cyclopium* tend to be shorter; the spores of *P. cyclopium* are mostly globose and of *P. expansum* elliptical; *P. expansum*, when inoculated into sound apples, produces large areas of soft brown rot in 7–10 days, whereas *P. cyclopium* has little or no effect.

P. crustosum Thom (Lat. *crustosus*, crusty). Colonies dull yellow-green to grey-green, becoming brown, almost velvety with fasciculation limited and not very evident, in old cultures forming continuous crusts of conidia which break off in irregular pieces when the dish is tapped; reverse colourless; penicilli with coarsely roughened conidiophores but otherwise

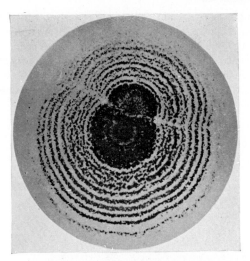

FIG. 156.—*P. expansum*—colony in Petri dish,
showing well-marked zonation with rings of
small coremia. × 0·75.

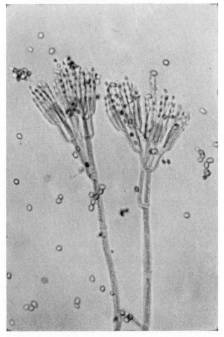

FIG. 157.—*P. expansum*—typical penicilli. × 500.

very much like those of *P. expansum*. Found fairly frequently on various materials and slowly produces a brown rot of pomaceous fruits.

P. italicum series includes only the one species.

P. italicum Wehmer (Lat. *italicus*, Italian). Colonies somewhat restricted, pale grey-green, usually showing obvious fasciculation and often producing curious prostrate coremia at the edge of established colonies (Fig. 158); odour fragrant; reverse pale tan to yellowish brown; penicilli normally with three stages of branching but sometimes more complex; conidiophores smooth, 4–5μ diam.; phialides few in number, mostly 8–12 × 3μ; conidia at first cylindrical, almost *Oidium*-like, becoming elliptical, smooth, 4–5 × 2·5–3·5μ. The chains of cylindrical conidia, which often make it difficult to decide where the phialides end, are highly characteristic (Fig. 159). Most strains deteriorate on continued cultivation, becoming floccose and non-fasciculate. Found frequently on all kinds of citrus fruits, and can always be obtained when required from such fruits which show blue-green patches of mould. It is distinguished from *P. digitatum* not only by colour and texture of the colonies on fruit, but also by the type of rot produced. Fruits attacked by *P. digitatum* shrivel and dry up, whereas *P. italicum* produces a soft rot which rapidly reduces the fruit to a slimy pulp. Even a very small colony of *P. italicum* will render the whole fruit nauseous. This species is found occasionally on other substrates but is of little importance outside the citrus fruit industry.

P. patulum series also includes but a single species.

P. patulum Bainier (Lat. *patulum*, spreading, open). Bainier also described as a distinct species *P. urticae* (of *Urtica*, nettle), but this is to be regarded as synonymous with *P. patulum*. Raper and Thom use the epithet *urticae* quite unjustifiably since *P. urticae* was described a year later than *P. patulum*, and, if the description of the latter is sufficiently precise for Raper and Thom to decide that it is identical with *P. urticae*, then there is no reason for adopting the later name, particularly as the earlier name is beautifully descriptive. Colonies restricted with abrupt margins, pale grey-green to almost pure light grey, thick, with prominent fascicles; drops colourless, large; reverse pale dull yellow to brownish; penicilli loosely divergent with three to four stages of branching; conidiophores smooth, sinuate, 3–4μ diam.; phialides short, 4·5–6μ long, crowded; conidia elliptical to subglobose, smooth, 2·5–3μ in long axis, borne in divergent chains. Common in soil and fairly frequently isolated from other sources. Produces the antibiotic variously known as "patulin", "expansin", "claviformin", etc.

P. granulatum series. Characterized by the production of colonies with a majority of the conidiophores aggregated into fascicles or small feathery coremia. There are 2 species but only one is of industrial importance.

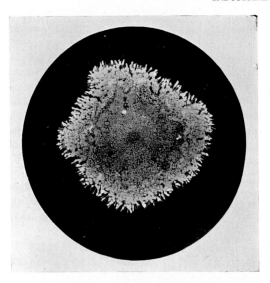

FIG. 158.—*P. italicum*—colony in Petri dish, showing
prostrate coremia around the edge. Natural size.

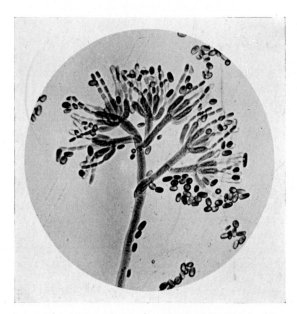

FIG. 159.—*P. italicum*—characteristic penicilli. × 500.
Note the *Oidium*-like young conidia.

P. corymbiferum Westling (Lat. ex Gr. *corymbus*, a cluster of flowers; *fero*, to bear). Colonies spreading, yellow-green to olive-green, with broad white margin, markedly fasciculate; drops blood red (usually on Czapek and not on malt agar); reverse deep reddish brown with the colour diffusing somewhat; penicilli compact, with three stages; conidiophores rough, $3 \cdot 5 - 4 \cdot 5 \mu$ diam.; phialides $9 - 12 \times 2 - 2 \cdot 5 \mu$; conidia globose, smooth, variable in size but mostly about 3μ diam., borne in tangled chains. Not uncommon, especially in soil. Like *P. cyclopium*, is an active parasite of various species of bulbous plants.

P. claviforme series. The conidiophores are mostly aggregated in definite large coremia. There are 2 species, of which one, *P. claviforme* Bainier (Lat. *claviformis*, club-shaped), is occasionally met with and, with its prominent coremia with pinkish stalks and large green heads, is unmistakable. It produces the same antibiotic as *P. patulum*.

The Biverticillata-Symmetrica

P. luteum series. Characterized by the production in the perfect state of perithecia with soft or rudimentary walls, mostly ripening quickly, yellow-white or pale buff. As in the *Aspergillus glaucus* series, shape and markings of the ascospores form the main basis of classification. Most of the species tend, on continued cultivation, to become entirely conidial or even completely sterile. Derx (1926) stated that *P. luteum* is heterothallic, and that the loss of ascospore production is due to the segregation of the haplont forms during subculturing. Emmons (1935), however, has shown that none of the species in this series is heterothallic, but that, for reasons not at present understood, most isolates show a sudden or gradual loss of fertility in laboratory cultures. Most strains retain their characteristics best on corn-meal agar. The series includes 11 species, but most of these are rare, or at least have so far been reported very infrequently. There are, however, 2 species which are common in soil, and hence on materials which are at some time in their history in contact with, or contaminated with, soil.

P. wortmannii Klöcker (J. Wortmann, German botanist). [*Talaromyces wortmannii* (Klöcker) Benjamin]

P. vermiculatum Dangeard (Lat. *vermiculatus*, like a little worm; refers to the perithecial initials). [*Talaromyces vermiculatus* (Dang.) Benjamin] The two species are very similar and are best considered together. Thom (1930) regarded them as synonymous. They both form soft yellow perithecia which are without a true wall, being bounded by a soft hyphal web. The ascospores are almost identical, at the same time differing from those of all other species, being ovate, without furrow, finely spinulose all over, $4 - 4 \cdot 5 \times 3 - 3 \cdot 5 \mu$. The points by which they may be distinguished are: colonies of *P. vermiculatum* are spreading and of soft texture, those of *P. wortmannii* restricted and consisting of tough felts; the perithecial initials are quite different, those of *P. vermiculatum* consisting of long, thick, deeply staining, club-shaped hyphae, around which thinner hyphae are tightly coiled (Fig. 160), those of *P. wortmannii* being much smaller, and, owing to the texture of the colonies, more difficult to find, and consisting of irregular knots of somewhat thickened hyphae. Fig. 161 shows two of the many different patterns which can be observed. Note that Fig. 161 is at twice the magnification of Fig. 160.

Some of the other ascosporic species have been reported in this country and are always to be found occasionally. They are distinguished as follows:

P. spiculisporum Lehman (Lat. *spiculus*, a little spine). [*Talaromyces spiculisporus* (Lehman) Benjamin] Perithecia white; ascospores elliptical, finely spinulose all over, $3-3\cdot5 \times 2\cdot2-2\cdot8\mu$.

P. luteum Zukal (Lat. *luteus*, golden yellow). [*Talaromyces luteus* (Zukal) Benjamin] Perithecia yellow, without definite walls; spores elliptical with 3–4 transverse bands, approximately $4\cdot5 \times 2\cdot2-2\cdot8\mu$.

P. helicum Raper and Fennell (name refers to perithecial initials). [*Talaromyces helicus* (Raper and Fennell) Benjamin] Perithecia yellow, soft, without definite walls, arising from helically coiled initials; ascospores elliptical, finely spinulose, $2\cdot5-3 \times 1\cdot4-1\cdot8\mu$.

P. duclauxii series. Abundant coremia produced. There is only one very distinctive species.

P. duclauxii Delacroix (E. Duclaux, French microbiologist). Different strains vary in appearance, some growing as a forest of slender erect coremia, 4–5 mm. high, others developing irregular spiky clumps of coremia. Surface colour grey-green; reverse yellow then purplish red, with colour diffusing somewhat; penicilli typically biverticillate with acuminate phialides; conidia elliptical to subglobose, rough. A fairly abundant species, reported from many sources.

P. funiculosum series. Colonies with ropes of hyphae from which the conidiophores mostly arise. Reverse mostly in reddish shades. Of the 5 species 4 are fairly common, the other of no importance.

P. funiculosum Thom (Lat. *funiculus*, a thin rope). Colonies spreading, with tough basal felt of mycelium and aerial growth as ropes or tufts of hyphae, sporing in irregular patches; reverse usually pink to reddish, in some isolates becoming very deep red, in others colourless or almost so; penicilli typical; conidiophores short, mostly arising from funicles (Fig. 162), smooth, about 3μ diam.; phialides $10-12 \times 2-2\cdot5\mu$; conidia elliptical, with thick walls, smooth or slightly rough, $2\cdot5-3\cdot5 \times 2-2\cdot5\mu$, borne in tangled chains. Many strains deteriorate in culture, growing as a sodden felt, sporing tardily and sparsely. Very common in soil and on decaying vegetation. It has also been reported as producing a core-rot of *Gladiolus* corms (Jackson, 1962).

P. verruculosum Peyronel (Lat. *verruculosus*, warted). Colonies floccose funiculose, rather loose in texture, a mixture of yellow-green and yellow;

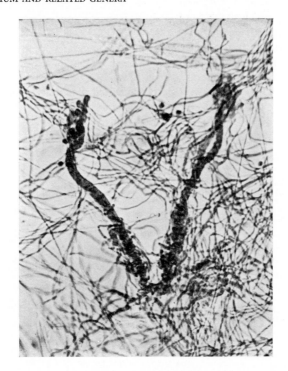

FIG. 160.—*P. vermiculatum*—perithecial initials. × 500.

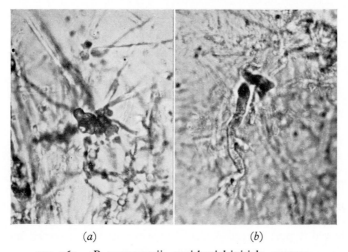

(a)   (b)

FIG. 161.—*P. wortmannii*—perithecial initials. × 1000.

reverse greenish or pale dull brown; penicilli rather short and broad; conidiophores smooth, 2·5–3μ diam.; phialides 8–10 × 2·5μ, more abruptly tapered than is usual in this section; conidia globose or nearly so, very rough, 2·8–3·5μ diam.

P. islandicum Sopp (Lat. *islandicus*, Icelandic). Colonies very striking in appearance, restricted, fairly thick and tufted, a mixture of orange, red, and dark green; reverse dull orange to red, becoming dull red-brown;

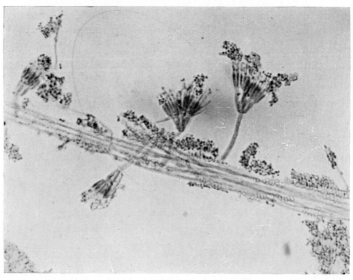

FIG. 162.—*P. funiculosum*—conidiophores arising from hyphal rope. × 500.

penicilli typical but rather short; phialides somewhat abruptly tapered, 7–9 × 2μ; conidia elliptical, smooth, 3–3·5 × 2·5–3μ, borne in tangled chains.

P. piceum Raper and Fennell (Lat. *Picea*, the spruce. Name suggested by the resemblance of the conidial heads to miniature spruce trees, *Picea excelsa*. The name is unfortunate since Lat. *piceus* = pitch black). Colonies grow fairly rapidly, fairly thick matted, yellow at first then producing irregular patches of dull yellowish green, with reverse dull orange to brown. The mature heads are very characteristic, the chains forming solid masses which are more or less conical (Fig. 163). Penicilli are typically biverticillate, and the conidia rough, 2·5–3 × 2·2–2·8μ.

P. purpurogenum series. Colonies velvety or floccose, not funiculose, with hyphae encrusted with yellow, orange, or red pigment, and reverse in bright red to purplish red shades. The series includes 4 species and one variety. Three of the species are not uncommon.

P. purpurogenum Stoll (Lat. *geno*, to cause; hence producing purple colour). Colonies usually velvety, more rarely somewhat floccose, sporing heavily, deep green on a yellow mycelium, becoming very dark green; reverse and agar intense blood red or purplish red (on malt agar almost colourless); penicilli typical; conidiophores fairly short, smooth, about

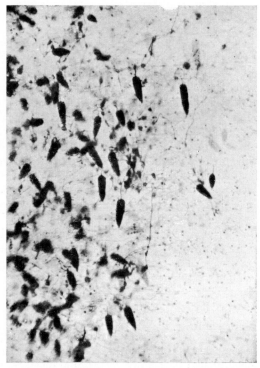

FIG. 163.—*P. piceum*—showing characteristic shape of spore-heads.
× 50.

3μ diam.; phialides 10–12 × 2–2·5μ; conidia elliptical to subglobose, with thick walls, in most strains rough, occasionally smooth, 3–3·5 × 2·5–3μ. Common in soil and reported as occurring on a variety of substrata.

P. rubrum Stoll (Lat. *ruber*, red). Very similar to the preceding species, but differs in its lighter grey-green colour, and in the conidia being smooth and almost globose.

P. variabile Sopp (Lat. *variabilis*, variable). Colonies growing moderately well, velvety or almost so, sage green, grey-green or almost grey, often sporing in patches with sterile areas white, yellow, or orange, and with margin white, cream, or yellow; reverse yellow, brownish or greenish; penicilli typical; conidiophores varying in length, smooth, 2·5–3μ diam.;

phialides 10–12 × 2μ approximately; conidia elliptical, often with somewhat pointed ends, smooth, 3–3·5 × 2–2·5μ. This species includes most
strains which, before the publication of the Manual, would have been
identified as belonging to Thom's "P. luteum series, non-ascosporic".
As the name implies, different isolates vary much in appearance, especially in the amount of pigmentation of the mycelium.

P. rugulosum series. Colonies show restricted growth, with reverse
colours orange, brownish or greenish, not red or purple. Conidia are
elliptical and mostly rough. There are 3 species and one variety, of which
2 species are of importance.

P. rugulosum Thom (Lat. *rugulosus*, wrinkled). Colonies restricted, almost velvety to definitely floccose, rich green becoming greyer; reverse
at first colourless, slowly becoming deep yellow to orange in spots and
patches, in slope cultures particularly along the edges; penicilli mostly
typical but not infrequently irregular, with metulae of different lengths;
conidiophores smooth, 2·5–3μ diam.; phialides 10–12 × 1·8–2μ; conidia
elliptical, markedly roughened, 3–3·5 × 2·5–3μ, borne in tangled chains.
An interesting occurrence of this species, and, less frequently, of some of
the other biverticillate species, is as parasites of the black Aspergilli.
Hyphae of the *Penicillium* twine up the stalks of the *Aspergillus* and
smother the black heads with fruiting masses of a dark olive-green
colour (Fig. 168, p. 275). Such parasitic species cause trouble from time
to time in factories where citric acid is manufactured by the mould fermentation process, attacking the mycelial mats of *A. niger* and causing
portions to sink in the solution.

P. tardum Thom (Lat. *tardus*, slow). Colonies on Czapek agar growing
very slowly, matted floccose 1–2 mm. thick, slowly becoming grey-green,
with submerged margin, and with reverse yellowish in the centre but
otherwise colourless; on malt agar growth is more rapid, velvety or
slightly floccose, pale dull bluish green becoming greyer, and with reverse pale dirty brown; penicilli often typically bivert. but sometimes
incomplete; conidiophores smooth, 2–2·5μ diam.; phialides mostly
8–10 × 1·8–2·2μ; conidia elliptical, fairly thick-walled, rough, 3–3·5 ×
2–2·5μ, in loosely parallel or tangled chains. Occurs somewhat infrequently but on a wide range of materials.

AIDS TO THE STUDY OF THE GENUS PENICILLIUM

1. Habitat. Most of the common species of *Penicillium* grow on a wide
range of substrata. Nevertheless, certain species, although not confined to
a definite habitat, can be found when required on certain definite substrata. The beginner is strongly advised to isolate some of these and study
them with a view to getting a correct conception of the criteria used for
identification.

Apples rotting in storage . . .	*P. expansum*
Citrus fruits—dull olive-green . .	*P. digitatum*
blue-green . . .	*P. italicum*
Toadstools—grey-green . . .	{ *P. brevi-compactum* { *P. stoloniferum*
blue-green . . .	*P. cyclopium*
Blue-veined cheese	*P. roqueforti*
Camembert cheese—white species . .	*P. caseicola*
pale bluish . .	*P. camemberti*

2. Chemical tests. There is little hope of there ever being a complete biochemical classification of the Penicillia. Different isolates which are almost indistinguishable morphologically, and belong, therefore, to the same species, may differ in the types of chemical products formed under any given conditions. On the other hand, substances for which there are specific tests may be produced by more than one species. Also there are

Colour with FeCl₃	Inference	Comments
Intense blackish green .	*P. frequentans*	Citromycetin produced.
Pale brown precipitate; dissolves in excess to give solution the colour of iodine solution	*P. citrinum*	Citrinin. Culture fluid gives yellow cryst. ppt. with HCl.
Brown at 7 days; crimson to purple at 21 days	*P. brevi-compactum* series	Mixture of acids. Solution giving good purple gives white ppt. of myco-phenolic acid with HCl.
Bright green . . .	*P. italicum*	Reaction not given by all strains.
Scarlet	*P. charlesii*	Mixture of tetronic acids produced.

many species which do not produce any substances for which there are simple tests. Chemical tests can be useful for confirming identifications and, in certain instances, are in themselves strong presumptive evidence of identity.

The most useful reagent is aqueous ferric chloride (about 2 per cent). The method is to grow the mould on small quantities of liquid, in test-tubes, and, at intervals of 7 days, test the filtered culture fluid by adding ferric chloride until no further visible reaction occurs. It is best to use both Czapek and Raulin-Thom media in all cases.

3. Bacterial spectrum test. A number of antibacterial substances have been obtained from species of *Penicillium*. They vary in their range of activity and, by determining which types of bacteria are inhibited, it is possible to obtain confirmatory evidence of the identity of the particular species.

H

The medium recommended by Raper and Thom has the following composition:

Yeast extract	2·0 g
Peptone	3·0 g
Dextrose	2·0 g
Sucrose	30·0 g
Corn steep solids	5·0 g
$NaNO_3$	2·0 g
$K_2HPO_4.3H_2O$	1·0 g
$MgSo_4.7H_2O$	0·5 g
KCl	0·2 g
$FeSO_4.7H_2O$	0·01 g
Agar	20·0 g
Water	to 1000 ml.

The medium is adjusted to pH 7·0 before sterilizing. After sterilization it is poured into Petri dishes, 20 ml. per dish, and the agar surface allowed to dry for one or two days. A loopful of spores of the *Penicillium* is streaked along one side of the dish and the dish incubated 4 days at 24° C. Suspensions of the test organisms are then streaked at right angles to the mould colony, and the dish further incubated 24 hours at 30° C.

Inhibited	Antibiotic	Species
St. aureus; B. subtilis	Penicillin	*P. chrysogenum* series
St. aureus; B. subtilis; B. cereus	Citrinin	*P. citrinum*
All bacteria; *Candida* to limited extent	Patulin	*P. patulum*
		P. expansum
		P. claviforme

The organisms used by Raper and Thom include three gram-positive bacteria—*Staphylococcus aureus*, *Bacillus subtilis*, and *B. cereus*—of which the first two are sensitive to penicillin, whilst *B. cereus* produces a penicillinase, and hence is insensitive; two gram-negative bacteria—*Salmonella schottmuelleri* and *Proteus mirabilis*; also one yeast-like fungus—*Candida albicans*.

The three species producing patulin are readily distinguished in culture, *P. claviforme* by its large coremia, *P. patulum* by its thick, very greyish colony, and *P. expansum* granular and definitely green, not grey.

REFERENCES

Barnett, H. L., and Lilly, V. G. (1962). A destructive mycoparasite, *Gliocladium roseum*. *Mycologia*, **54**, 72–7.

Benjamin, C. R. (1955). Ascocarps of *Aspergillus* and *Penicillium*. *Mycologia*, **47**, 669–87.

Biourge, P. (1923). Les moisissures du groupe *Penicillium* Link. *Cellule*, **33**, 1re fasc., Louvain.

Birkinshaw, J. H., and Raistrick, H. (1931). On a new methoxydihydroxy-toluquinone produced from glucose by species of *Penicillium* of the *P. spinulosum* series. *Phil. Trans.*, Ser. B, **220**, 245–54.

Brown, A. H. S., and Smith, G. (1957). The genus *Paecilomyces* Bainier and its perfect stage *Byssochlamys* Westling. *Trans. Brit. mycol. Soc.*, **40**, 17–89.

Derx, H. G. (1926). Heterothallism in the genus *Penicillium*. *Trans. Brit. mycol. Soc.*, **11**, 108–12.

Emmons, C. W. (1935). The ascospores in species of *Penicillium*. *Mycologia*, **27**, 128–50.

Hetherington, A. C., and Raistrick, H. (1931). On the production and chemical constitution of a new yellow colouring matter, citrinin, produced from glucose by *Penicillium citrinum* Thom. *Phil. Trans.*, Ser. B, **220**, 269–95.

Hughes, S. J. (1953). Conidiophores, conidia and classification. *Canad. J. Bot.*, **31**, 577–659.

Jackson, C. R. (1962). *Penicillium* core-rot of *Gladiolus*. *Phytopathology*, **52**, 794–7.

Langeron, M. (1922). Utilité de deux nouvelles coupures génériques dans les Perisporiacés. *Diplostephanus* n.g. et *Carpenteles* n.g. *C.R. Soc. Biol. Paris*, **87**, 343–5.

Moore, W. C. (1941). New and interesting plant diseases. *Trans. Brit. mycol. Soc.*, **25**, 206–10.

Raistrick, H., and Smith, G. (1941). Anti-bacterial substances from moulds. Part I.—Citrinin, a metabolic product of *Penicillium citrinum* Thom. *Chem. & Ind. (Rev.)*, **60**, 828–30.

Raper, K. B. (1957). Nomenclature in *Aspergillus* and *Penicillium*. *Mycologia*, **49**, 644–62.

Raper, K. B., and Thom, C. (1949). *A manual of the Penicillia*. Baltimore: The Williams and Wilkins Co.

Shear, C. L. (1934). *Penicillium glaucum* of Brefeld (*Carpenteles* of Langeron) refound. *Mycologia*, **26**, 104.

Smalley, E. B., and Hansen, H. N. (1957). The perfect stage of *Gliocladium roseum*. *Mycologia*, **49**, 529–33.

Thom, C. (1930). *The Penicillia*. Baltimore: The Williams and Wilkins Co.

Zaleski, K. (1927). Über die in Polen gefundenen Arten der Gruppe *Penicillium* Link. *Bull. int. Acad. Cracovie (Acad. pol. Sci.)*, Ser. B, *Sci. nat.*, 417–563.

Chapter XI

Laboratory Equipment and Technique

Studiosis tamen valde commando ut suas species non sub uno genere inquirant, sed et sub affinibus omnibus.

P. A. Saccardo, *Sylloge*, I. Introduction

The first essential in any laboratory experimental work involving micro-fungi is to obtain pure cultures; that is, to grow each species on a sterile substratum, known as the culture medium, free from admixture with any other organism and with suitable precautions to prevent subsequent con-tamination. Many fungi fail to grow characteristically in presence of other organisms and identifications are therefore rendered difficult. It is obvious also that studies of nutrition, of metabolic activity, or of the controlling effect of antiseptic substances, are without much value unless they are carried out with individual moulds and not with mixtures of species.

Methods of isolation, of culture, and of study are, in general, similar to those used by bacteriologists, but differ from them in detail. The methods described here are only a selection of those available but should be suffi-cient for preliminary essays in mycological studies, and the advanced worker and specialist can add to them as experience directs.

GENERAL EQUIPMENT

Cultures for morphological study, and for many other purposes, are grown either in plugged test-tubes or in Petri dishes. Cultures for storing in a collection are also usually grown in plugged tubes, but nowadays a number of mycologists prefer screw-top bottles for this purpose.

Tubes. The tubes may, for most purposes, be ordinary test-tubes, but the special bacteriological test-tubes are much to be preferred. They are made of glass which has a high degree of resistance to chemical action and will withstand repeated sterilization. At the same time they are sufficiently strong mechanically to resist a certain amount of rough handling. It is an advantage to have two sizes, the smaller without, the larger with rims. The rimless tubes, which are commonly 6 by $\frac{5}{8}$ inches but are preferably 5 by $\frac{5}{8}$ inches, pack into baskets and storage boxes much better than the rimmed variety and are used for "slopes" (see below). The larger tubes with rims, 6 by $\frac{3}{4}$ inches or 6 by $\frac{7}{8}$ inches, are used to hold medium for plating out, when the rim is a distinct help in pouring cleanly, and the greater capacity makes thorough mixing of the contents easier.

Plugs. Tubes of culture medium, and tubes containing pure cultures, are always plugged with cotton-wool so that any air which enters will be filtered from all contaminating organisms. The cotton-wool should be of the non-absorbent variety. It may be obtained in a number of different colours and these are useful for distinguishing the various culture media. A plug should project into the tube about an inch and should have a tuft outside the tube by which it can be extracted; it should fit accurately and tightly, but not so tightly that it cannot be extracted when gripped between any two fingers of one hand; and it should retain its shape so that, after withdrawal, it can readily be reinserted. There are several methods of making plugs by rolling and shaping before insertion, but an easy and quite satisfactory method is as follows: a strip about $2\frac{1}{2}$ inches wide is torn from the sheet of cotton-wool; from this a rectangular piece is torn, of a size that can be determined only by trial, and the edges are folded in to give a piece $2\frac{1}{2}$ inches long and of a width twice the diameter of the tube; this is laid across the mouth of the tube and its centre gently pushed in by means of a glass rod or the blunt end of an aluminium needle-holder. If the plug is definitely a tight fit when thus pushed in, it will be of the correct fit when the rod is withdrawn. The steaming which it receives during sterilization sets the plug more or less permanently to the shape of the tube.

Rectangular baskets of iron wire, galvanized or tinned, are used to hold tubes during sterilization and may be obtained of any convenient size.

Screw-top bottles. The usual size is 3 by 1 inches and the screw caps are fitted internally with rubber pads to effect an airtight seal. Such bottles have certain advantages over plugged tubes. During sterilization the caps must, of course, be somewhat loosened, but as soon as the bottles come out of the sterilizer the caps may be screwed down tightly until the medium is required. It is thus feasible to make culture medium in large batches and to store it almost indefinitely without any danger of its drying out. Cultures in these bottles are easy to store, since they stand upright in drawers or filing cabinets, and they can be labelled on the caps so that all particulars are in full view. The bottles are mechanically stronger than tubes and are thus very suitable for sending cultures by post. On the other hand, they are more difficult to manipulate when being inoculated, since it is virtually impossible to hold three bottles between thumb and first finger, as is done when inoculating duplicate cultures in tubes (see above).

Another respect in which tubes are more convenient than bottles is in the examination of living cultures. A tube is easily laid across the microscope stage, and the growth therein can be examined with a 16 mm. objective, or even with an 8 mm., and may be photographed if desired. The glass of the screw-top bottle is too thick to allow of any objective of shorter focal length than 16 mm. being focused, and the glass is usually too uneven to transmit a useful image. Another disadvantage is that, once the cultures have been sown, the caps must be unscrewed sufficiently to admit air, and it is not easy to do this without running a grave risk of infection getting in. In the author's experience not all moulds grow equally well in tubes and bottles. Many do, but some species definitely benefit from the freer aeration of the plugged tube.

Petri dishes. These are flat, circular, shallow glass dishes with perpendicular sides, provided with covers of the same shape, but of slightly larger diameter, so that they fit loosely over the dishes. The air which enters, and it must be remembered that fungi require a continual supply of air, is not filtered as in the case of plugged tubes. Owing, however, to the tortuous path it must take to enter the dish, suspended dust and spores are deposited outside and cultures are reasonably safe from contamination, provided that the dishes are handled with care and that the air of the laboratory is not too heavily charged with infection. As an extra precaution Petri dishes should, whenever possible, be stored and incubated upside down. Dishes may be obtained in various sizes but, for most purposes, the most convenient size is 10 cm. by 1·5 cm. They are sometimes made with plane, polished bottoms, but this is a refinement too costly for ordinary use.

The greatest disadvantage of the Petri dish is that it is easily broken, so it is not surprising that various modifications have been suggested for reducing the rate of breakage. Fisher (1958) designed a dish with a recessed lid, and with a ridge cast on the lower surface of the dish. This arrangement prevents slipping when dishes are piled on top of one another, thus avoiding what is probably the commonest cause of breakage. Unfortunately, although the author states that the dishes can be obtained, all attempts to purchase some of them have failed.

Some laboratories use dishes made of polystyrene, a transparent plastic. They are virtually unbreakable, and, in any case, have very little tendency to slip when piled together. Sterilization of plastic dishes by heat is difficult, owing to the low softening point of the material. Also, of course, these dishes are easily scratched. They are now obtainable ready sterilized, packed in tens in polythene bags, and, owing to their low cost, they mostly are used once only and then discarded.

In some institutions, particularly bacteriological laboratories, glass dishes with nickel lids are used. They are obviously not so suitable for mycological work, since the surface of the colonies cannot be seen without opening the dish. In America, dishes made of a special type of paper are popular. They are used once and then scrapped.

General apparatus. A certain amount of the necessary equipment is such as is to be found in any chemical laboratory. This includes beakers and flasks of various sizes, measuring cylinders, pipettes, funnels, bunsen burners and tripods, balances and weights. It is assumed that the reader is sufficiently familiar with this type of apparatus for it to require no further mention.

A supply of needles and loops will be required for inoculations and the best are made from short lengths of stiff nichrome wire (20-gauge is suitable), permanently fixed into long aluminium handles. Platinum wire cannot be set satisfactorily into aluminium as it becomes loose after a short period, owing to the difference in coefficients of expansion of the two metals. Adjustable needle-holders, which grip the needle in a small brass chuck, are preferred by some workers, but they have too many hidden surfaces and require much heating to ensure perfect sterilization, particularly when working with liquid media. Occasionally a very stiff needle is required and a convenient form is triangular in section with fairly sharp edges that can be used for scraping. Fine-pointed scissors and scalpels are useful for cutting up infected material preparatory to making cultures, and the ones made entirely of stainless steel (if obtainable) are worth the extra cost, as they stand up to repeated sterilization without corroding. The list of tools should also include three pairs of forceps, one strong and blunt-ended, one with fine points, and the third of the special type for handling cover-glasses.

Some kind of support is necessary for holding a sterile needle whilst both hands are occupied in such tasks as manipulating plugs and labelling culture tubes. The porcelain saw-tooth type of rack sold for the purpose is not really suitable, as it supports the needle-holder at the wrong end. The simplest support is a block of wood, 1–2 inches thick, or a large rubber bung, with a hole, just large enough to take the handle of the needle comfortably, bored through it. The needle stands upright in this, without serious danger of picking up infection. Such a support is invaluable when sowing large numbers of cultures. Other tools, which are used much less frequently, may be supported, after sterilization, in any way which ensures that the tips of forceps or the cutting edges of scissors and scalpels are not touching anything, e.g. resting on the lid of a Petri dish or a tripod.

For labelling culture tubes and Petri dishes grease pencils, specially made for writing on glass, are obtainable in several colours. There are also special inks for marking glass, and these give a more permanent label than the grease pencil. Even ordinary ink will mark glass quite readily, provided that the surface is dry.

Incubators. The majority of moulds will grow reasonably well at the temperatures prevailing in most laboratories. However, in order to induce maximum rate of growth, and, in some cases, to promote the formation of certain types of spores, higher or lower temperatures are essential. The containers used for this purpose are known as incubators, these being provided with means of maintaining automatically and continuously a predetermined temperature. Also, in physiological work it is often desirable to study the effect of temperature on growth, and this, of course, unless the work is to be unduly prolonged, involves the use of several incubators set for different temperatures.

The usual type of incubators obtainable commercially are designed to work at temperatures above that of the laboratory, and are of two types, for gas and electricity respectively. Gas incubators comprise a chamber which is water-jacketted on all sides except the front, and is usually constructed entirely of copper, the whole surrounded by a cabinet of wood and asbestos sheet. It is heated by a bat's-wing burner, the inflow of gas being controlled by a flat copper capsule containing a volatile liquid, which, at a temperature somewhat above the boiling-point of the liquid, expands and pushes on a lever which cuts off the gas supply. A by-pass, which can be adjusted independently, ensures that a pea-sized flame is left burning when the main gas supply is cut off. A sliding weight on a rocking lever, which controls the cut-off point by putting pressure on the capsule, allows of an adjustment of the temperature over a range of about 5° C. This type of incubator was originally designed, for bacteriological laboratories, to work at 37°. When used at 24–25°, which is the most favourable temperature for a great many fungi, excessive corrosion of the bottom of the water-jacket occurs, and, apart from the gradual eating away of the copper, the scale formed frequently drops on to the burner and extinguishes the by-pass flame.

Electric incubators are usually not water-jacketted, but consist of a single-walled copper container, insulated by dry packing. They are controlled by the same type of expandable capsule as is used for the gas incubators, this controlling a simple off-and-on switch. They are less trouble to maintain than the gas incubators, but tend to show more variation in temperature within the usable space.

Both types of incubator have one serious fault for mycological work, in that they necessitate incubation in the dark. Light is often more important than warmth for characteristic growth of moulds. Many species spore much better in the light than in the dark. For maintenance of stock cultures it is often best to dispense with incubators, and grow cultures at room temperature in the light. However, the Commonwealth Mycological Institute has carried out successful trials of specially designed incubators which obviate the difficulty. So far as is known, such incubators are not available commercially, but they are not difficult to construct. The top, back and sides are of glass, set in a light wooden frame, and the front consists of sliding glass doors. The shelving is also of glass. Along both ends and the bottom plastic-covered resistance wire is laid zig-zag fashion, stretched between supports of plastic-covered curtain wire. The heating wire is connected to the mains through a "Sunvic" control, which can be adjusted to give a short range of constant temperatures. Above the incubator is a "daylight" fluorescent tubular lamp, which can be turned on at night if required, or used to provide extra illumination on dull days.

Whilst the maintenance of temperatures above that of the laboratory is easy, the attainment of constant temperatures below 20°C presents some difficulty. Some fungi grow best at comparatively low temperatures, and, in any case, it is often useful to find the effect on growth of a considerable range of temperatures. The domestic type of refrigerator can be set to give temperatures over the range 0° to about 8° C. In fact there is usually an appreciable variation in temperature, at any setting, in different parts of the machine. The difficulty is to maintain a steady temperature within the range 10–20°. Some form of cooling is essential, and one method of achieving this is to place a small incubator (or more than one) inside a large refrigerator, when it is not difficult to balance the heat input against the cooling effect of the ambient atmosphere.

Sterilizers. Culture media are usually sterilized by steam, either at atmospheric or higher pressure, so as to avoid change of concentration by evaporation. Whilst much may be done by judicious adaptation of domestic utensils, the special pieces of apparatus made for the purpose have distinct advantages when cultural work on a large scale is undertaken.

The Koch type of steamer, designed for sterilization at ordinary pressure, is a tall copper vessel, cylindrical in the smaller sizes, the lower portion of which serves as a water-bath and is fitted with a gauge and sometimes with a constant level attachment. A perforated shelf, fixed a little distance above the water level, serves to hold apparatus. The lid is provided with a tubulure to take a thermometer and to act as a steam outlet, and the whole outside of the steamer, except the bottom, is lagged with felt. It is normally supplied with a stand of sheet iron and a ring burner. The internal height of the Koch steamer should be sufficient to accommodate a litre flask and a funnel of 7 inches diameter supported above it. Larger sizes of steamers, to hold several baskets of tubes or a number of flasks, are usually rectangular in shape. In some laboratories very large steamers, constructed on the same principle, are used for the sterilization of large numbers of culture flasks or bottles.

Some culture media are difficult to sterilize completely at atmospheric pressure and it is necessary to use an autoclave, in which steam is generated under pressure, usually 10 to 20 pounds per square inch. An autoclave, if used properly, can also be employed, in place of the steamer, for the sterilization of the more usual media, and the time required for sterilization thereby much shortened. Full particulars of autoclaves, as well as of Koch steamers, may be obtained from the catalogues of laboratory furnishers.

STERILIZATION

When working with pure cultures it is necessary, in order to avoid contamination, to sterilize all tools, utensils, containers, and culture media. The usual method of sterilization is by heating to a sufficiently high temperature, and for a sufficiently long time, to kill all fungus spores and bacteria.

Tools and glassware. Metal tools are sterilized by heating in a bunsen flame, needles until just red hot, cutting tools at a somewhat lower temperature to avoid loss of temper. Dry glassware, such as tubes, flasks, and Petri dishes, are sterilized by dry heat, and a capacious air-oven should be available capable of being maintained at a temperature of about 160° C.

Three hours' heating at this temperature is sufficient to sterilize anything. Petri dishes are preferably packed in boxes before heating, and allowed to cool and remain therein until required for use. Special boxes of sheet iron or copper are made for the purpose, but they are expensive, and the rectangular tins in which biscuits are sold serve the purpose just as well.

Sterile graduated pipettes are frequently required for sowing liquid culture media and for making adjustments to sterile media. They are plugged at the mouthpiece ends, pushing the plugs well into the tubes, wrapped separately in brown paper so that they are completely enclosed, and sterilized by dry heat, at a temperature not exceeding 130° C, for five to six hours. At a higher temperature the paper is charred and becomes useless as a protection.

Tubes for use in aeration experiments are fitted with the necessary rubber bungs, ready for insertion into flasks or tubes, every open end is plugged with cotton-wool, and the whole is wrapped in grease-proof paper and sterilized in the autoclave.

Culture media. The sterilization of culture media is apparently a simple matter but is actually beset with a number of pitfalls for the inexperienced. Very few media are completely stable to heat, and it is not unusual for deleterious changes to take place during sterilizing. Most sugars are altered to some extent and may form products which are toxic to fungi. Of the common sugars, glucose is the most and sucrose the least altered. (For a detailed account of the effect of heat on sugars, see Davis and Rogers, 1940.) Many media made from vegetables owe their particular value to their content of vitamins and kindred substances, many of which are destroyed to some extent on heating. Also, agar-agar, and more particularly gelatine, which are used for solidifying media, lose their power of setting when overheated and agar, like some of the sugars, may give rise to toxic products (see Robins and McVeigh, 1951). Therefore, whilst it is essential to heat sufficiently to destroy infections, it is equally essential to avoid over-sterilization. A most instructive experiment is to make up a quantity of any culture medium, divide it into several portions, and sterilize these by autoclaving respectively for 15 minutes, 30 minutes, 1 hour, 1½ hours, 2 hours, etc.; then sow all the batches of medium with the same species of mould. The differences in rates of growth, and often in types of growth, are usually most striking.

Steaming at atmospheric pressure needs to be very prolonged in order to kill all possible contaminants, the spore-bearing bacteria being particularly resistant. It is usual, therefore, to use an intermittent process, the medium being steamed for 30–60 minutes on each of three successive days. The theory of the process depends on the fact that vegetative structures are more readily killed by heat than are the spores. The first short cooking destroys most of the vegetative growth but may not kill the spores. These, however, being in a favourable situation for growth, and being already swollen by the moist heat, germinate rapidly. The second heating destroys the new growth and the third day's treatment accounts for any spores whose germination has been delayed.

Bacterial spores which can withstand boiling for a long time are killed comparatively rapidly at somewhat higher temperatures. The autoclave is therefore used for sterilization of most vegetable media, which are more likely than synthetic media to be contaminated with spore-bearers. It is, of course, easier with the autoclave than with the steamer to over-sterilize and damage constituents of the medium. To avoid trouble see that the pressure-gauge is reading correctly, and time the duration of the process strictly, reckoning from the time the pressure reaches the required value to the time the source of heat is shut off. Very few materials require more than 30 minutes at 15 pounds, and most media require less. The size of the auto-clave is also important (see Langeron, 1945). A large autoclave takes too long to heat and to cool down, with the result that some portions of the contents are almost sure to be overcooked. It is much better, in a large laboratory, to have several small autoclaves than one big one.

Cold sterilization. There are two methods by which certain culture media can be sterilized without running the risk of damage by heat. Filtration through a Seitz filter will remove all contaminants from a liquid medium, but the process is slow and is seldom used in the mycological laboratory. The second method is described by Hansen and Snyder (1947) and is used for the sterilization of seeds, leaves, portions of stems, and thin slices of vegetables. By its means it has been possible to grow in pure culture a number of fungi which have never been grown on heat-sterilized media. The material to be treated is placed in a screw-top fruit-preserving jar together with a small quantity of propylene oxide, at the rate of approximately 1 ml. per litre space. The lid is screwed down tightly and the jar allowed to stand overnight. The screw is then slackened somewhat to allow the fumigant to escape, and after a few hours the material is ready for use. The most convenient way of handling the sterilized material is to remove portions as required with sterile forceps and transfer to melted plain agar in Petri dishes. The warm agar rapidly drives off the last traces of propylene oxide and, as soon as the agar has set, the material can be inoculated.

Sterilization of rooms. It is sometimes necessary to sterilize the air of a laboratory when it becomes so heavily charged with infection that cleanly work is difficult. A quick way of partial cleansing of the air is to spray thoroughly with a 2 per cent solution of thymol in spirit, using an instrument which gives a fine mist-like spray. Thymol is a very efficient fungicide and the alcohol ensures that any spores with which the mist comes in contact are properly wetted. Another excellent spray is the proprietary antiseptic "Dettol". It owes its fungicidal properties mainly to its content of terpineol.

A thorough method of sterilization, to be used preferably during a week-end, is to charge the air of the room with formaldehyde. A heap of sodium permanganate (the potassium salt serves just as well but is more expensive) is placed on an iron tray on the floor and commercial formalin is poured on in sufficient quantity to wet the heap through. A rapid evolution of gaseous formaldehyde occurs and cleanses the air quickly and efficiently. It is necessary to see that all windows and ventilators are closed before commencing operations, and also, when the formalin has been added, to remove oneself quickly, shut the door, and see that no one enters the room for at least twenty-four hours.

Another good method of purifying the air is to seal up all openings and blow live steam into the room but, obviously, this practice can be adopted only when the fittings of the laboratory are made to withstand it. In some institutions, where the special requirements of microbiological work have been considered at the time of building, the culture laboratory has walls of glazed brick, benches of stone or tiles, drained floor, and metal doors and window-frames which can be made airtight, the air being steamed periodically as a matter of routine. In this method a complete kill of all airborne organisms is not aimed at, since this would involve a steaming sufficiently prolonged to raise all the contents of the room to a temperature approaching 100° C. The efficiency of a comparatively short steaming is due to the fact that spores and particles of dust act as foci around which the steam condenses. The particles are carried down to the floor in the water-drops and drained away with the rest of the condensate.

In some laboratories the air of the culture chamber is sterilized by means of ultra-violet radiation. Special lamps for the purpose are readily obtainable. Such lamps, of course, must be switched off before the chamber is entered.

CULTURE MEDIA

A great many different culture media are used for the propagation and study of fungi, but the majority have been designed for some special purpose, such as the securing of optimum growth of a particular species or for determining the availability of specific substances for mould growth. The only ones which will be noted here are such as are in general use for isolation, propagation, and morphological studies. They can be added to or altered as experience suggests.

Mycological media differ in several respects from the media used by bacteriologists. The latter are usually slightly alkaline in reaction, whereas most fungi prefer a slightly acid medium and many species can tolerate a fairly high acidity. Bacteriological media commonly contain protein as a source of carbon and nitrogen, whilst the majority of the substrates used by the mycologist have carbohydrates as a source of carbon, and nitrogen is supplied in an inorganic form, as nitrates or ammonium salts. The chemical elements which are known to be essential for the growth of fungi are carbon, hydrogen, oxygen, nitrogen, sulphur, phosphorus, potassium, magnesium, and iron. Minute amounts of a number of other elements have been shown to be necessary for growth of certain fungi, and are probably essential for many species whose requirements have not yet been investigated. Most natural media, and synthetic media made up with crude chemicals, normally contain sufficient of these so-called "trace elements", but it is sometimes necessary to add them deliberately when media are compounded with chemicals of "analytical reagent" quality.

Many vegetables and vegetable extracts are eminently suitable as culture media without any addition, or by addition of sugar only, and a few such are very useful for isolation of moulds and maintenance of stock cultures. For some types of work—nutritional studies and biochemical investigations —synthetic media are used almost exclusively, since it is necessary to have media which are of known composition and which can be duplicated as required. In taxonomic work on some groups of fungi synthetic media are advantageous for the same reason. The type of growth shown by any particular species may vary considerably according to the medium used and, in consequence, it is of great importance, when describing new species, to record observations of cultures made on media which can be prepared in identical form by any other worker.

Culture media may be either solid or liquid. The term solid media is used in two different senses, to mean, on the one hand, actual solid substances, usually portions of roots, stems, or seeds of plants, and, on the other, aqueous solutions made into jellies by the addition of gelatine or agar. Agar is usually considered to be a solidifying agent pure and simple. This is not strictly true because even the best samples of agar contain minute amounts of nutrients, and quite a number of species can make limited growth on a medium made with agar and water only. Nevertheless, when agar is used to solidify culture media, the amounts of nutrients which it contributes are such a small proportion of the whole that their effect is virtually nil. The most useful property of agar is the great difference in temperature between its melting-point and setting-point. At the concentrations commonly used, agar media do not melt till the temperature exceeds 95° C, and they can therefore be used for incubating cultures at high temperatures. The molten medium does not resolidify until the temperature falls below 40° C, so that infected material or portions of mould growth may be mixed with the medium, before pouring into Petri dishes, without any danger of killing the spores. Gelatine, on the other hand, serves as an excellent culture medium for many moulds without the addition of any other nutritive substance and cannot, therefore, be considered as an inert support when used to solidify any other medium. Gelatine media are readily liquefied on warming and cannot be used at temperatures above 30° C at the outside. They are also liquefied by the enzyme action of many fungi, and the chief use of such media is for diagnostic purposes based on the presence or absence of this liquefying property.

Solid media are easier to handle than liquid media and are to be preferred whenever requirements permit. Liquids are used chiefly in biochemical work, when it is often necessary to determine the course of metabolism by analysis of the medium or to isolate some specific product of fungal activity.

Preparation of agar media. Agar-agar is a carbohydrate obtained from certain species of seaweeds. It may be purchased as strips of a pale brownish colour, or as a fine greyish powder. The strip form is preferred by some workers, as it is said to give a more transparent medium, but it requires prolonged soaking before it can be dissolved, whereas the powder dissolves very readily and is more convenient to use. The amount of agar to use varies with the type of culture fluid. Media which are not strongly acid seldom require more than 1·5 per cent, as this amount gives a jelly which is firm without being too solid, and having little tendency to crack. With media which have a distinctly acid reaction it is necessary to use 2 per cent or even more, since agar is readily hydrolysed by heating with acid and thereby rendered incapable of setting.

The method of preparation is very simple if powdered agar is used. The required amount of powder is added to the culture fluid, well distributed by shaking, and the mixture heated to the boiling-point. Heating may be carried out in the steamer but is quite safely performed over a bunsen burner if the liquid is constantly stirred to prevent sticking on the bottom of the vessel. The agar is completely dissolved by the time the temperature reaches boiling-point. For most purposes it is unnecessary to filter the agar medium, the slight cloudiness, which is nearly always present, being no disadvantage, but for some purposes, such as making single spore cultures, involving microscopical examination of agar plates, it is undoubtedly preferable to have a perfectly clean medium. Filtration through ordinary filter paper is impracticably slow, even with the aid of a hot-water funnel or by carrying out the filtration in the Koch sterilizer. The "Chardin" type of paper, a very thick soft variety, permits of more rapid filtration but does not give a perfectly clear medium. One of the best methods is to tear up sheets of "Chardin" paper into small pieces, place these in very hot water and shake vigorously until a pulp is obtained, pour this on to a Buchner funnel and suck dry to form a fairly thick pad, wash with hot water until the funnel and flask are heated through, and immediately filter the hot medium with suction. Filtration is rapid and a beautifully clean product results. When the agar is dissolved, and after filtration if this is carried out, the medium is filled into tubes, which are then plugged and sterilized, either by autoclaving for 20 minutes at 15 pounds pressure, or by steaming for 30–60 minutes on three successive days.

Preparation of gelatine media. A good quality sheet gelatine is cut into small pieces, covered with the required amount of water or liquid medium, and allowed to soak for several hours or overnight. The vessel is then heated until solution is complete and the medium is tubed and sterilized by steaming for thirty minutes on three successive days. Gelatine media must not be autoclaved. The strength of gelatine media needs to be varied somewhat according to the season. A 10 to 12 per cent solution will give a firm jelly in all but very hot weather, and 15 per cent is sufficient for such occasions.

VEGETABLE MEDIA

1. Potato plugs. Large, healthy potatoes are scrubbed and washed in water. In some laboratories they are next sterilized by soaking for about twelve hours in a 0·1 per cent solution of mercuric chloride or in a dilute permanganate solution, but if the potatoes are sound this is unnecessary. The tubers are peeled and again washed, and then cylinders, $2\frac{1}{2}$ to 3 inches long, are cut out with a cork borer. The diameter of the plugs should be only very slightly less than the internal diameter of the culture tubes. The cylinders are cut in two, lengthwise and diagonally so as to give wedge-shaped pieces, and the halves are put into the tubes, with the thick ends downwards and resting on pieces of wet cotton-wool. The latter is to prevent the surface of the potato from drying out too rapidly. The tubes are plugged and sterilized by autoclaving at 20 pounds pressure for 20 minutes on two successive days.

Potato plugs are sometimes useful for inducing obstinate cultures to produce spores, but, like other similar media, they have the disadvantage that only the surface of the colony is seen and characteristic features, which are noted when the reverse side of a Petri dish is examined, are missed.

2. Other solid vegetables. Carrots may be used as plugs, made in the same way as potato plugs, or may be cut into slices. Beans and other seeds are used whole, whilst stems of various plants are cut into short lengths. Otherwise, the details of preparation are the same as for the potato medium.

3. Disintegrated vegetables. The nutritive value of whole vegetables can be combined with the convenience of agar media by using one of the modern high-speed disintegrators, of which the American "Waring Blendor" is the prototype. Some vegetables can be treated in the raw state, but those, such as carrot, which are somewhat tough are preferably lightly cooked. The water in which they are cooked is put into the machine with the vegetables, so as to lose none of the water-soluble constituents. Vegetables are rapidly reduced to a creamy consistency resembling an emulsion. After treatment the mixture is diluted to the required volume, 1·5 per cent of agar is added, and the medium tubed and sterilized in the autoclave.

4. Vegetable decoctions. Extracts of potatoes, beans, carrots, and prunes are the decoctions most commonly used. Fifty grams of prunes or dried beans, 100 g of carrots, or 200 g of peeled potatoes cut into small pieces, are boiled for an hour in 1 litre of water. The liquid is strained through fine muslin, or filtered, and solidified if required with 1·5 to 2 per cent of agar. The medium is tubed and autoclaved for 20–30 minutes at 15 pounds.

The amounts given are average figures and may be varied according to requirements.

The prune medium, without additions, gives good growth of most moulds, but the bean, carrot, and potato extracts lack soluble carbohydrate and are improved by the addition of 1·5 to 3 per cent of glucose or cane sugar. Potato-dextrose agar, as used in many mycological laboratories and commonly referred to as P.D.A., contains 2 per cent of glucose. Carrot extract, with addition of 2 per cent cane sugar, is an excellent medium for the majority of the Dematiaceae and for species of *Fusarium*, and has been found to induce sporulation of certain Ascomycetes which grow scarcely at all on other media in common use. For some species of moulds it has been found that a mixture of equal parts of carrot and potato extracts is better than either extract alone.

5. Wort. Unfermented sweet wort from a brewery is diluted (if necessary) to a specific gravity of 1·05, heated in the autoclave for half an hour at 10 pounds pressure and filtered hot. If no autoclave is available the wort may be heated for an hour in the steamer, but in this case it will be found that subsequent sterilization causes a further precipitate to be thrown down, whereas, if the wort is first heated to a temperature above 100° C and sterilization is carried out in the steamer, this does not occur. Wort agar is made by the addition of 2 per cent of agar and the medium, being distinctly acid in reaction, should not be over-sterilized.

In some districts wort is difficult to obtain and recourse must be had to commercial malt extract. This is not a complete substitute, since it is not practicable to make a solution which is as rich in nutrients as natural wort. A medium which has been found to be completely satisfactory is made by adding 5 per cent malt extract, 3 per cent cane sugar, and 2 per cent agar to Czapek's solution (see p. 233).

A modified malt medium is recommended by Yuill (private communication) for species of the *Aspergillus glaucus* series and the *A. restrictus* series. It contains 2 per cent malt extract, 8 per cent sodium chloride, and 1·5 per cent agar. Most strains of the above series grow just as well on this as on Czapek agar with 20 per cent cane sugar (see p. 233).

Wort agar gives vigorous and characteristic growth of the great majority of common moulds and is very useful for isolation of species and for stock cultures of some of the more delicate and slow-growing species. Some species, however, tend to produce an undue amount of mycelium at the expense of spore-bearing structures, and these are best kept on a less rich medium. A still richer medium, much favoured by some mycologists, is made with 10 per cent of gelatine instead of agar. One interesting feature of wort is that it is heavily buffered and has no tendency to become alkaline, as many synthetic media do, when the food material is becoming exhausted. In many species of *Penicillium* the colouring matter of the spores acts as an indicator, being green with acid and turning greyish or brownish as the pH rises. Such species frequently become dirty grey on Czapek agar but retain their green colour for months on wort.

6. Corn-meal agar. Consists of corn meal (ground maize), 60 g; agar, 15 g; water, to 1 litre. Mix the meal to a smooth cream with water, simmer for one hour, filter through cloth, add the agar and heat till dissolved, make up to original volume, and sterilize, preferably in the autoclave for 30 minutes at 10 pounds pressure.

This medium is in regular use in many laboratories. It is very valuable for stock cultures of many fungi, particularly of most of the dark-coloured moulds. Although it contains a fair concentration of nutrients there is little soluble carbohydrate and, in consequence, moulds must hydrolyse the starch before they can obtain the glucose they require and there is never any excess of the latter to stimulate over-rapid development of mycelium.

7. Wheaten flour agar. This is made in the same way as corn-meal agar. It contains more nitrogen than the latter and a somewhat different assortment of accessory factors (vitamins and kindred substances). Some species which grow only very slowly on corn-meal make satisfactory and typical growth on the flour medium.

8. Plant-stems. Various stems of plants have been, from time to time, suggested as culture media. At the Centraalbureau voor Schimmelcultures, Baarn, Holland, *Lupin* stems are used regularly, and have been found valuable for inducing spore production when cultures on the usual agar media remain sterile. The flowering stems, of such diameter that they will fit easily into culture tubes, are cut into lengths of 5–6 cm., packed into fruit-processing jars, covered with water, and sterilized in the autoclave. They remain in good condition almost indefinitely in the unopened jars.

SYNTHETIC MEDIA

9. Plain gelatine. Simply a 10–15 per cent solution of gelatine in water, prepared as described above.

10. Sugared gelatine. The addition of 1 to 3 per cent of cane sugar to plain gelatine gives a medium which usually induces more vigorous growth than the unsweetened medium. It serves for determination of liquefying power in many species which grow very poorly in the absence of sugar. The method of preparation is obvious.

11. Raulin's medium. The first attempt to compound a rational synthetic medium was by Raulin (1869), who analysed the ash of *Aspergillus niger* and, on the basis of his analysis, made a medium which he used for biochemical studies of this species.

The only difficulty in making up this solution is to dissolve the silicate, but it is hard to see what useful purpose it serves and it can certainly be

omitted without detriment. The reaction is strongly acid, pH 2·9, and it is
almost impossible to make an agar medium which will set. Growth of
Aspergillus niger, for which the medium was designed, and of a few other
species is good, but a great many moulds fail to grow characteristically or
at all. Raulin's solution is of interest chiefly because it forms the basis of
many other media, some of them of great value.

Sugar candy	70 g
Tartaric acid	4 g
Ammonium nitrate	4 g
Potassium carbonate	0·6 g
Ammonium phosphate	0·6 g
Magnesium carbonate	0·4 g
Ammonium sulphate	0·25 g
Zinc sulphate (crystals)	0·07 g
Ferrous sulphate (crystals)	0·07 g
Potassium silicate	0·07 g
Distilled water	to 1500 ml.

12. Raulin-Thom solution. Thom and Church, in *The Aspergilli* (1926,
p. 40), make a curious mistake in quoting Raulin's medium, and the mistake is repeated in Thom's *Penicillia* (1930, p. 36). Although it is mentioned
in the text that the solution contains ammonium nitrate, the formula given
omits this salt and substitutes ammonium tartrate. The result is an extra-
ordinarily interesting medium for the purpose of biochemical work and one
which gives good growth of most common moulds. It is appreciably less
acid than the original Raulin's medium, the pH being about 3·9.

The medium is not easy to make up, as magnesium tartrate tends to
separate and does not readily re-dissolve. The following is a satisfactory
method. The tartaric acid, finely powdered, is dissolved in about 500 ml.
of warm water and the $MgCO_3$ is dissolved in this solution. The ammonium
tartrate is added and the liquid stirred till it is all in solution. Next the
sugar is dissolved in the mixture. The potassium carbonate is dissolved
separately in 100 ml. of water and added to the rest. The ammonium, zinc,
and ferrous sulphates are also separately dissolved and added. Finally, the
phosphate is dissolved in 500 ml. of water and added to the main solution
with constant stirring, and the medium is made up to the correct volume.
If properly compounded, the medium should be perfectly clear and of a
faint yellow colour which becomes somewhat deeper on sterilization.

13. Neutral Raulin-Dierckx medium. This is another interesting
modification of Raulin's medium, designed by Dierckx for use with species
of *Penicillium* and strongly recommended by Biourge (1923, p. 43). It
certainly stimulates pigment production in many species of *Penicillium* and,
in addition, often gives typical growth, when other media fail, of the core-
miform species of this genus. The method of making up the solution is
given by Biourge as follows:

1. Dissolve 0·40 g of $MgCO_3$ with 0·71 g of tartaric acid in 100 ml. of water.
2. In 800–900 ml. of distilled water dissolve saccharose, 46·6 g; NH_4NO_3, 2·66 g; ammonium phosphate, 0·40 g; K_2CO_3, 0·40 g; $(NH_4)_2SO_4$, 0·16 g; $ZnSO_4.7H_2O$, 0·04 g; $FeSO_4.7H_2O$, 0·04 g.
3. Add to solution (2) 66·7 ml. of solution (1) and make up to 1000 ml.

In a note on another page (p. 37) Biourge says: "Take 0·27 g $MgCO_3$ and 0·40 g tartaric acid. Bring together in a small mortar until clear, then dilute at once considerably to stop crystallization"; the rest of the formula being as above. He adds: "The small amount of precipitate can be ignored if distributed equally in containers." For a solid medium Biourge recommends the addition of 100 g of gelatine.

14. Czapek's solution. Many workers consider that Raulin's solution and its various modifications are unnecessarily complicated. Czapek's solution is an attempt to supply all the elements necessary for growth with the minimum of duplication. The basal salt solution is here given as modified by Dox (1910) and Thom and Church (1926).

Sodium nitrate	2·0–3·0 g
Potassium chloride . . .	0·5 g
Magnesium sulphate, $MgSO_4.7H_2O$.	0·5 g
Ferrous sulphate, $FeSO_4.7H_2O$.	0·01 g
Potassium phosphate, K_2HPO_4 .	1·0 g
Distilled water	to 1000 ml.

The basal solution may be used with various amounts of sugar, with different sugars, or with other sources of carbon. The most usual addition for general taxonomic work is 30 g of sucrose, for optimal growth of species of the *Aspergillus glaucus* group 200–400 g sucrose, and for many biochemical studies 50 g of glucose. If glucose is used it is necessary to make up the solution without phosphate and add the latter in concentrated solution after sterilization; otherwise the medium becomes brown and turbid on heating. The reaction is approximately neutral and quite an appreciable amount of magnesium phosphate is precipitated, indicating that the medium might with advantage be slightly modified. Many workers make up the solution with acid potassium phosphate, KH_2PO_4. The reaction is then definitely acid, the pH being about 4·2, but, even so, there is still a small amount of precipitation on sterilization. With the acid phosphate there is no browning of glucose on heating. The acid medium gives rather better growth of most moulds than the neutral form.

In compounding the medium, it is best to dissolve all the salts except the phosphate in about half of the water, add the sugar, dissolve the phosphate separately and add to the rest, finally making up to correct volume.

Czapek agar, one of the most generally useful of all solid media, is made by the addition of 1·5 per cent of agar. It is recommended by various authors for taxonomic studies: for *Aspergillus* by Thom and Church (1926) and Thom and Raper (1945), for *Penicillium* by Thom (1930) and Raper and Thom (1949), and by Waksman for Actinomycetes (1931).

Smith (1949) states that many species of *Penicillium* fail to grow characteristically on Czapek agar if this is made up with only the purest ingredients. Typical growth and, in certain cases, increased yields of metabolic products are obtained if traces of zinc and copper are added to the medium. The amounts recommended are $ZnSO_4.7H_2O$, 0·01 g, and $CuSO_4.5H_2O$, 0·005 g per litre.

ANTI-BACTERIAL MEDIA

15. Littman's medium (1947).

Dextrose	1·0%
Peptone	1·0%
Oxgall	1·5%
Agar	2·0%
Crystal violet	1 : 100,000
Streptomycin	30 units/ml.

The medium is made up without the streptomycin and sterilized in the usual way. It is cooled to 46° C and the streptomycin, in sterile saline, is added just before pouring. The oxgall in this medium is to limit the growth of "spreaders", that is such species as *Rhizopus stolonifer* and *Trichoderma viride*, which otherwise tend to swamp slow-growing moulds.

16. Smith and Dawson's medium (1944).

Glucose	10 g
$NaNO_3$	1 g
K_2HPO_4	1 g
Agar	15 g
Rose bengal	0·067 g (1 : 15,000)
Soil extract	1000 ml.

The soil extract is made by autoclaving 500 g of loam in 1200 ml. water for one hour. The extract is filtered through paper, and should have a final volume of 1 litre.

17. Cooke's medium (1954).

Dextrose	10 g
Phytone or peptone	5 g
KH_2PO_4	1 g
$MgSO_4.7H_2O$	0·5 g
Agar	20 g
Water	to 1000 ml.
Rose bengal	0·035 g
Aureomycin	35 μg

The aureomycin must be added to a sterilized and cooled medium, just before pouring.

Some moulds grow very feebly, or fail to grow, on synthetic media but are strongly stimulated by the addition of small amounts of preparations which are rich in accessory factors. Other species which grow well on ordinary media are stimulated by the same means to grow still more luxuriantly, and often to give increased yields of desired metabolic products. There are many sources of these accessory factors but, of these, the four mentioned here are the most widely used and easily obtainable.

1. Yeast extract. Pressed (bakers') yeast, 200 g, is autoclaved with a litre of water. The suspension is filtered hot, and again after cooling, and made up to the original volume. The best amount of extract to use in any particular case can only be found by experiment.

2. Corn steep liquor. This contains most of the accessory factors of raw maize. The liquid, at any rate in this country, is of variable composition and its use in biochemical work makes reproducibility of results very uncertain. However, additions of quantities of the order of 0·1–0·2 per cent to any of the usual synthetic media will frequently result in increased rate of growth and more abundant sporulation.

3. "Marmite." This has been widely used, as a source of the B group of vitamins, in nutritional experiments on animals. It can be used also, as a stimulant, for growing fungi which do not thrive on synthetic media. The amount required is small, of the order of 0·1 g per litre of culture medium.

4. "Panmede." This is a digest of ox-liver, manufactured by Messrs Paines & Byrne, Ltd., Greenford, Middlesex. It is claimed that it induces vigorous growth of organisms which are difficult to cultivate on other media. The medium recommended for moulds contains 1 per cent Panmede, 4 per cent crude dextrose, and 1·5 per cent agar, adjusted in the usual way to any desired pH.

Slopes. Cultures which have to be kept for some time for study, or which are to be stored as part of a culture collection, are almost invariably made on agar slopes in tubes. For making slopes in tubes of $\frac{5}{8}$ inch diameter the amount of agar medium should be 5–6 ml. The medium is sterilized in the tubes in the ordinary way and then, whilst the agar is still hot and molten, the tubes are inclined at such an angle that the medium forms a layer of decreasing thickness from the bottom of the tube to within about an inch of the plug. Laying the tubes on the bench with the plugged ends supported on a glass rod half an inch in diameter gives about the right amount of slope. The object of sloping the agar is to provide a relatively large surface, on which the progress of growth can be watched far better than on a level surface of agar in a comparatively narrow tube. Incidentally, the varying thickness of the layer of medium often reveals interesting cultural characteristics, the type of growth at the shallow end showing marked differences from that at the deep end.

The method of sowing slopes (assuming that the worker is right-handed) is to hold the tubes by their lower ends between the thumb and first finger of the left hand, with the plugs on the palm side of the hand and pointing slightly downwards. Three is the maximum number of tubes which can be conveniently held, allowing of two transfers being made in one operation from a tube culture, or three cultures being made from infected material. A needle is sterilized by heating to redness in a bunsen flame. Whilst it is cooling the plugs are removed, one at a time, by the right hand, using a twisting motion, and placed between the other fingers of the left hand so that they are held by their tops only. The bunsen flame is played round and into the mouths of the tubes, an operation known as "flaming", until the glass is too hot to touch. The tip of the needle is plunged into the agar in one of the unsown tubes in order to make sure that it is cool and to wet it slightly, and is then used to pick up a few spores or a fragment of mycelium from the parent culture. This is deposited, as rapidly as possible and without being allowed to come in contact with the hot glass, on to the fresh agar surface. When both tubes have been thus sown the mouths of the tubes are again flamed and the plugs re-inserted. Some workers flame the tubes again, after insertion of the plugs, but this is not necessary unless the plugs have been out of the tubes for an appreciable time, and it causes the tops of the plugs to become charred and messy to handle.

A slightly different method of handling tubes is claimed by some to be safer, in the sense that it gives more security against contamination. The tubes are held between thumb and first finger but with the palm of the hand pointing directly downwards. When the plugs are placed between the other fingers they are protected by the hand and there is little risk of their picking up infection. Some people find it very difficult to acquire this technique and these should stick to the more usual method described above.

When large numbers of cultures of a single species are required the best method, provided that the parent culture is sporing freely, is to use a spore suspension. With the usual precautions a fairly large mass of spores is picked up on the needle and transferred to about 1 ml. of sterile water contained in a very small test-tube. The tube is shaken to distribute the spores and is then supported, as nearly horizontal as possible, in such a way that the mouth of the tube can be flamed periodically during the sowing. The inoculations are made with a wire loop, a tiny drop of suspension being transferred to each fresh tube of medium. Loops of standard size for bacteriological work, made of either platinum or nichrome wire and either unmounted or fixed in handles, can be purchased but are readily made from the ordinary wire needles by winding the ends round a metal or glass rod of about 1 mm. diameter.

One other precaution should be taken when sowing slopes. The most characteristic growth is obtained when a slope is sown at a single point near the centre. The bacteriologist inoculates in a wavy line from bottom to top of the slope, but this method is quite wrong when handling moulds. With a single spot inoculation not only is the growth more typical but the surface of the slope is actually more quickly covered.

Cultures on liquid media. Such media in tubes are sown in the same way as slopes except, of course, that the tubes cannot be held with mouths pointing downwards. They should be held as nearly horizontal as possible without getting the liquid on to the hot glass near the mouths, in order that air-borne spores cannot fall directly on to the surface of the medium but will fall on the hot glass and be killed.

Plates. For morphological studies microfungi are grown in Petri dishes (commonly known as "plates"), as well as on slopes. Dishes planted with single colonies are useful for determining rate of growth and colony characteristics such as zonation and sectoring, whilst plates containing several colonies are more suitable for microscopical examinations. Many species form dense, opaque felts of mycelium, so that the only part of an isolated colony which can be examined by transmitted light is the extreme edge, and this usually shows no ripe fruiting structures. In a dish containing several colonies it is usually found that, along the edges where these approach each other, narrow sterile zones are left and, in the more mature portions of the colonies, spore-bearing organs can be clearly viewed as they hang over these gaps (Figs. 164, 165).

The usual 10-cm. Petri dish requires approximately 12 ml. of medium to give a layer of adequate thickness, and, if comparative cultures of different species, or different strains of the same species, are required, the amount of medium per dish should be fairly accurately standardized. The medium is filled into tubes in correct amounts and sterilized therein. Before pouring the plates the tubes should be allowed to cool down to about 45° C, best by leaving them for a few minutes in a water-bath maintained at this temperature, as very hot medium gives off water vapour, which condenses on the cool lid of the dish and then drips back on to the medium. Each tube, as it is lifted from the water-bath, is first quickly wiped free from adhering water. If this precaution is omitted there is a danger of drops of this water, usually far from sterile, getting into the plates. Next the tube is held in a sloping position whilst the plug is removed and the mouth flamed, and then the medium is poured gently into the dish whilst one edge of the cover is raised as little as is necessary for the purpose. After replacing the lid the dish is carefully tilted to spread the medium and then left to stand on a level surface until the agar has set. Whenever possible it is best to store Petri dishes in the inverted position, both before and after sowing, as this minimizes the chance of infection, and when sowing plates, or handling

FIG. 164.—*P. expansum*—colonies in Petri dish,
showing the absence of a sterile edge where the
colonies approach one another.   ×0·6.

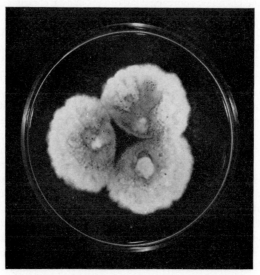

FIG. 165.—*P. nigricans*—floccose, tardily sporing
but showing the same effect.   ×0·6.

them for examination, every care must be taken to avoid exposure of the medium more than is absolutely necessary, since protection by flaming, as in the case of tubes, is impossible. With dishes in the normal position it is not easy to sow colonies in predetermined positions, the inoculating needle often leaving a trail of spores right across the surface of the medium. If the dish is inverted, then lifted out of its cover, and the needle approached and receded from directly below, it will be found that very few stray colonies will appear. Another method of sowing colonies exactly where required is to inoculate with loopfuls of a spore suspension.

Slide cultures. For the study of some species which produce very small and fragile sporophores the only satisfactory methods of making preparations for microscopical examination are by means of slide cultures and cultures on cellulose film. There are a number of simple ways of making cultures on microscope slides.

(1) If the fungus tends to spread close to the substratum it is grown on a very thin layer of agar spread on the slide. The agar medium should preferably be filtered till it is clear and transparent. A piece of glass rod, or better, a strip of aluminium sheet not more than 1 cm. in width, is bent twice at right angles, in such a way that it will fit into a Petri dish and support a slide clear of the bottom. Shallow slots in the two opposite sides of the bent strip will prevent the slide from slipping about. The dish, with the support and slide in position, is sterilized by dry heat. A drop of melted agar is poured on the centre of the slide whilst the latter is still warm and spread as evenly as possible by means of a bent glass rod. If it is desired to study the germination of spores, a little spore material is mixed with the agar before pouring on to the slide, but if spore production is to be observed, it is better to plant the medium at two or three points. About 10 ml. of a sterile 20 per cent solution of glycerol is poured into the bottom of the dish. This keeps the thin layer of agar moist but not wet, as it would become if pure water were used. The slide is sufficiently near to the lid of the dish to allow observation of growth with a low-power objective without removing the lid; if the height of the support has been properly adjusted a $\frac{2}{3}$-inch objective may be used. When the desired stage of growth has been reached the slide is lifted out and placed for a few hours in a similar dish (which need not be sterilized) containing a little formalin, or a solution of osmic acid, in order to kill the fungus and partially fix the structures. The slide may be examined dry, as it is, or, if the growth is too decidedly aerial, with a cover-glass laid on very gently. It is often possible, in this way, to make fairly flat preparations without breaking down fragile structures unduly. With some species which are not quite so fragile permanent mounts may be made in lacto-phenol, afterwards cutting away the agar round the cover-glass and sealing with cement (see below, under Microscopical Methods).

(2) Another way of making slide cultures is described by Henrici (Skinner et al., 1947), utilizing a shallow cell, built up on the slide with sealing-wax

and a large rectangular cover-glass. With the plane of the slide vertical agar medium is run in to about half the depth of the cell, and the fungus is planted on the narrow surface thus provided. All stages of growth are readily observed owing to the spread of the mould being confined approximately to one plane, and the effect is as if a thin section through a colony were being examined. With moulds of vigorous habit and with large fruiting structures, such as many of the Mucoraceae, a larger cell may be built up on the same lines, using sheets of thin glass, such as old photographic plates stripped free from gelatine, clamped together in pairs with separators made from narrow strips of cardboard. The photograph of *Thamnidium elegans*, Fig. 23, was taken from a culture made in this way.

(3) An elegant and most useful method, particularly for the examination of fragile sporing structures, has been perfected by the late Dr. J. T. Duncan and has been used successfully for many years for the examination of dermatophytes in the Department of Medical Mycology, London School of Hygiene and Tropical Medicine.

Agar medium is poured into a Petri dish to a depth of about 2 mm. When it has set completely, a small block, about 1 cm. square, is cut out with a sterile tool and transferred to the centre of a sterile slide. The block is inoculated on all four *edges*, then covered with a large sterile cover-glass. The slide is incubated in a moist chamber, such as is described above. The fungus spreads out from the agar block and tends to attach itself to the two glass surfaces. The slide may be taken out of the moist chamber from time to time for examination, without much danger of contamination. When growth has reached the desired stage the cover-glass is carefully stripped from the agar block and carefully lowered, fungus side down, on to a drop of mounting fluid (with or without stain) on another slide. Alternatively the preparation may be stained by any standard procedure before mounting. Next, the agar block is carefully loosened and removed, leaving a second preparation on the original slide. The photograph of *Sporothrix schenckii*, Fig. 83, was obtained from such a slide culture.

Still another method of making slide cultures is used at the Commonwealth Mycological Institute, and described in their Handbook (p. 64). Agar medium is poured into a Petri dish and allowed to set. Using a sterile knife or scalpel, two diametrical slits are made, at right-angles to one another, in the agar. Each section in turn is partially raised, from the central cross, and a flamed cover-glass placed under the agar, which is then allowed to fall back into place. The dish is inverted and the positions of the cover-glasses marked with grease pencil. A small square, about $\frac{1}{4}$ inch side, is also drawn over the centre of each cover-glass. The dish is turned right side up, and pieces of agar, corresponding to the marked squares, are cut out. The medium is now inoculated. In most cases the growth will eventually extend over the bare areas of cover-glass. When such growth is satisfactory the cover-glasses are removed by cutting away the surrounding agar, and are mounted in the usual way.

LABORATORY EQUIPMENT AND TECHNIQUE 241

Culture on cellulose film. Instead of glass slides, and in some ways more satisfactory, transparent viscose film may be used as a support and has some distinct advantages. Cellulose acetate film is quite unsuitable for culture work as it is non-absorbent, but the pure regenerated cellulose film readily absorbs water and substances in aqueous solution. Cellulose itself is utilized by comparatively few moulds, so it is necessary to add suitable nutrients. There are two simple ways of doing this.

(1) Small pieces of the material (about ½ inch square is a convenient size) are soaked in dilute culture medium, blotted to remove excess liquid, and sterilized by steaming in any suitable receptacle. Using sterile forceps a number of pieces, six to eight are usually sufficient, are laid in a sterile Petri dish, or, if incubation is to extend for more than a few days, in a dish lined with wet blotting-paper. Each piece is then inoculated at one or more points as desired and the dishes incubated.

(2) In the method described by Fleming and Smith (1944) small pieces of film are immersed in water in a short test-tube and sterilized by boiling or autoclaving. A number of pieces are picked out by means of sterile forceps, laid on the surface of an agar culture medium in a Petri dish, and inoculated at one or more points. Soluble food material diffuses through the film and normal growth occurs. For the study of spore germination almost any culture medium may be used, but for examination of later stages of growth and spore production, it is better to use a medium of 10 per cent or even 1 per cent the usual strength, so as to obtain thin transparent colonies.

With either method single pieces of film may be removed at intervals for examination of various stages of growth, and may be handled in a number of different ways. They may be examined without any preliminary treatment, being simply laid on a slide or first bent sharply across the centre of the mould colony so that fruiting structures may be examined where these overhang at the fold. They can be mounted directly in lacto-phenol, or they may be first fixed in formalin vapour and stained with any reagent which does not stain cellulose. For permanent preparations they may be fixed, stained, dehydrated, cleared, and mounted in balsam, the methods being fully described in a number of books on microtechnique. Two modern books which can be recommended are by Johansen (1940) and Sass (1941).

A further use of viscose film, described by Fleming and Smith (*loc. cit.*), is for the preparation of museum specimens of mould colonies. Circles of film, or of black or white paper which is not too highly glazed, are cut so that they just fit comfortably in a Petri dish. A number of circles are placed between sheets of wet filter-paper in a dish and sterilized in the autoclave. Each piece is carefully laid on the surface of a culture medium in a Petri dish and inoculated at its centre. After a colony has grown sufficiently the film or paper is lifted out of the dish, floated on 10 per cent formalin and left overnight to kill and fix the colony, dried in the air and finally mounted between a pair of watch-glasses. Various modifications of the proceduer are described in the original paper.

Kondo *et al.* (1959) have described a variation of this technique, for the purpose of preparing herbarium specimens in such a way that these can be used to maximum advantage by subsequent students. The original paper should be consulted for details. An excellent book which discusses the principles of preparing specimens for microscopical examination, is by Baker (1958).

Cultures for biochemical studies. A number of methods have from time to time been used for growing moulds on comparatively large volumes of liquid media, for studies of biochemical activity and isolation of metabolic products. The widespread interest in such studies at the present time, and the difficulties which have been encountered in some laboratories, are the reasons for describing here the methods which have been found satisfactory in the Department of Biochemistry at the London School of Hygiene and Tropical Medicine.

Containers. It is necessary, if moulds are grown as surface felts, to use shallow layers of liquid, in order to secure complete utilization of all the nutrients. Trays take up little room but are difficult to handle and to protect from contamination. Conical flasks, although taking up more space in the incubator, have been found to be in every way more satisfactory for total volumes up to 50 or even 100 litres. Cultures can be examined during the period of incubation for uniformity, purity, and vigour; distribution in a number of small containers localizes chance contamination; and the course of metabolism is more satisfactorily followed by examination of the whole contents of a single flask than by drawing off samples from a large container.

Flasks are of good-quality resistance glass and are used solely for this kind of work. For isolation of products 1-litre flasks are used, but smaller sizes, from 100 ml. to 750 ml., are used for special purposes and for pilot experiments. The volume of liquid in each flask is such as will give a depth of not more than 4 cm. A normal 1-litre flask holds 350–400 ml., whilst a 250-ml. flask usually contains 100 ml.

Inoculation. Flasks are sown with an aqueous spore suspension (or suspension of mycelium if the fungus is sterile), giving a heavy sowing so as to ensure the rapid establishment of a continuous felt of mycelium. It has been found that the most satisfactory method is to start with a number of cultures on agar slopes, in $\frac{3}{4}$- or $\frac{7}{8}$-inch rimmed tubes, allowing one slope for each 4 or 5 flasks if the mould spores well, and one slope for 3 or even 2 flasks if spore production is scanty or the fungus sterile. A similar number of tubes of sterile water, 12–15 ml. per tube, are also prepared. The contents of one tube of water are poured, with of course the usual precautions, into one of the culture tubes, the surface of the agar is lightly scraped with a stiff tool, working as rapidly as possible, and the suspension is poured directly into the flasks. The aim should be to distribute the spores evenly, and not necessarily to add the same volume of liquid to each flask. There is usually no need to flame the mouths of the flasks, since the plugs are out for only a few seconds. Finally the flasks are shaken to distribute the spores over the surface of the liquid. The whole process is carried through as rapidly as possible, consistently with efficiency, so as to avoid contamination. Rapid work also ensures that most of the spores are not completely wetted, and therefore float on the surface of the medium in the flasks.

A non-sporing mould is not so easy to handle as a species which spores freely. Flasks must then be sown with a suspension of bits of mycelium, with the disadvantage that most of the inoculum usually sinks, and the formation of a surface felt does not begin until hyphae from the submerged mycelium have extended sufficiently to reach the air, often a matter of several days. It is therefore essential to work in a clean atmosphere and to take every possible aseptic precaution. A method adopted with fungi which will not grow when submerged is to use a long needle to transfer bits of dry mycelium from a slope to the surface of the liquid medium, but to do this successfully requires considerable practice.

In some laboratories adaptations of bacteriological technique are used. Cultures are grown in Roux bottles or flat culture flasks, spore suspensions are then made by scraping off the growth into comparatively large volumes of liquid, and flasks are sown with measured volumes of the suspension delivered from a pipette. This procedure does not usually give such satisfactory results as the method outlined above. Sowing with a pipette certainly means that each flask receives the same volume of suspension but seldom ensures equal distribution of spores. In the rare cases in which the spores of a mould are readily wetted equal distribution is of course achieved, but then most of the spores sink in the culture fluid and the establishment of a surface felt is delayed. If the spores are not easily wetted, as is usually the case, most of them float to the top of the pipette and remain there when the liquid is run out. There is also much greater risk of contamination when many flasks are sown from one suspension than when only a few flasks are sown from one culture.

In certain fermentations which are carried out on a factory scale, such as the manufacture of penicillin, the mould is actually grown submerged in the culture medium, the latter being artificially aerated. In these cases the inoculum should be wetted as completely as possible and, to ensure this, the liquid used for making the spore suspension contains a wetting agent.

Special Media. Most of the culture media used for moulds are slightly acid in reaction, the pH being usually between 4·0 and 5·0. If the mould grows better at a higher initial pH it is necessary to make up and sterilize the medium at a low pH and adjust to the final reaction after sterilization, in order to avoid decomposition of the sugar. The contents of one flask are used for a titration, to determine the requisite amount of alkali to be added.

This amount of sterile alkali, usually NaOH, is added to each flask by means of a sterile pipette. It is advisable to have an assistant for this task, to flame the necks of the flasks and hold them in an inclined position whilst the alkali is being run in.

Media containing both glucose and ammonium salts cannot be sterilized without serious discoloration if the pH is higher than about 4·0. In such cases it is usual to dissolve the ammonium salts separately in a volume which is equivalent to an easily measured volume per flask, say 10 ml., and to dissolve the remaining ingredients in a correspondingly reduced volume. Both solutions are sterilized and, when cold, the solution of ammonium salts is measured into the flasks by means of a sterile pipette. An easier way, and one sufficiently accurate for most purposes, is to pipette the solution of ammonium salts, before sterilization, into a series of plugged tubes. Then, after sterilization, the contents of one tube are poured into each flask.

METHODS OF ISOLATION AND PURIFICATION OF MOULDS

The method used to isolate a particular mould from a natural substratum, and to obtain a pure culture, depends somewhat on circumstances. If the fungus is growing more or less luxuriantly, and typical aerial fruiting structures can be clearly seen, it is usually easy, working with a fine sterile needle and with the aid of a good hand-lens or dissecting microscope, to pick off a few spores, or a single spore-head, and transfer to a suitable culture medium. Very often a pure culture results from this first transfer. It is seldom, however, that a mould is found growing, under natural conditions, entirely free from other organisms, and there is always a danger that direct transfers will carry a contaminant. When this happens to be a slow-growing species its presence may not be detected for some time and, therefore, cultures made in this way must be watched carefully over a period of several weeks, and purified if at any time there is reason to suspect contamination.

When mould occurs on industrial products, such as leather, textiles, and cereals, the presence of the fungus is often betrayed only by a stain or discoloration instead of the more familiar furry growth which one associates with mould. Even when the stain is due entirely to the presence of coloured spores it is difficult, and often impossible, to demonstrate the presence of spore-bearing heads. In such cases direct cultures can seldom be made without introducing gross contamination. A few adventitious spores of a very rapidly growing species, in presence of a much larger number of those of the causal organism, may result in cultures being completely swamped with a mould which has nothing whatever to do with the damage. With material of this type the best way of isolating the dominant organism is by making a series of dilution cultures, an operation commonly termed "plating out".

Dilution cultures. A number of tubes of agar medium, 10–12 ml. in ach tube, are heated till the agar is melted and then placed in a water-bath naintained at 45° C until required. Whilst the tubes are cooling a small ortion of the mouldy material is reduced to as fine a state of subdivision s possible. A few fragments are dropped, by means of sterile forceps, into ne of the tubes of medium, taking the usual precautions in handling the ube. The plug is then reinserted and the tube is rotated between the palms f the hands in order to mix the contents. The tube should not be shaken n the usual way, as this introduces numerous air bubbles which are very ersistent in the viscous fluid. The plug is removed, the mouth of the tube lamed, and the contents poured into a sterile Petri dish. The medium from nother tube is now poured into the first, mixed by rotation with the small mount of agar remaining after pouring the plate, and then poured into a econd Petri dish. The contents of a third tube are now poured into the ame tube, mixed and poured as before, and this process continued for a umber of plates which can be determined only by experience with the articular material; usually five or six plates are sufficient. As soon as the gar has set the plates are inverted and incubated.

The rationale of the process is simple. Thorough mixing of the infected naterial with the agar in the first tube serves to disseminate the spores of he fungus, or, in some cases, fragments of mycelium, throughout the nelted medium. The small amount of agar left in the tube after pouring ontains relatively few spores and these are again distributed throughout considerable volume of medium when the contents of the second tube re added, and so on. Each plate, after the first, will contain only a fraction f the number of spores contained in the previous one and, on incubation, he successive plates will show fewer and fewer colonies. If the spores of a articular mould are very numerous in the infected material they are likely o persist through all the dilutions and give rise to colonies in the later lates, whereas a few purely adventitious spores, which are to be found on lmost any material, are eliminated in the first two or three dilutions. The inal plate, if the amount of material and the number of plates have been udged correctly, should not show more than nine or ten well-separated olonies. If these are all alike, a pure culture has automatically been ob- ained and nothing is required but to make transfers to slopes or fresh lishes as desired. If more than one mould is present on the final plate, ncubation should be continued only sufficiently long for the colonies to be lifferentiated, and transfers should immediately be made from a colony of ny species which it is desired to study or retain. It is advisable to plate out eparately each of the moulds isolated, as soon as the transfers are showing pores, in order to check their purity and effect further purification if this s necessary.

It is policy, in most cases, to plate out infected material on two or three lifferent media, for an unimportant mould may grow so well on a rich nedium, such as wort agar, that it swamps a slow-growing species, or, on he other hand, an important species may grow very poorly or not at all on

a synthetic medium such as Czapek agar and be completely missed unless
a more suitable medium is used as well.

A slightly different method of making dilution cultures is often advo
cated. Instead of pouring all the plates from one tube, which is refilled for
each plate, a little of the thoroughly mixed contents of the first tube is
poured into the second tube and the remainder poured into a dish, a little
of the second is poured into the third tube, and so on. The method given
above is better, unless an exact dilution ratio is necessary (and this involves
making the dilutions with sterile graduated pipettes), since it ensures an
approximately uniform degree of dilution at each stage, and it is also much
easier.

For the purification of an impure culture the procedure is the same as
for the isolation of a mould from infected material. The most usual method
is by making a series of dilution cultures, exactly as described above. It is
necessary to use a very small amount of inoculum to mix with the first tube
of agar and, in cases where the spores are small and happen to be easily
wetted, it is advisable to make one or two preliminary dilutions in sterile
water or saline (a 0·9 per cent aqueous solution of sodium chloride). However, in some cases of infection of cultures the growth of the contaminant
is clearly confined to a small area of the slope. It is then frequently possible
to obtain a pure culture by picking off a few spores from a portion of the
slope remote from the invader.

Hyphal tip cultures. This is a method which is particularly valuable
for the purification of a sterile fungus, or one which produces spores only
very sparingly. It can be used only when the initial rate of growth of the
mould to be purified is greater than that of the contaminant. It happens
not infrequently that cultures, particularly when mites have had access to
them, become contaminated with *Cephalosporium*. In such cases plating
out serves only to isolate the invader, since *Cephalosporium* produces myriads of small, easily wetted spores. Fortunately species of this genus grow
very slowly for the first few days, being easily outpaced by very many
species of other genera, and can therefore be eliminated by the hyphal tip
method.

A single colony is planted in the centre of a Petri dish containing a
medium on which the mould to be purified will grow well. When the
colony is about 1 cm. in diameter a small piece of agar, containing the tip
of one of the radiating hyphae at the edge of the colony, is cut out and
transferred to a fresh slope. It is usually possible to find a growing tip
which is well separated from other hyphae and which can be cut off cleanly.
The most suitable tool for the purpose is the dummy microscope objective
described below.

The problem of bacterial contamination. When plating out some
materials, particularly soils and industrial effluents, it is not unusual to find
all the plates on a series heavily infected with bacteria, often to such an
extent that sub-cultures made from colonies of moulds are more likely

han not to be contaminated. Another complication is that some common soil bacteria strongly inhibit the growth of moulds, so that, if such are present, species of fungi present in the infected material do not grow at all.

It is possible to prevent the growth of all but a few species of bacteria, without impeding the development of the fungi, by the use of selective antiseptics in the culture medium. The compositions of three such antibacterial media are given on pp. 234–5. Tests made by the author, using ordinary malt agar and potato agar, showed that there is little advantage to be gained from using antibiotics along with the dyestuffs. Media containing 0·035 g per litre of rose bengal have been found to be very satisfactory for plating out a wide variety of soils, and they involve no additions after sterilization. The amount of rose bengal may be increased, if necessary to 0·067 g per litre. At this concentration it exerts a distinct anti-spreading effect. Above this concentration some species of moulds are inhibited.

For the purification of a culture which is infected with bacteria, but is free from contaminating moulds, several methods are useful. Raper (1937) recommends the use of a small glass cylinder (the so-called van Tieghem cell) to one edge of which are fused three small glass beads, $\frac{1}{3}$ to $\frac{1}{2}$ mm. in diameter. This is placed in a Petri dish, resting on the beaded edge, and the whole sterilized. Melted agar is poured in to a depth about half-way up the cylinder. When cool, the contaminated fungus is inoculated in the centre of the ring. The fungus grows underneath the free edges of the ring, within the agar, whereas the bacteria cannot spread. Hyphal-tip cultures can be made from hyphae which have spread well outside the ring. In a modification of the method by Ark and Dickey (1950) small pellets of modelling clay are used instead of glass beads for raising the glass cylinder. Clay has the advantage that it sticks the cylinder to the bottom of the dish, so that the latter can be inverted if desired.

Another method is to plate out the fungus on a rose-bengal medium. If the infected fungus is not producing spores, a fragment of mycelium can be sown on a medium containing the maximum amount of rose bengal. Usually the bacteria, if not suppressed completely, are prevented from spreading, and it is possible to obtain pure hyphal-tip cultures of the mould.

Single spore cultures. A number of microbiologists regard a culture as pure only when it has been obtained by germination of a single spore. It is true that in some types of studies on fungi single-spore cultures are a *sine qua non*. For example, investigations on the genetics of species of *Neurospora*, and yeasts, necessitate not merely the making of very numerous single-spore cultures but also the separation in serial order of all the spores in a single ascus. Again, in order to prove the connection between an Ascomycete and its conidial state it has usually been necessary to make cultures derived from both single ascospores and single conidia. However, in industrial work, biochemical studies, and the maintenance of culture collections, the value of single-spore cultures has been much overrated.

Assuming that single-spore cultures are a necessity, there are quite a

I

number of methods available. However, it should be noted that, whilst any
method can be used successfully for isolation of large and/or highly col
oured spores, all methods, with the exception of those involving the use of
a micro-manipulator, are difficult when applied to moulds which have very
small, lightly coloured or colourless spores.

(1) The most elegant method of isolating single spores is by means of a
good micro-manipulator. A number of types are marketed with which it is
possible to pick up any particular spore and transfer it to a fresh substrate
all under a high power of the microscope. Such machines are virtually a
necessity in genetical investigations and in any work involving the making
of very large numbers of single-spore cultures. Unfortunately they are all
very expensive and, for the kind of work done in most mycological laboratories, the outlay is not justified.

A book on the subject by El-Badry (1963) gives details of most types of
machines, and discusses their application in a number of scientific dis
ciplines.

(2) A series of dilution plates is prepared in the usual way, using a clear
filtered agar. The plates are incubated only for sufficient time for the spores
to put out the primary germ-tubes and must, therefore, be examined under
the microscope at frequent intervals during the first forty-eight hours. A
few spores are found which have just germinated and which are individually
well separated from all other spores. These are marked, with a glass pencil,
whilst the dish is on the stage of the microscope, then the agar is cut round
them with a sterile scalpel, the flattened end of a needle, or the tool described below, and the tiny pieces of agar are transferred to tubes of fresh
culture medium. The great disadvantage of this method is that the spores
do not all lie in one plane and examination of the plates can be extremely
tedious. The advantage is that, in a set of dilution plates, there is usually
one plate with a particularly favourable distribution of spores, not too few
to be difficult to find, and sufficiently spread to make isolation of individual
spores fairly easy.

(3) Another method is to flood the surface of the agar in a Petri dish
with a dilute spore suspension in sterile water, using not more than about
2 ml., allow to stand for a few minutes and then pour off the water. It will
be found that quite a number of spores have stuck to the agar and, as these
lie approximately in one plane, they are fairly easily found. Otherwise the
procedure is the same as in (1).

A very convenient tool for cutting out minute blocks of agar enclosing
single spores was originally described by La Rue (1920). It consists of a
dummy microscope objective in which the front lens is replaced by a sharp-
edged metal tube, about 5 mm. long and 1·5 mm. in diameter (this being
approximately the diameter of the field using a $\frac{2}{3}$-inch objective and a × 10
eyepiece). This is screwed on to the nosepiece of the microscope in place
of one of the objectives, and, if possible, should be fairly accurately centred
with the $\frac{2}{3}$-inch. The plate is examined with the latter lens and a field is

found which shows a single spore. The cutter is swung into position and lowered carefully until it just touches the glass at the bottom of the dish. A small circular block is cut out of the agar, but is left behind when the cutter is raised and may be re-examined with the $\frac{2}{3}$-inch to ensure that all is well. The little block is then lifted out with a needle. If the cutter is not perfectly centred with the objective it is only necessary to make a few trial cuts, in order to determine the direction of the error, and thereafter to move the dish the determined amount before making the cut.

A modified form of the cutter, designed particularly for easy sterilization and replacement of the cutter is described by Keyworth (1959).

(4) A dilute spore suspension is made by shaking vigorously a small mass of spores in a tube of sterile water or saline. A series of dilutions, in water or saline, are made from this until a suspension is obtained such that single loopfuls contain usually one spore, otherwise none. This is determined by spreading a series of loopfuls on a slide and examining under the microscope. Single loopfuls are then transferred to agar plates and the presence of not more than one spore in each is confirmed by direct examination. The positions of the drops which do contain a spore are marked on the bottoms of the dishes, so that the spores may be readily located after the drops have evaporated, and so that accidental infection may be recognized by its position on the plate. The plates are incubated, and in those in which the spores are viable pure cultures are obtained.

(5) A very ingenious method is described by Hansen (1926) which works well with species which have large coloured spores, such as species of *Alternaria*, *Helminthosporium*, etc., but which is difficult to apply to most species of *Penicillium* and many of *Aspergillus*. A dilute spore suspension is made in melted agar medium. This is sucked up into a number of fine glass capillary tubes, of bore slightly greater than the diameter of the spores, and the medium is allowed to set therein. Microscopic examination of the capillaries should show short lengths containing each a single spore. These are broken off, sterilized externally with alcohol, and planted in fresh medium. The spores germinate and the germ-tubes emerge from the ends of the capillaries and form typical colonies.

An excellent review of methods of single-spore isolation is given by Hildebrand (1938). The paper contains a detailed description of the author's technique and there is an extensive bibliography.

MICROSCOPICAL METHODS

Equipment. A microscope is essential for any work on micro-fungi. Identification of species involves the use of the highest powers and, whilst much good work has been and may be done by skilful manipulation of indifferent equipment, a well-made and properly equipped instrument makes for ease, speed, and accuracy. However, what is much more important than elaborate and expensive equipment is knowledge of how to use the microscope properly, so as to get the best out of such instrument as may be available. Some notes on the proper use of the microscope and on choice of equipment are given in the Appendix.

In addition to the microscope proper a good hand-lens is very useful, particularly for the preliminary examination of cultures and for observation of fungi growing on natural substrata. The best kind is the type known as "Aplanatic", made in various powers by all the manufacturers of microscopes. It consists of three lenses cemented together and, although comparatively expensive, is worth the difference in price, compared with the singlets and doublets, in that it gives a much flatter field and is almost entirely free from chromatic aberration. Suitable powers for examination of fungi are $\times 12$ and $\times 15$. Some people have difficulty in using the high-power aplanats because the short working distance renders it far from easy to illuminate the specimen adequately. In such cases the "Illuminator Magnifier", made by Messrs. R. and J. Beck, in which the light from a small electric lamp is focused on the specimen, may be found preferable.

Apart from optical equipment there are a number of small pieces of apparatus used particularly for microscopic work. At least two needles are required for mounting specimens. The nichrome needles used for sowing cultures are unsuitable, since they are not sufficiently stiff for teasing out material. Fine sewing needles are satisfactory if mounted in wooden or metal handles. If still finer tools are required entomological pins (with the heads cut off) may be used and, for mounting these, special metal holders with very small chucks are obtainable.

A simple but very useful piece of apparatus is a 6-inch square tile, half white and half black. The black half is used for manipulating colourless specimens, which are often invisible in the mounting fluid if viewed against a white surface.

Several small dropping-bottles are required for stains and mounting media. The ordinary type of dropping-bottle usually delivers a drop which is far too big, and it is better to take a short length of glass tubing, with one end drawn out to a fine jet, and insert this through the cork of an ordinary bottle.

The requirements include a supply of glass slides, 3×1 inches, and of cover-glasses. The latter should be of No. 1 thickness and may be circular or square as preferred, convenient sizes being $\frac{3}{4}$ inch diameter or $\frac{3}{4}$ inch square. If larger than this it is difficult to apply a ring of cement all the way round. Both slides and cover-glasses should be cleaned before use by soaking in chromic acid, then thoroughly washing in water and finally in alcohol.

Examination of living cultures. In the study of moulds for purposes of identification a great deal of information can be obtained from observations made on dry living cultures under the compound microscope. Examination in this way should always precede the preparation of slides. Slopes can be placed across the stage and the edges examined by transmitted light with objectives of $\frac{2}{3}$ inch and low power. Petri dishes can be laid flat on the stage, either side up, and examined by incident or transmitted light. As mentioned earlier, when a mould forms dense matted growth the planting of several colonies in one dish will often result in there being narrow sterile zones, where the colonies approach each other, and mature fruiting structures can be observed partially overhanging these clear spaces (Figs. 164, 165). In the case of species of *Penicillium* and *Aspergillus* the shape of the spore-heads, the disposition of the chains of spores, and the origin of the conidiophores, whether from submerged or aerial hyphae, have considerable diagnostic value, and the information required for identifications can be obtained only from the study of undisturbed living cultures. In the same way, determination of species of *Mucor*, and other members of the same family, is possible only when study of slide preparations is combined with examination of living material in Petri dishes. Many other moulds produce conidial structures which fall to pieces at the least touch and, with these, the value of direct observation is apparent.

In the examination of living cultures it is important to remember that a good deal of water is given off by fast-growing species, and some of it appears on the hyphae as small droplets, invisible to the naked eye. It is not uncommon to see the sporangiophores of species of *Mucor* thickly studded with these droplets, and by the beginner they may readily be mistaken for chlamydospores. In tube cultures an appreciable amount of water often collects on the walls of the tube. Sporangia coming in contact with this usually burst and liberate irregular masses of spores. Similarly, false spore-heads held together by mucus, such as those of *Verticillium* and *Cephalosporium*, are broken up. Spores often germinate in such damp situations, even whilst still attached to phialides or in sporangia, and give rise to structures which are puzzling the first time they are seen under a low power of the microscope.

Preparation of slides. For study of fine detail, and for accurate measurements, slide preparations must be made. With most species of moulds it is difficult, if not impossible, to fix, stain, and mount specimens, as is done for botanical and zoological material, and at the same time preserve structure. In all but a few cases it is general practice to use fluid mounts made with as little manipulation as possible. The method to be adopted often depends on whether the slide is required merely for a temporary examination, or whether it is intended for a permanent collection of typical slides. In the descriptions of methods given here, those which are unsuitable for making permanent slides are indicated.

Success in making slides depends largely on the age of the culture. If this is too young there will be few or no fully developed fruiting structures or ripe spores. If the culture is too old the fruiting structures, particularly in the Hyphomycetales, drop to pieces. For this reason Petri-dish cultures are more satisfactory than slopes. So long as the culture is still spreading it is possible to pick off small portions, for making slides, from areas of exactly the right age. If tubes must be used, the right age of culture for the majority of moulds is 5–7 days. Exceptions to this are the species which form balls of spores. In very old cultures which are drying out, the mucilage in which the spores are enveloped acts as a cement, and most of the balls remain intact when mounted in fluid medium. If young cultures are examined the spores are dispersed all over the slide, but, on the other hand, the conidiophores are still turgid, whereas in old cultures they usually collapse.

Water has a limited application as a mounting fluid (see below) but for the majority of moulds is quite unsuitable. It evaporates rapidly, it often causes swelling of hyphae by osmosis, and it usually causes the parts of the specimen to adhere together as a tangled mass of hyphae, spores, and air bubbles. Alcohol wets efficiently and makes a fairly satisfactory mountant for a brief and rapid examination, but is too volatile for general use. Easily the most generally useful medium is that known as "lactophenol".

A number of formulae for this medium have been published but, for regular use, the original recipe of Amann (1896) is to be preferred.

Lactophenol

Phenol (pure crystals)	10 g
Lactic acid (syrup, sp. gr. 1·21) . .	10 g
Glycerol (pure)	20 g
Distilled water	10 g

It is readily prepared by warming the phenol with the water until dissolved and then adding the lactic acid and glycerol. The refractive index is 1·45.

The method of mounting is to place a *small* drop of mounting fluid in the centre of a clean glass slide, with a sterile needle pick off from the culture a very small portion of typical material, place this in the drop of fluid and very gently tease it out with a pair of needles until it is well wetted, then lower on to it a cover-glass in such a way as to avoid air bubbles as far as possible. It is difficult to make slides which contain no air bubbles without teasing out the specimen to such an extent that structure is destroyed and, within reason, their presence does not seriously interfere with observation. Some moulds are wetted by lactophenol only with great difficulty and do not give presentable slides unless some means is taken to accelerate the process. One way which is sometimes effective is to warm the slide over a small flame before putting on the cover-glass. Another way of expelling air, which has been found particularly useful for species of *Penicillium*, is to place the specimen in a drop of alcohol on the slide, wait until most of the liquid has evaporated, then add a fairly large drop of lactophenol, tease out the specimen as required and apply a cover-glass. If even this fails to give a satisfactory slide, as may happen with thick structures such as perithecia, the specimen should be placed in a small watch-glass, completely covered with mounting fluid and the whole placed in any vessel from which the air can be exhausted. A vacuum desiccator, if available, will take several specimens at once and is admirable for the purpose. To make sure of the removal of all air from the specimen, air is admitted to the vessel after 10–15 minutes and then suction is again applied for about another 10 minutes. The specimen may now be transferred to a slide and a cover-glass applied.

It is usually impossible to make good slides by using mounting fluid in such quantity that it just fills the space underneath the cover-glass (this is the practice when Canada balsam is used as mounting medium). Generally an appreciable amount oozes out, especially if the cover-glass is gently pressed down with a pair of forceps, and must be absorbed on bits of filter paper carefully applied to the edge, taking care that none of the fluid gets on to the upper surface of the cover-glass.

The advantages of lactophenol are that it rarely causes either shrinkage or swelling of the cells of most fungi, and that it has no tendency to evaporate, so that preparations are reasonably permanent, if carefully handled, without any further manipulation. It is also sufficiently viscous to allow of the use of an oil-immersion lens without undue movement of parts of the specimen, provided that the immersion oil has not become oxidized. It should be noted that, since lactophenol is a non-swelling medium, measurements of objects in this fluid will not necessarily agree with measurements of the same objects mounted in water, which often causes appreciable swelling of fungal hyphae and spores. Many published diagnoses of fungi include measurements recorded from specimens mounted in water (although there has been for some years an increasing tendency to use lactophenol), and it is necessary to bear this in mind when trying to identify moulds.

In a very few cases it has been found that lactophenol is not a suitable mounting medium. One of its characteristics is that its refractive index is very close to that of fungal hyphae, with the result that colourless structures appear very faint, owing to lack of contrast with the background. The majority of colourless and pale coloured moulds can be stained to give adequate contrast, but the yeasts are difficult to stain without causing shrinkage and distortion of the cells. These are best mounted in plain water, in which the cells are clearly visible. Water is also the best medium for mounting a few species of dark coloured moulds whose spores shrink to a serious extent in lactophenol. Slides made with water as the mounting medium may be protected from evaporation, so that they remain in good condition for periods up to about two days, by sealing the edges of the cover-glass with shellac cement (see below).

Locquin (1952) has described a mounting medium of very high refractive index, which he claims to be suitable for mounting fungi directly, that is without any previous manipulation. It is made by saturating pure glycerol alternately with potassium iodide and mercuric iodide until no more of either will dissolve. It is unfortunate, but not surprising, that this medium causes very serious shrinkage of fungal hyphae and spores, for its refractive index, 1·7–1·8, would otherwise make it valuable for observation of fine structures.

An interesting and completely different method of making slides is described by Skerman (1946). A portion of the mould colony containing sporing structures of suitable age is wetted with a drop of a mixture of three parts ether and one part alcohol. Before this has completely evaporated a drop of 10 per cent collodion in the same mixed solvent is placed, by means of a 4-mm. loop, on to the moistened area. The collodion dries in 1–3 minutes to give a film in which mycelium and conidiophores are embedded. The film is removed, as soon as it is sufficiently firm, to a slide. It may be mounted directly in lactophenol containing stain, or it may be stained first and then mounted in plain lactophenol. Alternatively, if much mycelium is present, the collodion may be removed with warm solvent and the mycelium then teased out as required. The original paper contains some fine photomicrographs of species of *Penicillium*, showing spore chains *in situ*.

Butler and Mann (1959) have described another useful method of removing portions of colonies without disturbance. A piece of cellulose self-adhesive tape, about 3 inches (8 cm.) long, and ½ to ¾ inch (1·3–1·9 cm.) wide, is lightly pressed on to a mould colony, in a Petri dish or on a natural substratum, then placed, sticky side down, in a drop of mounting fluid on a slide. The tape is stretched tight and the ends stuck down to the glass to hold it in position. If it be necessary to wet the colony first with alcohol, in order to remove air bubbles, care should be taken to avoid getting any of the spirit on to the ends of the tape. If the ends are wetted with any fluid they will not stick to the glass. The amount of pressure to be applied when the centre of the tape is placed on the colony can be found only by practice. If pressed too hard the tape picks up a dense mass of mycelium and sporing structures. However, one or two trials, with any particular type of colony, should be sufficient to determine the correct procedure.

The method is excellent for making temporary mounts, but there does not seem to be any immediate prospect of making permanent slides by this procedure. Commercial varieties of tape all tend to become brittle with age, so that, even if one could seal the edges of the tape, or use a small piece which would lie under a cover-glass, it is improbable that the slides would have any real degree of permanence. Also, lactophenol appears to react with the sticky material on the tape, with the result that the background becomes spotty after the slide has been kept for a few days.

The original paper includes photomicrographs of chains of *Alternaria* spores. The method works very well with such sporing structures, where the spores are not too readily deciduous, but with very fragile structures, as for example the branched spore chains of *Cladosporium*, the production of a slide showing unbroken structures is a matter of luck.

Staining. In general, moulds tend to take up stains somewhat unevenly, due to the fact that their affinity for colouring matters, which in young structures is considerable, decreases rapidly with age.

The simplest, and often the best, method of staining is to mix the dye with the mounting fluid, but only comparatively few colouring matters can be used successfully in lactophenol, the chief being cotton blue, picric acid, orange G, picro-nigrosin, acid fuchsin, and trypan blue. Cotton blue (0·05 g in 100 ml. lactophenol) is probably the most widely used of all stains for moulds. It colours very young structures fairly deeply, but, with most specimens, acts very unevenly, a single hypha often showing irregular patches of colour. Another peculiarity is that staining proceeds slowly over a long period, so that a specimen which, when first mounted, shows a satisfactory depth of colour, will often become hopelessly overstained after a few months storage. The picric acid medium is made by saturating lacto-phenol with the dye, filtering off any excess before use. It stains com-paratively evenly, but the yellow colour appears to many people to be of insufficient intensity. Contrast may be enhanced, especially for photo-micrography, by using a blue light filter on the microscope lamp. Most of the photographs in this book were taken from specimens mounted in this medium. Orange G is used as a one per cent solution in lactophenol. It usually stains well and reasonably evenly, and gives adequate contrast, which, as in the case of picric acid, may be increased by the use of a blue light filter. Acid fuchsin is used as a 0·1 per cent solution (see below for comments). Picro-nigrosin was originally recommended to be used as a 0·4 per cent solution of water-soluble nigrosin (bacteriological stain) in picro-lactophenol. Nigrosin alone, in aqueous solution or in lactophenol, does not stain fungi, and is actually used in the so-called relief or negative staining, in which the specimen is mounted in a 5–10 per cent aqueous solution of the dye, and appears as a colourless image on a blue-black background. In combination with picric acid it stains readily from lacto-phenol, the colour being a neutral grey. Picro-nigrosin-lactophenol has, unfortunately, one grave fault. On keeping, often for only a short time, it forms an exceedingly fine precipitate, which cannot be completely re-moved by filtration. The trouble may be avoided by keeping two solutions, picro-lactophenol and 2 per cent aqueous nigrosin. When required, single drops of each of the two solutions are quickly mixed on a slide, and the specimen immersed in the drop immediately. An even better method is first to mordant the specimen with picro-lactophenol, remove the excess as completely as possible, first by drainage and finally by careful applica-tion of bits of filter paper, wash with two or three successive drops of water, stain with 0·4 per cent nigrosin solution, remove excess of this, and mount in plain lactophenol. This method may appear to be complicated, but all the operations are easy and rapid, and, since the stain is a neutral shade and very fast, the results are worthwhile.

Carmichael (1955) recommends a slightly different combined staining and mounting medium, termed lacto-fuchsin. This is made by dissolving 0·1 per cent acid fuchsin in pure lactic acid. It stains rapidly and evenly, and, unlike most other stains, colours the spores of many moulds. The contrast given by the stain is enhanced by the fact that the refractive index of lactic acid is slightly lower than that of lactophenol. This medium, whilst one of the best for making temporary mounts, is, unfortunately, quite unsuitable for the preparation of permanent slides, owing to the instability to light of the dyestuff. Slides exposed to daylight gradually fade, and, if exposed to a powerful light such as is necessary for micro-projection, lose their colour entirely within a few minutes. Another disadvantage is that, whilst the stain gives adequate visual contrast, it is very difficult to obtain sufficient contrast in photomicrography, owing to the peculiar absorption spectrum of the dye.

Another combined stain and mounting medium, which is stated to give a permanent blue-black, is described by Isaac (1958). It is of complicated constitution, containing haematoxylin, bismarck brown, methyl green, ferric and chrome alums, chloral hydrate, glycerol, and gum arabic, and, at one stage in its preparation, the use of an ultra-speed centrifuge is necessary.

All the dyes mentioned so far colour cell contents, but have no or little effect on cell-walls. Boedijn (1956) describes the use of a dye, trypan blue, which he claims will colour the cell-walls of all the species tested. He recommends the use of a 0·1–0·5 per cent solution of the dye in either 45 per cent acetic acid or in lactophenol, accelerating the action by heat if necessary. In the Author's experience 0·05 per cent of the dye in lactophenol stains rapidly, and gives adequate depth of colour. Staining is very even if the specimen be well teased out in the drop of stain, and the cell-walls are coloured as deeply as the cell contents. The stain appears to be stable to light.

With all these combined staining and mounting fluids most of the stain is absorbed by the specimen, and, if the mount is reasonably thin, that is with the specimen well teased out and the cover-glass pressed well down, the "background" should be almost colourless. With thicker specimens, such as whole perithecia, there is often too much background colour. In such cases it is advisable to allow the stain to act for a few minutes, drain off as much as possible, and mount in a drop of plain lactophenol.

A method of staining which is preferable with some dyes, and which often gives very good results, is to treat the specimen with a solution of the stain in acetic acid or alcohol, and then replace the fluid with plain lactophenol. Boedijn, as stated above, suggested using trypan blue as a solution in 45 per cent acetic acid, but staining by this method is extremely rapid, and it is difficult to avoid overstaining. In any case the dyestuff stains well as a solution in lactophenol. Alcorn and Yeager (1937) recommend Orseillin BB, used as a 0·25 per cent solution in 3 per cent acetic acid. The stain is allowed to act for 3–5 minutes, then the specimen is examined under the microscope, and, if too deeply coloured, is treated with successive drops of very dilute acetic acid until satisfactory. This gives a pleasant red shade, which gradually turns brownish, particularly if the slide be exposed to strong light. The dye has little affinity for fungal hyphae if applied as a solution in lactophenol. Another stain which works well with young structures, and, incidentally, shows up the nuclei in the cells, is Chlorazol Black E (= New Black D.E. in some catalogues). This is used as a one per cent solution in 90–95 per cent alcohol, and is allowed to act for about one minute, using alcohol to differentiate if staining has been too rapid.

Permanent slides. Any slide which is to be kept as a permanent record should be protected by sealing the edge of the cover-glass with suitable cement. Lactophenol mounts can, it is true, be kept without sealing, but there is always a danger of the cover-glass being accidentally moved. In addition, it is almost impossible to protect slides completely from dust, and dust on an unsealed cover-glass is difficult to remove without disturbing the specimen. For the application of fluid sealing cements a turntable is almost a necessity if circular cover-glasses are used. If square covers are preferred, the sealing must of course be done freehand.

There are three cements which are of use for sealing slides of moulds mounted in lactophenol. The first is brown shellac cement, obtainable from any of the dealers in microscopic sundries. The second is nail varnish, and of this the clear unpigmented varieties are best, since the pigments usually incorporated are soluble in lactophenol and often stain the specimens on long standing. Both cements are useful but neither is completely dependable, as sometimes, even when every care is taken in the application, there is some leakage under the cover-glass. The following are the main conditions for success: (1) No cement should be applied to a slide until it is certain that the cover-glass is not under strain. If a slide is allowed to stand for a few days, air bubbles often appear at the edge, showing that the cover-glass has been bent slightly when making the slide, and has gradually recovered on standing. Such air spaces must be filled in with mounting fluid before applying cement. (2) The cement must be of suitable consistency, i.e. it should run smoothly, but not too readily, from a brush. The shellac cement, if it is too thick, may be brought to the right consistency by the very gradual addition of alcohol. Nail varnish may be thinned by the cautious addition of acetone. (3) Using a turntable, a very thin coat of cement should be put on first, using a fine brush. When this is quite dry, a second thicker coat is added to overlap the first. (4) A necessary precaution with nail-varnish is to ensure that the portion of the slide immediately around the cover-glass is clean and dry. This is not so essential when using shellac cement. (5) If the layer of mounting fluid is thick, due to the presence of thick structures which it is unadvisable to squash, any attempt to apply cement with the aid of a turntable will cause movement of the cover-glass. In these cases the first layer of cement must be put on freehand. When this is quite dry the slide may be tidily finished by ringing on the turntable in the usual way. With these precautions the majority of preparations sealed with either of the cements will keep in good condition for many years. Another purpose for which shellac cement (but not nail varnish) is very suitable is, as noted above, for sealing temporary mounts in water, for the purpose of retarding evaporation.

The third cement, recommended by Dade (1960) is Araldite, a synthetic epoxy resin. There are several varieties which are cold-setting, of which the one sold in the shops as a 2-tube pack is not suitable for ringing slides. The preferred variety is X 83/4, and is obtainable only from Ciba (A.R.L.) Ltd., Duxford, Cambridge. Resin and hardener are supplied separately, and are mixed in small amounts as required. Both are syrups, and the mixture is too viscous to apply directly to a slide. It is therefore recommended that the cover-glass be first anchored by the application of a narrow ring of nail-varnish, and that the first revolutions of the turntable be very slow, to avoid the resin streaking across the cover-glass. If the resin is still too viscous it may be thinned by the addition of a small quantity of toluene. The cement takes 2–3 days to harden completely. This is probably the most permanent finish available for fluid mounts, but it is more troublesome to apply than the other cements. It is therefore advisable to wait until a number of slides require ringing, and deal with these all together. The brush used to apply the cement must be thoroughly cleaned immediately after use, best by means of acetone.

A different method of sealing was described originally by Diehl (1929). The object is placed in a drop of mounting fluid on a 22-mm. No. 0 cover-glass and is then covered with a 12-mm. No. 2 cover-glass. Any excess fluid is carefully cleaned away from the edge of the small cover, a large drop of Canada balsam is placed in the centre of it and a slide is gently lowered on to it. When the balsam has spread to the edge of the large cover-glass the slide is inverted and allowed to stand until the balsam has set. Linder (1929) claims that this method can be used with lactophenol provided that the mount is thoroughly dehydrated in a desiccator before the final mounting in balsam. However, most workers who have experimented with the method have found that lactophenol and balsam are not compatible, the balsam gradually creeping in between the cover-glasses and eventually ruining the slide. In theory this is an ideal method of making permanent slides and it is to be hoped that someone will discover either a clear cement which does not react with lactophenol, or a mounting medium which is as convenient as lactophenol and is, at the same time, compatible with balsam.

It is unfortunately true that some slides deteriorate in spite of every care in preparation. As stated above, spoilage may occur because of the leakage of ringing cement under the edge of the cover-glass, but there is another cause of trouble which is entirely unconnected with the method of sealing. The mycelia of many species of fungi contain large amounts of fat and there is a tendency for this to be displaced by mounting fluid, with the consequent appearance of numerous unsightly globules all over the slide.

In such cases it is not possible to prepare slides which will keep in good condition unless the mould to be mounted will withstand much more drastic treatment than the simple methods described here.

Whatever the method of preparation, all slides which are to be kept should be labelled as soon as made, giving the name of the species, its number in the culture collection, the stain (if any), the mounting fluid, and the date of preparation.

Artefacts. In the examination of slides beginners are often puzzled by certain artefacts. Perhaps the commonest are very small air bubbles. These are usually circular if the layer of medium is thick, but may be of any shape if the cover-glass has been well pressed down. The circular ones have a very thick dark outline and look, as they are, highly refractive. The thin ones have dark edges, but the most noticeable point about them is that portions of hyphae or spores lying inside the bubbles are not in focus when the rest of the slide appears sharp, and also show a greater degree of contrast than similar structures surrounded by fluid.

Another common appearance is a series of circular bodies of very varying size, somewhat refractile and always with smooth edges. These are oily substances exuded from the cells of the fungus, or squeezed out during the mounting process. As mentioned above, these oily drops sometimes increase in number when a slide is stored and may entirely ruin the preparation.

It is not uncommon, when making slides from cultures of moulds which form thin tough colonies on agar, to carry over bits of culture medium to the mounting fluid. These become flattened out when the cover-glass is pressed down. The first thing noticed about such patches of agar is that they do not take the stain from coloured lactophenol. They usually contain submerged hyphae and often have a few spore-heads pressed into them. The latter always appear somewhat shrivelled and are, like the agar, unstained.

IDENTIFICATION OF SPECIES

Methods of obtaining pure cultures and general methods of examination having been discussed, it now remains to describe the routine to be followed when an unfamiliar species of fungus has to be identified.

The first necessity is to prepare a full and accurate description of the fungus, as grown on standard culture media. It is assumed that some idea of its general behaviour on one or two media has already been gained during isolation and purification. If the species grows satisfactorily on Czapek agar this should now be used, but if the Czapek medium induces only poor and stunted growth, or if sporing structures are lacking or tardily produced, some more suitable medium must be selected. Several cultures should be made in Petri dishes. One or two of the dishes may be planted with single colonies for measurement of growth rate and for recording the general appearance of colonies, the rest with three colonies spaced about an inch apart. It is necessary to make a number of cultures because some of them will be examined, probably with the dishes uncovered before maturity, and are likely to develop infections on continued incubation. Cultures should be examined at frequent intervals and the following details recorded:

1. Rate of growth; described as slow, very slow, moderate, rapid, etc. Some diagnoses give actual diameters of colonies after specified numbers of days.

2. Colony colour and colour changes; whether uniform or in zones or patchy. Evanescent colours which are often to be observed at the edges of growing colonies should be recorded. Where a standard colour index is available records can be made with a precision which is unattainable by the use of ordinary names of colours.

3. Colour and colour changes of the reverse of the colonies.

4. Colour changes in the medium; whether confined to the area covered by the colony or diffusing.

5. Texture of surface; whether loose or compact; plane, wrinkled or buckled; velvety, matted, floccose, hairy, ropy, gelatinous, leathery, etc.

6. Odour, if any. Odours are usually very difficult to describe. Many species have a smell which can only be described as "mouldy", but quite a number have characteristic and sometimes fragrant odours, and, of course, many have no odour.

7. Character of drops of transpired fluid often found on aerial growth.

8. Character of the submerged hyphae; colour, presence or absence of septa, approximate diameter, characteristics of special structures if any present.

9. The stage at which fruiting structures develop.

10. The character and disposition of the mature fruiting organs; whether sporangia, perithecia, pycnidia, sporodochia, coremia, or detached conidiophores; whether borne in the substratum, on the surface, or on the aerial mycelium. The presence of more than one type of sporing structure should be particularly noted.

11. Colour, size, and shape of mature fruiting organs or fruit-bodies.

12. Details of structure of the fruiting organs, including measurements of essential parts and disposition of the spores thereon.

13. Full details of spores; colour, shape, septation, surface markings, size (including both average and extreme measurements).

Data numbered 1–7 are obtained by examination of cultures with the naked eye or with a hand-lens, 8–11 by observations on living cultures with the aid of a low or moderate power of the microscope. Numbers 12 and 13 necessitate the preparation of slides and the use of the highest powers of the microscope.

The information recorded under 8 and 10 should be sufficient to place the species in its correct Class and Order, and consideration of the rest of the data will lead to the Family and then the genus. The determination of the correct specific epithet is, except in a few genera, a matter of some difficulty. If there is a good monograph of the genus it is usually a question of careful and patient observation, along the lines indicated by the particular authority, then repeated consideration of the data until the unknown fits into its proper place. In absence of an authoritative treatment of the genus the usual procedure, and often the quickest in the long run, is to consult Saccardo's *Sylloge*, look up all the recorded species in the genus, and follow up references to any which seem to be near the one to be identified. Unfortunately, many species have been inadequately described and there are even numerous genera which are ill-defined. In some cases the accepted conception of a genus is more a matter of tradition than of adequate diagnosis, and it is difficult to find any published data sufficiently exact for recognition. However, the genera which are of commonest occurrence are just the ones which have been most studied and concerning which there is an extensive literature.

Perhaps the greatest difficulty in identifications is to find that, whilst most of the data from the unknown fit a published description, there is a discrepancy in spore dimensions. It has already been pointed out (p. 253) that the mounting fluid may affect apparent size of spores. It has also been shown (Williams, 1959) that environmental conditions during growth of the fungus can affect spore size. In general, when the fungus is grown in the dark the spores are larger than when cultures are kept in the light. Other factors, such as temperature and nutrition, affect different species in different ways. What this amounts to is that reasonable differences in spore dimensions should not necessarily preclude identity, if other characteristics correspond.

One thing the student should guard against. It is doing a great disservice to other mycologists to assume too hastily that an unrecognized fungus is a new species and to publish a description under a new name. The literature is cumbered with a mass of generic names and specified epithets which are nothing more than synonyms of well-known forms, and which ought never to have been bestowed.

REFERENCES

Alcorn, G. D., and Yeager, C. C. (1937). Orseillin BB for staining fungal elements in Sartory's fluid. *Stain Tech.*, **12**, 157–8.

Amann, J. (1896). Conservirungsflüssigkeiten und Einschlussmedien für Moose, Chloro- und Cyanophyceen. *Z. Mikroscopie*, **13**, 18–21.

Ark, P. A., and Dickey, R. S. (1950). A modification of the Van Tieghem cell for purification of contaminated fungus cultures. *Phytopathology*, **40**, 389–90.

Baker, J. R. (1958). *Principles of biological microtechnique*. London: Methuen.

Biourge, P. (1923). Les moisissures du groupe *Penicillium* Link. *Cellule*, **33**, 1re fasc., Louvain.

Boedijn, K. B. (1956). Trypan blue as a stain for fungi. *Stain technol.*, **31**, 115–16.

Butler, E. E., and Mann, M. P. (1959). Use of cellophane tape for mounting and photographing phytopathogenic fungi. *Phytopathology*, **49**, 231–2.

Carmichael, J. W. (1955). Lacto-fuchsin: a new medium for mounting fungi. *Mycologia*, **47**, 611.

Cooke, W. B. (1954). The use of antibiotics in media for the isolation of fungi from polluted water. *Antibiot. & Chemother.*, **4**, 657–62.

Dade, H. A. (1960). On mounting in fluid media, with special reference to lactophenol. *J. Quekett microscr. Club*, Ser. 4, **5**, 308–17.

Davis, J. G., and Rogers, H. J. (1940). The effect of sterilization upon sugars. *Z. Bakt.*, Abt. II, **101**, 102–10.

Diehl, W. W. (1929). An improved method for sealing microscopic mounts. *Science*, **69**, 276–7.

Dox, A. W. (1910). The intracellular enzymes of *Penicillium* and *Aspergillus*. *U.S. Dept. Agric. Bur. anim. Ind. Bull.*, 120.

El-Badry, H. M. (1963). *Monographien aus dem Gebiete der qualitativen Mikroanalyse. Band* **3**. *Micromanipulators and micromanipulation*. Wien: Springer-Verlag.

Fisher, C. B. (1958). An improved Petri dish. *J. med. Lab. Tech.*, **15**, 282–4.

Fleming, A., and Smith, G. (1944). Some methods for the study of moulds. *Trans. Brit. mycol. Soc.*, **27**, 13–19.

Hansen, H. N. (1926). A simple method of obtaining single-spore cultures. *Science*, **64**, 384.

Hansen, H. N., and Snyder, W. C. (1947). Gaseous sterilization of biological materials for use as culture media. *Phytopathology*, **37**, 369–71.

Hildebrand, E. M. (1938). Techniques for the isolation of single microorganisms. *Bot. Rev.*, **4**, 627–64.

Isaac, P. K. (1958). A haematoxylin staining mountant for microorganisms. *Stain Technol.*, **33**, 261–4.

Johansen, D. A. (1940). *Plant microtechnique*. New York: McGraw-Hill.

Keyworth, W. G. (1959). A modified La Rue cutter for selecting single spores and hyphal tips. *Trans. Brit. mycol. Soc.*, **42**, 53–4.

Kondo, W. T., Graham, S. O., and Shaw, C. G. (1959). Modifications of the cellophane culture technique for photographing and preserving reference colonies of microorganisms. *Mycologia*, **51**, 368–74.

Langeron, M. (1945). *Précis de mycologie*. Paris: Masson et Cie.

La Rue, C. D. (1920). Isolating single spores. *Bot. Gaz.*, **70**, 319–20.

Linder, D. H. (1929). An ideal mounting medium for mycologists. *Science*, **70**, 430.

Littman, M. (1947). A culture medium for the primary isolation of fungi. *Science*, **106**, 109–11.

Locquin, M. (1952). Nouveau réactif pour l'étude des structures fines chez les champignons. *Bull. Soc. mycol. Fr.*, **68**, 172–4.

Raper, J. R. (1937). A method of freeing fungi from bacterial contamination. *Science*, **85**, 342.

Raper, K. B., and Thom, C. (1949). *Manual of the Penicillia*. Baltimore: The Williams & Wilkins Co.

Raulin, J. (1869). Études chimiques sur la végétation. *Ann. Sci. nat. (Bot.)*, **11**, 201.

Robbins, W. J., and McVeigh, I. (1951). Observations on the inhibitory action of hydrolyzed agar. *Mycologia*, 43, 11-15.

Sass, J. E. (1941). *Elements of botanical microtechnique*. New York: McGraw-Hill.

Skerman, V. B. D. (1946). Simple techniques for the preparation of mould mounts. *Aust. J. exp. Biol. med. Sci.*, 24, 319-20.

Skinner, C. E., Emmons, C. W., and Tsuchiya, H. M. (1947). Henrici's *"Molds, Yeasts and Actinomycetes"*. London: Chapman and Hall.

Smith, G. (1949). The effect of adding trace elements to Czapek-Dox culture medium. *Trans. Brit. mycol. Soc.*, 32, 280-3.

Smith, N. R., and Dawson, V. T. (1944). The bacteriostatic action of rose bengal in media used for the plate counts of soil fungi. *Soil Sci.*, 58, 467-71.

Thom, C. (1930). *The Penicillia*. Baltimore: The Williams & Wilkins Co.

Thom, C., and Church, M. B. (1926). *The Aspergilli*. Baltimore: The Williams & Wilkins Co.

Thom, C., and Raper, K. B. (1945). *A manual of the Aspergelli*. Baltimore: Williams and Wilkins.

Waksman, S. A. (1931). *Principles of soil microbiology*. 2nd Ed. London: Baillière, Tindall & Cox.

Williams, C. N. (1959). Spore size in relation to culture conditions. *Trans. Brit. mycol. Soc.*, 42, 213-22.

Chapter XII

Physiology of Mould Fungi

> The fungi, in common with other living organisms, possess
> tools or reagents far more specific, more delicate, and more
> powerful than those available in the laboratory.
>
> Lilly and Barnett, *Physiology of the Fungi*, 1951

The growth of fungi, like that of all other living things, is profoundly affected by environment. Variations in external conditions may not only affect the rate of growth of a mould, but, in many cases, can bring about differences in type of growth. Ignorance of this fact led many of the older mycologists to describe as different species, or even different genera, what were in reality physiological variants of one and the same species.

The various topics discussed below are treated only in broad outline, since it would be impossible, in one chapter, to do justice to the mass of published work on the subject. Two books which can be recommended are by Hawker (1950) and Lilly and Barnett (1951), each of which discusses a wide range of relevant subjects. A more recent book by Cochrane (1958) is more in the nature of a review. It covers a wide field with a strong biochemical bias.

One thing which soon becomes obvious to a student of the physiology of fungi is that very few generalizations can be made about these organisms. In their food requirements, response to external stimuli, and tolerance of adverse conditions they show the utmost diversity.

Food requirements. Water and oxygen are both absolutely necessary for growth of fungi and, in addition, the following elements are known to be required: carbon, nitrogen, potassium, phosphorus, sulphur, and magnesium. Iron also is needed, albeit in small amounts, by many species and possibly by all, and, in consequence, is included in most synthetic culture media. It has been shown that quite a number of other metals are required in minute amounts by some species (most of the published work refers to *Aspergillus niger*) and, in view of the fact that most of the same elements have been shown to be necessary for both plants and animals, it seems probable that traces of a whole range of metallic elements are needed by most if not all fungi. It is because all the essential trace elements are required by plants that culture media, for fungi, made from parts of plants are rarely deficient in this respect. Some of the elements are required in such minute amounts that they occur in adequate concentrations in A.R. chemicals, or are extracted, during sterilization, from the glass of culture vessels. This being so, it is apparent that very elaborate methods of purification are necessary in studying the effects of trace elements. On the other hand, it has been shown by Smith (1949) that Czapek agar made up with A.R. chemicals and glass-distilled water gives abnormal growth of many species of *Penicillium*, and that amounts of zinc and copper comparable with the amount of iron can be added with advantage to Czapek solution and Czapek agar in order to induce normal growth and abundant production of spores.

Most mould fungi can utilize inorganic sources of all the essential elements except carbon, but a few are unable to use inorganic nitrogen. Others, such as *Rhizopus stolonifer*, grow well on media containing ammonium salts or amino compounds, but cannot utilize nitrogen from nitrates. However, the great variety of natural substrata which are regularly colonized by numerous species of moulds indicates that these organisms can tolerate wide variations in concentrations of essential nutrients, and can utilize both carbon and nitrogen combined in very diverse forms. An interesting observation by Gupta and Nandi (1957) is that the production of perithecia in a strain of *Penicillium vermiculatum* was dependent mainly on the concentration of nitrogen. The favourable concentrations corresponded to 0·05–0·1 per cent of $NaNO_3$.

In addition to essential elements fungi can utilize, or at least take up from the medium, a few other elements whose presence or absence seems to have no effect on growth. Raulin analysed the ash of *Aspergillus niger* and, on the basis of his analyses, devised his well-known culture medium. One of the ingredients, potassium silicate, is certainly not essential, but, when it is used, silica is found in the ash of the mould. Chlorine is not necessary for growth, the potassium chloride often added to culture media (as in Czapek-Dox solution) being used as a convenient source of potassium. A number of fungi can, however, utilize the chlorine contained in such media, building up complex organic compounds containing chlorine. Raistrick and Smith (1936) showed that one strain of *Aspergillus terreus* could utilize over 95 per cent of the chlorine contained in Czapek-Dox solution, most of it being found in two compounds, "geodin", $C_{17}H_{12}O_7Cl_2$, and "erdin", $C_{16}H_{10}O_7Cl_2$. Clutterbuck *et al.* (1940) have published a survey of the chlorine metabolism of a large number of species of moulds, and shown that the majority can utilize this element to a small extent, whilst a few metabolize over 25 per cent of the total chlorine in the medium, an outstanding species being *Caldariomyces fumago* Woronichin. This species utilizes about 95 per cent of the chlorine in the medium, producing a compound named "caldariomycin", $C_5H_8O_2Cl_2$. Two other interesting chlorine-containing metabolic products of moulds are "griseofulvin", $C_{17}H_{17}O_6Cl$, from *Penicillium griseofulvum* (Oxford, Raistrick and Simonart, 1939), and "sclerotiorin", $C_{21}H_{22}O_5Cl$, from *Penicillium sclerotiorum* and *P. multicolor* (Curtin and Reilly, 1940; Birkinshaw, 1952).

It has been found that some fungi require, in addition to the substances listed above, very small amounts of complex organic compounds, some of which are vitamins, better known as essential metabolites in animal nutrition, and all of which resemble these vitamins in producing effects on growth which cannot be attributed to their value as nutrients in the usual sense. Many common moulds grow well on synthetic media made up with glucose and pure salts, and obviously these do not need to be supplied with any of the accessory growth-substances. However, it is not true to say that they do not require such substances. They synthesize sufficient for their own requirements (sometimes more than sufficient) and are therefore independent of outside sources. Other moulds grow very poorly or not at all on Czapek agar but spread rapidly on malt agar or other "natural" medium, e.g. *Penicillium digitatum*, *Sordaria fimicola*, and many other Ascomycetes. Some Ascomycetes grow rapidly on synthetic media and form perithecia, but these are abortive, containing no asci, and only when the medium is supplemented with certain growth substances are fertile perithecia formed.

The first fungus proved to require one of the vitamins which were already known to students of human and animal nutrition was *Phycomyces blakesleeanus*, which requires thiamin (aneurin, vitamin B_1). Since then it has been found that most of the fungi which need external supplies of growth-substances require thiamin, with or without other substances. Some of them require the complete molecules, whilst others, like *P. blakesleeanus*, can synthesize the vitamin if supplied with the pyrimidine and thiazole components, whilst still others need to be supplied with only one of the two components of the thiamin molecule and hence can synthesize the other. Substances other than thiamin which are required by some species are biotin, inositol, and pyridoxin, but for details of such one of the books mentioned above should be consulted.

The knowledge of fungal requirements of growth substances has been put to good use by employing fungi as test organisms for various vitamins. Such methods are both much less expensive and much quicker than animal feeding experiments. *Phycomyces blakesleeanus* is used for estimating thiamin; a yeast-like fungus, *Nematospora gossypii*, is used to estimate biotin and inositol; and a number of deficient strains of *Neurospora sitophila* and *N. crassa* are available for estimating several vitamins and a number of amino-acids. Up to the present no fungus has been found to be deficient for riboflavin (vitamin B_2), but this can be estimated by means of species of *Lactobacillus*.

The actual quantity of food material required to support mould growth is very small. Fungi are frequently to be found growing on polished furniture woods, or even on old iron, where the total amount of available nutrients must be exceedingly small; textiles made of cellulose—cotton, linen, and some rayons—may be extensively mildewed without the fabrics showing the least sign of weakening or tendering, the only available food material being the small amounts of substances other than cellulose which are contained in all such fabrics; solutions of many inorganic salts often develop, on standing, slimy growths of mould mycelium; and it is common experience that many species can grow and produce spores on plain agar made up with tap-water, a medium which is, since the agar is not utilized, extremely poor in nutrient material.

Mould growth on the lenses and prisms of optical instruments, often resulting in the glass being etched, has been a nuisance to scientific workers in the tropics for a very long time, and became a serious problem during the second world war. There seems to be little doubt that certain moulds can grow on the most carefully cleaned glass, provided that the relative humidity of the air is suitable, but how they obtain necessary nutrients is still a puzzle. Ohtsuki (1962) has isolated two species of *Aspergillus*, which he claims to be the true "glass moulds". One species belongs in the *A. restrictus* group, and was described as new as *A. vitricola* (= *A. penicilloides, fide* Raper and Fennell). It is stated to be unable to grow in a saturated, or nearly saturated atmosphere, but spores germinate readily in air of 80–90 per cent R.H. The other species is a member of the *A. glaucus* group, and is described as *Eurotium tonophilum* (recognized by Raper and Fennell as *A. tonophilus*). The results obtained by Ohtsuki on germination in various concentrations of nutrient salts, and in atmospheres of varying relative humidity, confirm the findings of Galloway (p. 296), that atmospheric humidity is more important than moisture in the substrate for growth of moulds.

Respiration. All fungi require oxygen for growth. Many of the yeasts develop characteristically when completely submerged in liquid, where the amount of available oxygen is small, but, whilst fermentation of sugar may proceed rapidly under these conditions, the amount of new cell substance formed is very small compared with the amount of growth under truly aerobic conditions. Some other fungi produce atypical structures, sometimes yeast-like, at other times slimy, when growing submerged in liquid, but in all such cases growth is slow and, so far as is known, depends on the oxygen dissolving in the fluid. The spores of a great many filamentous fungi will germinate when immersed in a liquid medium, but grow exceedingly slowly until some hyphae reach the surface, after which growth is rapid and typical until the surface is covered.

Normal metabolism results in the breakdown of some of the organic food material to CO_2 and, unless this is continually removed and replaced by a fresh supply of air, growth ceases, whether on account of the diminution of oxygen tension alone or by actual poisoning is not known. That being so, it is rather surprising that the means taken to exclude contaminants from cultures do not have more effect on growth and sporulation. Moulds are commonly grown on liquid media in plugged test-tubes 6 inches long, the depth of liquid being about 1 inch, and it would be reasonable to assume that, with the tube in a vertical position, CO_2 would accumulate above the surface of the fluid. In a culture flask or in a tube of sloped agar, with a much greater active surface relative to the plugged opening, conditions should be decidedly worse, yet analyses of air drawn from flasks containing actively growing moulds invariably show only very small differences from analyses of ordinary air, proving that there must be a very rapid removal of CO_2 and intake of fresh air.

It is readily shown, by passing measured volumes of air through a culture vessel (properly stoppered, of course, and not plugged), or through a number of similar vessels in series, that lack of an adequate supply of air not only influences rate of growth but, in many cases, alters the type of growth. One common effect of very restricted aeration is to suppress normal colour production in both mycelium and conidia. A striking instance of this is mentioned by Thom in *The Penicillia* (1930, p. 83) where he states that when Roquefort cheese is cut the fresh surface is often colourless, but turns green very rapidly on exposure to air.

Although fungi are unable to use CO_2 for photosynthesis, and normally liberate the gas, it has been shown that, with some fungi at least, a portion of the CO_2 produced by respiration can be reabsorbed and utilized for building up compounds containing four or more carbon atoms from the 3-carbon primary breakdown products of glucose. Experiments with labelled carbon have shown that some of the common acidic metabolic products of moulds are formed in this way (Krebs, 1943).

Little is known at present about the mechanism of respiration in the filamentous fungi. Most of the studies on heterotrophic respiration have been carried out with yeasts or bacteria, since these organisms lend themselves readily to manometric studies.

Reaction of medium. Most of the commonly occurring moulds will tolerate a wide range of hydrogen-ion concentration, provided other conditions are favourable, but, as might be expected, different species respond in various ways to changes in reaction. In general, a slightly acid medium is favourable to spore germination and rapid growth of the young colony, just as a slight alkalinity is preferred by the majority of bacteria. Once growth is established the reaction of the medium usually changes, owing to the accumulation of products of metabolism. Many species, particularly of *Aspergillus* and *Penicillium*, form fairly large quantities of organic acids, oxalic, citric, and gluconic acids being very frequently found, and many others more rarely. In absence of metabolic products of a strongly acid nature, alteration in reaction of the culture medium depends largely on the types of inorganic salts present. For example, the utilization of nitrogen from sodium or potassium nitrate tends to liberate base and increase the pH, and from ammonium sulphate to liberate acid and decrease the pH, whilst the presence of appreciable amounts of buffer salts tends, of course, to stabilize the reaction or at least minimize the pH drift. It is not ususual for appreciable quantities of organic acids to be produced during the early stages of growth and then to be utilized by the mould as other sources of carbon become exhausted. In such cases the pH of the medium will first decrease and then gradually increase as growth proceeds. Sometimes the

FIG. 166.—*Phycomyces blakesleeanus*—culture in Roux bottle, illuminated from one shoulder only during period of growth, showing well-marked phototropism. × 0·5.

appearance of the mould itself indicates changes in reaction of the medium. In many of the green species of *Aspergillus* and *Penicillium* the bright green of the young colony is very stable, often persisting for several months, on a buffered acid medium such as beer wort, whereas on Czapek's medium, which tends to become alkaline unless strong acids are synthesized, the initial green colour gradually changes to brown or grey shades. Many mould pigments, produced either in or on the mycelium or in the medium, act as indicators, showing very distinct colour changes according to alterations in reaction.

Influence of light. It is impossible to generalize concerning the effect of light on the growth of fungi, for, whilst many common species seem to grow equally well and characteristically in light or darkness, others are definitely stimulated or affected in other ways by light. For example, many of the Mucoraceae seem to be quite unaffected by moderate illumination whilst some strains of the common mould *Rhizopus stolonifer* grow appreciably faster in diffused light than in darkness. Some fungi are positively phototropic; that is, they grow towards the light or orientate their fruit-bodies in the direction of maximum illumination. The sporangiophores of *Phycomyces blakesleeanus* are comparatively short when a culture is exposed fully to the light, but attain a length of 20–30 cm. if the mould is grown on a layer of medium placed at the bottom of a tall vessel which allows light to enter only from the top. Fig. 166 shows a very characteristic growth of *P. blakesleeanus* in a Roux bottle which was covered with opaque material except for a patch on one shoulder, the sporangiophores being all turned towards the point of entry of the light. A number of other species exhibit similar tendencies. The sporangia of *Pilobolus* species, which are violently shot off at maturity, point in the direction of strongest illumination; the stipe of the wood-destroying fungus, *Lentinus lepideus*, grows towards the light and the cap develops only when the intensity of the light exceeds a certain minimum; the necks of the perithecia in many Pyrenomycetes, particularly the species which grow on dung, are regularly orientated towards the point of maximum illumination. In all such cases phototropism represents an adaptation for the more efficient dispersal of spores.

An interesting effect, described by Williams (1959), is on spore size. Spores of all species examined were appreciably larger when produced in the dark than when cultures were continuously illuminated. Other factors influencing size of spores are discussed, but it was found that the response of different species of fungi, to all the factors except light, varied considerably.

The effect of light on many moulds is to stimulate the production of spores or fruit-bodies, short exposures to sunlight or ultra-violet rays often being a useful method of inducing sporulation in cultures which have deteriorated to the point of sterility. Longer exposures to ultra-violet rays results in death of both mycelium and spores, dark coloured spores being more resistant than pale or colourless ones. Doses of ultra-violet rays just

insufficient to kill a culture completely often results in a high rate of mutation in the surviving spores. This method of inducing mutation has been
used to obtain strains with altered biochemical characteristics.

Another common effect of moderate illumination is to stimulate production of mycelial pigments.

Temperature relationships. Fungi show great differences in their
response to temperature changes and in their powers of resistance to heat
and cold. Application of heat is, of course, the usual method of sterilizing
culture media, vessels, and tools, and of killing unwanted cultures, there
being few moulds which can withstand the action of steam or boiling water
for any appreciable length of time. Thermal death points, however, cover a
fairly wide range, some strains of *Penicillium brevi-compactum*, for example,
being killed by prolonged exposure to a temperature of 33° C, whilst
Byssochlamys fulva, a fungus responsible for much trouble in the canning
industry, can survive the normal sterilization process for canned fruits in
which, for a short time, the temperature exceeds 90° (Olliver and Smith,
1933), and ascospores of *Neurospora sitophila*, the red bread mould, can
survive inside a loaf during baking, and actually will not germinate unless
previously heated to at least 60°. On the whole, spores are more resistant
than mycelium, and both are less affected by dry than by moist heat. In
the same way the spores of many moulds are killed by freezing in presence
of water, but when dry have their germinative powers unimpaired by
prolonged cooling in liquid air.

The limiting temperatures which will admit of growth, as distinct from
mere retention of vitality, are less extreme. Brooks and Hansford (1922)
and Brooks and Kidd (1921) have shown that *Cladosporium herbarum* will
grow, and its spores germinate, on meat in cold storage at a temperature
of −6° C, and several other species will grow at slightly higher temperatures which are still below freezing-point. Ingram (1951) states that osmophilic yeasts grow in concentrated orange juice at least down to −10° C.
McCormack (1950) claims that a pink yeast, isolated from frozen oysters,
will grow at temperatures down to −18°, whilst, according to Davis (1951),
a temperature of −23° is required to stop the growth of *Oospora lactis*
(= *Geotrichum candidum*). Numerous other workers have confirmed that
many moulds can grow at or just below 0° C. At the other end of the scale
some fungi will flourish at temperatures sufficient to inhibit the majority
of species and to kill some of them. *Aspergillus fumigatus* grows well at
50° C; La Touche has described a species of *Chaetomium* which grows at
temperatures up to 62° C, with optimum between 40° and 50°; and quite
a number of moulds, including species of *Mucor*, *Absidia*, *Aspergillus*, and
Penicillium, flourish at blood heat, 37° C. Cooney and Emerson (1964)
have published a monograph on the thermophilic fungi. They restrict the
term to fungi which can grow at 50° or over, and which cannot grow at
20° or below. Species which can grow below 20°, such as *Aspergillus fumigatus*, are described as "thermotolerant".

Each species is found to grow best at round about some particular temperature, known as the optimum temperature for that species. On the
whole, species of *Penicillium* are commonest in temperate countries, and
have optima between 20° and 25°, whilst species of the nearly related genus
Aspergillus grow best at about 30°, and hence are found most frequently in
warm regions. This is a very rough generalization, with a number of exceptions, but is often of value in industrial work, where 90 per cent of the
fungi encountered belong to these two genera. One particular section of
the Aspergilli, the *A. glaucus* group, deserve special mention, as some of
the common species have two temperature optima. Conidia and vegetative
growth are produced most freely at somewhat low temperatures and,
therefore, optimum temperatures, as determined from growth rate, are
low, for some strains as low as 10°. Perithecia, on the other hand, are
formed most readily and abundantly at higher temperatures, some species
being almost entirely perithecial above 30°, and there is thus a second
optimum relating to the perfect stage of the fungus. Although the majority
of Penicillia have optima which are comparatively low, there are a number
of species, mostly belonging to the section Biverticillata-Symmetrica, which
show maximum growth at temperatures near to 30° and which flourish up
to at least 37°.

A few species of the Mucoraceae have high optimum temperatures, in
particular *Absidia corymbifera* and *A. ramosa*, both of which are parasitic
to warm-blooded animals and show maximum development in culture
at 37°.

An interesting type of reaction to temperature change is shown by
Thamnidium elegans. Although it grows over a wide range of temperature,
its optimum for growth is 25–27°. Hesseltine and Anderson (1956) have,
however, shown that, in order to obtain zygospores, from paired cultures
of mating strains, it is necessary to incubate at low temperature, 6–7°.

Moisture requirements. A supply of water is absolutely essential for
the growth of all fungi and, although a few species can thrive in presence
of very small amounts of available moisture, it is fortunate that the great
majority of moulds require what may be termed damp conditions. Dry rot
appears on timber only when it has not been properly seasoned or when it
is exposed in a humid, stagnant atmosphere, and mould appears on wall-
paper only when joints in the wall are faulty and the paper itself has become
perceptibly damp. Many industrial products, paper, textiles, leather, and
foodstuffs of various kinds, readily absorb or adsorb water from a humid
atmosphere and are liable to attack by only a very limited number of
moulds except when stored under conditions where this intake of water is
excessive.

It is important to realize that the total amount of water contained in any
particular substance does not necessarily determine its liability to fungal
attack. Jam, for instance, contains a high percentage of water but, if properly compounded, does not become mouldy because the high concentra-

tion of soluble matter, chiefly sugar, renders the water unavailable to fungi, whereas materials consisting chiefly of cellulose, such as cotton or paper, are liable to attack if the moisture content exceeds about 8 per cent. It has often been stated that glycerol has antiseptic properties but, at low concentration, it is an excellent food for many fungi, and it inhibits growth only in concentrated solution, when the effect is mainly or entirely due to its desiccating action and not to any specific toxicity. Of all the fungi which can tolerate high osmotic concentrations of dissolved substances easily the most important are members of the *Aspergillus glaucus* series. These are frequently encountered on all types of materials which contain only very small amounts of excess moisture. They are commonly found, as pure cultures, on jams and similar products which contain slightly less than the normal percentage of sugar. Requiring only slightly more moisture for growth are *Aspergillus candidus* and *A. versicolor*, both found frequently on tobacco and nuts in the shell. Even more tolerant than *A. glaucus* of high concentrations of dissolved substances are the so-called osmophilic yeasts, which are commonly found as causes of spoilage of honey.

At the other end of the scale there are a number of fungi, chiefly plant parasites, which require the presence of liquid water for spore germination, and Ingold has made a special study of a large group of Hyphomycetes which complete their development submerged in the water of ponds and streams (see Ingold, 1954).

The question of moisture requirements is further discussed in Chapter XIV.

Poisons. The Fungi, like other classes of living things, are affected adversely by the presence of certain substances in their food supply. Our knowledge of what may be termed poisons has been gained chiefly through the demands of various industries for efficient antiseptics to combat the ravages of fungi amongst their raw materials and products. This topic is discussed in the chapter on "Control of Mould Growth" (Chapter XIV) and will not be elaborated here.

It is an interesting fact that the use of certain poisonous substances, in very small doses, as general tonics for the human system is paralleled by the use of small quantities of antiseptics for stimulation of fungal growth. Zinc salts are probably the best known of such substances, and are regularly incorporated in a number of culture media, but many other substances show the same reversal of effect on particular species or groups of species. Copper, although used extensively as an antiseptic in agriculture, is tolerated in high concentrations by some moulds. Raper and Thom (1949) report that *Penicillium lilacinum* is commonly found in laboratory reagents, and has also appeared in nickel and copper electroplating baths. *P. ochrochloron* is also commonly found in strongly acid solutions containing high percentages of copper sulphate. A culture in the Author's collection was isolated from an electrotyping bath containing 15 per cent copper sulphate and 5 per cent sulphuric acid. It has resisted all attempts to induce sporulation, and has not, therefore, been identified. Kendrick (1962) investigated the fungi in the soil of a swamp which contained varying amounts of copper. Some species were found exclusively in samples of soil containing over 7500 ppm of copper, the dominant species being *P. ochrochloron*.

Influence of other fungi. If several species of fungi are growing together on the same substratum they may affect each other's growth in various ways. In the simplest case there is competition for the available food material, and the ultimate result is that some species are prevented from spreading, whilst others flourish. It is impossible to predict, from knowledge of the growth rates of the individual species, which will be dominant in mixed cultures, since rapid growth in pure culture does not necessarily mean ability to acquire food in face of competition. In many cases of mixed infections it is found that some of the moulds grow abnormally, producing freak or dwarfed sporing structures, or remaining sterile,

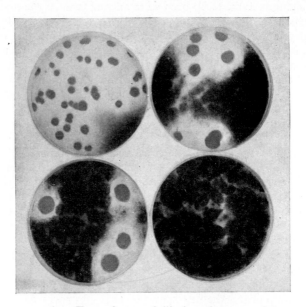

FIG. 167.—Four of a set of dilution plates containing colonies of *Penicillium* and *Stachybotrys.* × 0·35.

and it is therefore very unsafe to attempt to base identifications on examination of such material. Fig. 167 is a photograph of the first four of a set of dilution plates made from a mixed culture of *Stachybotrys atra* and a species of *Penicillium*. The fact that the fourth plate still contains scores of colonies of the *Stachybotrys* (black) shows that the fruiting of this species has been completely suppressed in the vicinity of the *Penicillium* colonies (grey).

The next case to be considered is not just competition for food but active antagonism. Some species, when growing in proximity to other moulds, produce more or less circular colonies surrounded by clear zones in which growth of the competing organisms is inhibited. This means, of course, that a toxic substance is being produced and liberated into the culture medium. Such substances, commonly known as antibiotics, are more suitably described in connection with the uses of fungi (Chapter XV).

In contrast to the interreactions described above are many cases in which the growth of one fungus assists that of another. A common state of affairs,

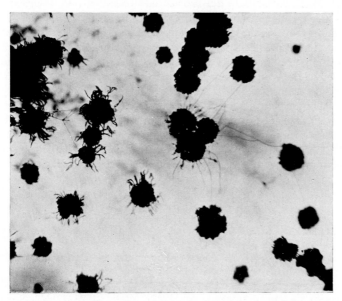

FIG. 168.—*Penicillium* disease of *Aspergillus niger*—infected heads as seen in Petri dish. ×25.

when a suitable substrate is exposed to chance infection, is for there to occur a well-marked progressive change in the fungal population, correlated with the different enzymatic activities of the various species. One group of species breaks down complex organic material to provide suitable food for another group, whilst a third group will utilize the metabolic products of the second. Cellulosic materials in nature, particularly lignified cellulose, are usually broken down in stages, and the same applies to dung.

Another case of frequent occurrence is for one species to synthesize, in excess of its own requirements, a growth-substance essential for another species. The first such case to be described was by Heald and Pool (1908) and, in consequence, the phenomenon is often termed "the Heald-Pool effect". They found that *Melanospora pampeana* grows very poorly in pure culture and produces no perithecia, but thrives and gives abundant perithecia when grown in mixed culture with *Fusarium moniliforme*. *Melanospora* (now *Sordaria*) *destruens*[1] does not produce perithecia on a medium containing only salts and sugar, but is stimulated by growing in mixed culture with any one of a large number of moulds and bacteria. Hawker (1939) has shown that *S. destruens* requires both thiamin and biotin for production of fruit-bodies and is stimulated by any organism which produces both these growth-substances.

True parasitism amongst the fungi is by no means uncommon. The higher fungi, particularly mushrooms and toadstools, are attacked by a large variety of micro-fungi, and even a few of the species of moulds commonly grown in the laboratory are parasitized, usually by nearly related species. Thus crude cultures of *Mucor* species, particularly when isolated from dung, are often attacked by species of *Chaetocladium*, and quite a number of the Mucorales are attacked by species of *Piptocephalis*. The *Penicillium* disease of *Aspergillus niger* is another well-known case of parasitism. The parasites are members of the section Biverticillata-Symmetrica, usually *P. rugulosum* or closely related species, and they cover the *Aspergillus* heads with dense clusters of dark green penicilli, eventually killing the host (Fig. 168). *Gliocladium roseum* is reported by Barnett and Lilly (1962) to parasitize a number of other fungi, attacking chiefly the conidia.

REFERENCES

Barnett, H. L., and Lilly, V. G. (1962). A destructive mycoparasite, *Gliocladium roseum*. *Mycologia*, **54**, 72–7.

Birkinshaw, J. H. (1952). Metabolic products of *Penicillium multicolor* G.-M. and P. with special reference to sclerotiorin. *Biochem. J.*, **52**, 283–8.

Brooks, F. T., and Hansford, C. G. (1922). Mould growths upon cold-store meat. *Trans. Brit. mycol. Soc.*, **8**, 113–42.

Brooks, F. T., and Kidd, M. N. (1921). Black spot of chilled and frozen meat. *Special Report no. 6*, Food Invest. Board, D.S.I.R., London.

Clutterbuck, P. W., Mukhopadhyay, S. L., Oxford, A. E., and Raistrick, H. (1940). A.—A survey of chlorine metabolism by moulds. B.—Caldariomycin, $C_5H_8O_2Cl_2$, a metabolic product of *Caldariomyces fumago* Woronichin. *Biochem. J.*, **34**, 664–77.

Cochrane, V. W. (1958). *Physiology of fungi*. London: Chapman & Hall.

Cooney, D. G., and Emerson, R. (1964). *Thermophilic fungi*. London: W. H. Freeman & Co.

Curtin, T. P., and Reilly, J. (1940). Sclerotiorin, $C_{21}H_{22}O_5Cl$, a chlorine-containing metabolic product of *Penicillium sclerotiorum* van Beyma. *Biochem. J.*, **34**, 1419–21.

Davis, J. G. (1951). The effect of cold on micro-organisms in relation to dairying. *Proc. Soc. appl. Bact.*, **14**, 216–42.

[1] = *S. fimicola* (see p. 52).

Gupta, A. D., and Nandi, P. N. (1957). Role of nitrogen concentration on production of perithecia in *Penicillium vermiculatum* Dangeard. *Nature, Lond.*, **179**, 429–30.

Hawker, L. E. (1939). The nature of the accessory growth factors influencing growth and fruiting of *Melanospora destruens* Shear, and of some other fungi. *Ann. Bot.*, N.S. **3**, 657–76.

Hawker, L. E. (1950). *Physiology of the fungi*. London: Univ. London Press.

Heald, F. D., and Pool, V. W. (1908). The influence of chemical stimulation upon the production of perithecia of *Melanospora pampeana* Speg. *Rep. Neb. Agric. Exp. Sta.*, **96**, 185–99.

Hesseltine, C. W., and Anderson, A. (1956). The genus *Thamnidium* and a study of the formation of its zygospores. *Amer. J. Bot.*, **43**, 696–703.

Ingold, C. T. (1954). Fungi and water. (Pres. Address). *Trans. Brit. mycol. Soc.*, **37**, 97–107.

Ingram, M. (1951). The effect of cold on micro-organisms in relation to food. *Proc. Soc. appl. Bact.*, **14**, 243–60.

Kendrick, W. B. (1962). Soil fungi of a copper swamp. *Canad. J. Microbiol.*, **8**, 639–47.

Krebs, H. A. (1943). Carbon dioxide assimilation in heterotrophic organisms. *Annu. Rev. Biochem.*, **12**, 529–50.

La Touche, C. J. (1950). On a thermophilic species of *Chaetomium*. *Trans. Brit. mycol. Soc.*, **33**, 94–104.

Lilly, V. G., and Barnett, H. L. (1951). *Physiology of the fungi*. New York: McGraw-Hill.

McCormack, G. (1950). "Pink yeast" isolated from oysters grown at temperatures below freezing. *Conn. Fish Rev.*, **12**, 28.

Ohtsuki, T. (1962). Studies on the glass moulds. IV. On aerial germination of mould spores on glass. *Bot. Mag. Tokyo*, **75**, 221–7.

Olliver, M., and Smith, G. (1933). *Byssochlamys fulva* sp. nov. *J. Bot., Lond.*, **71**, 196–7.

Oxford, A. E., Raistrick, H., and Simonart, P. (1939). Griseofulvin, $C_{17}H_{17}O_6Cl$, a metabolic product of *Penicillium griseofulvum* Dierckx. *Biochem. J.*, **33**, 240–8.

Raistrick, H., and Smith, G. (1936). The metabolic products of *Aspergillus terreus* Thom. Part II. Two new chlorine-containing mould metabolic products, Geodin and Erdin. *Biochem. J.*, **30**, 1315–22.

Raper, K. B., and Thom, C. (1949). *Manual of the Penicillia*. Baltimore: The Williams & Wilkins Co.

Smith. G. (1949). The effect of adding trace elements to Czapek-Dox medium. *Trans. Brit. mycol. Soc.*, **32**, 280–3.

Thom, C. (1930). *The Penicillia*. Baltimore: The Williams & Wilkins Co.

Williams, C. N. (1959). Spore size in relation to culture conditions. *Trans. Brit. mycol. Soc.*, **42**, 213–22.

Chapter XIII

The Maintenance of a Culture Collection

The cultivation of micro-fungi is still almost wholly an empirical art.

H. A. Dade, *Herb. I.M.I. Handbook,* 1960

In most laboratories where serious work on micro-fungi is carried out it is necessary to maintain a collection of moulds. If the work in hand aims at controlling the harmful activities of fungi on some industrial product most of the preliminary work will have to be done on pure strains of moulds isolated from actual cases of damage, and every important species will have to be kept in pure culture so as to be available for numerous experiments extending perhaps over a period of years. Those who study the biochemical activities of fungi, with a view to isolating or manufacturing special products of metabolism, usually find it necessary to test many strains of the same or related species, and to grow them all on many different media and under varying conditions. Again a type collection is essential.

Requirements of different laboratories, as regards size and type of collection, will vary greatly, but there are a number of general methods and precautions whose consideration is applicable to all cases. The main essentials for successfully maintaining a culture collection may be briefly summarized thus:

1. All cultures must be kept alive as long as required.
2. Every strain must be maintained in a state of purity.
3. As far as possible, each species or strain must be kept true to type, that is, with all the characteristics it had when first added to the collection.

KEEPING MOULDS ALIVE

The main essential is transfer to fresh culture medium at sufficiently frequent intervals. Different species, even of the same genus, vary much in longevity, the spores of some common moulds remaining viable for years whilst others lose their power of germination after a few months. Some strains can withstand desiccation and may readily be sub-cultured even after the medium has dried up to a horny mass, whilst others, chiefly those which do not spore freely, must be kept on a moist medium and die off if allowed to become even approximately dry. The frequency with which sub-culturing of a collection is necessary depends very largely, therefore, on the peculiarities of the particular fungi, but it also depends to some extent on external conditions, especially on the temperature of the place of storage. If kept in a room where the average temperature is 18–20° C most of the common moulds may be kept alive by transferring at intervals of 6 months. If a refrigerator, kept at about 8° C, is available for storage, cultures will remain in good condition for a longer time, the majority of species remaining viable for at least 8–9 months. If the average temperature of the laboratory is above 20° the advantage of cold storage is, of course, still greater. In this case it is best to grow sub-cultures in the 25° incubator until grown sufficiently to be able to decide whether the transfers are entirely typical, and then to remove the cultures to the refrigerator, where growth continues more slowly. A few species, notably some belonging to the *Aspergillus glaucus* group, grow well at low temperatures, occasionally better than at 20°. Even in these cases the lower temperature is suitable for storage, since one of its chief advantages is that drying out of the medium is retarded. When special accommodation at a low temperature cannot be provided, experience will soon show how often the collection should be sub-cultured, but it is always better to err on the side of frequency, since the loss of a valuable culture may be very serious. Although transfer about twice a year is suitable for most moulds, almost every collection includes a few species which will not survive for 6 months even in the refrigerator. It is best to keep these separate from the rest of the collection and to transfer at, say, 3-monthly intervals.

As regards choice of media for stock cultures it is advisable to aim at encouraging growth which is characteristic but not too luxuriant. For many species, particularly Aspergilli and Penicillia, Czapek agar is ideal from this point of view. However, as pointed out in Chapter XII, many Ascomycetes, and a number of species belonging to other classes, grow very poorly and do not produce fruit-bodies on synthetic media. For these, and for the more delicate species of Mucorales, wort agar is usually satisfactory. Nevertheless there are some moulds, particularly species of *Rhizopus* and *Absidia*, which grow poorly on Czapek agar but grow too luxuriantly on wort agar, often completely choking up the culture tubes. For these agar media made with extracts of vegetables, such as potato, carrot, mixed carrot and potato, and prune, are very suitable, since they provide the necessary accessory growth factors and, at the same time, do not encourage rampant growth. For most of the dark coloured Hyphomycetes corn-meal or oat-meal agar are amongst the best media, usually inducing adequate spore production without excessive aerial mycelium. Gelatine media are unsuitable for stock cultures, since the gelatine is rapidly liquefied by very many species of moulds. In any collection it is advisable to keep at least two stock cultures of each mould as an insurance against loss by accident or contamination, and it is advisable that these should be on different

K

media, since a periodical change of diet helps to preserve vigour and character in many species.

It sometimes occurs, on account of delay in sub-culturing or even for no apparent reason, that transfers made in the usual way fail to grow. Even if three or four attempts have been unsuccessful it is unwise to discard the culture as dead. There are two methods of saving the species and obtaining fresh cultures even when there remain only a very few viable spores, and when the chance of picking up one of these by needle transfer is extremely remote. The first method is to flood the surface of the old culture with a small quantity of sterile liquid medium, or with a small amount of agar medium at 40–45° C. On incubation the fresh medium provides sufficient moisture and nutriment to bring about the germination of any spores which are still alive, after which sub-cultures may be made in the usual way. The second method is to scrape off as much as possible of the old growth in about 3 ml. of sterile water, then to flood this suspension over the surface of a suitable agar medium in a Petri dish. If one lot of water is insufficient to remove all the old colony a second lot may be used, and this should be poured off into a second dish.

In comparatively recent years a number of methods of preservation have been devised, mainly with a view to preventing loss of valuable biochemical properties, but also aiming at diminishing the frequency of sub-culturing.

Oil. Ordinary cultures in tubes are completely flooded with sterile pure mineral oil. They are then stored in the usual way.

Freeze-drying. This process, known also as the "lyophil method", depends on the fact that very rapid freezing does not impair the vitality of fungal spores. A spore suspension in serum is made; this is frozen almost instantaneously in a bath at –40° C and dried at this temperature in a high vacuum. Full details of the apparatus and *modus operandi* are given by Wickerham and Andreasen (1942), also by Thom and Raper (1945). The advantages of the lyophil method are, first, that the suspensions are contained in very small sealed tubes, which are easily stored, not readily broken, and absolutely proof against contamination; secondly, that spores of most species which have been tested remain viable for many years; thirdly, and most important, that the spores which germinate when the tubes are opened are the actual spores sealed up at the beginning of the period of storage and hence retain their original morphological characteristics and biochemical properties.

Cryogenic storage. During the past few years the technique of storing living material at very low temperatures has been applied to micro-organisms. Suspensions of spores or mycelium in a suitable medium are frozen at a carefully controlled rate and then immersed in or suspended over liquid nitrogen in specially constructed containers. The material is revived merely by removing it from the container and culturing it at normal temperature in the usual way. The biochemical processes which eventually lead to the degeneration and death of living cells at normal temperatures virtually cease at temperatures approaching that of liquid nitrogen (– 196° C). Long-term storage of microbial cells is thus possible, with all the concomitant advantages of the preservation of biochemical properties.

Soil cultures. Spores are distributed in sterile soil, contained in plugged tubes or flasks. The containers are stored in any convenient way. The great advantage of the method is that successive cultures can be obtained over a fairly long period, each of which will have all the characteristics of the original isolate, assuming, of course, that technique is faultless.

Sand cultures. Spores are distributed in dry sterilized sand, contained in plugged tubes. After inoculation of the sand it is further dried over P_2O_5. The tube is drawn out below the plug and sealed off. Spores preserved in this way usually retain their viability for periods of two to several years, and, as with soil cultures and lyophil cultures, retain their original characteristics.

The results of an extended test of the lyophil method are given by Raper and Alexander (1945) and Mehrotra and Hesseltine (1958). A comparison and appraisal of various methods of preservation are given by Fennell, Raper, and Flickinger (1950). A more recent reappraisal of the various methods is by Fennell (1960). She also includes in her paper an extensive and useful bibliography. At Ottawa in 1962 a conference on culture collections was held, the proceedings being subsequently published (Martin, 1963). This volume should be consulted by all who have charge of culture collections.

When a collection is constantly dipped into, and when any culture may be required at short notice, there is little doubt that agar slopes provide the most convenient method of storage, particularly if supplemented by the use of sand or soil cultures for preserving critical cultures. The lyophil method is best adapted to collections from which cultures are regularly sent out to other laboratories.

The method of storing under oil is used in many laboratories, since a number of species which cannot be lyophilized successfully, such as many of the Mucoraceae and some Hyphomycetes with thin-walled spores, preserve their vitality for long periods under oil. Sub-cultures can be taken at any time, almost as easily as from an agar slope. One disadvantage is that the culture tubes must always be kept vertical or nearly so, even when making sub-cultures. Another is the difficulty of removing the mineral oil completely from perhaps hundreds of tubes when the cultures are discarded.

MAINTAINING PURITY

The problem of keeping stock cultures in a state of purity is beset with more difficulties than is the task of merely keeping them alive. In the first place it is obvious that sub-culturing a number of species, one after the other, demands the utmost care and faultless technique. However, even when every care is taken, infection will sometimes occur if the work is done in a highly contaminated atmosphere. Sometimes it is unavoidable that cultures are handled in such a way that clouds of spores are discharged into the air. If this happens in a room where transfers of stock cultures have to be made, it is well, before starting to sow, to minimize the risk of infection by sterilizing the air, using one of the methods described in Chapter XI.

Even the best workers are troubled by occasional contamination of cultures and it is therefore very necessary to be able to recognize such when it does occur. Infection of a culture by a species of a different genus, say the invasion of a *Mucor* by a species of *Penicillium*, is easily detected on casual inspection, but infection by a closely related species of similar appearance may go unnoticed for a long time, and may even result in complete loss of the original and its replacement by the contaminant. The main essential, the importance of which cannot be over-emphasized, is that the worker in charge of the collection should keep full records of all species, noting gross appearance at various ages, rates of growth, and microscopical appearance, both of the living cultures and of mounted specimens. It is true that an enthusiastic mycologist in time gets to know all his moulds as well as the keen gardener knows all the plants in his garden, but, even then, records are more valuable than memory alone, and absolutely necessary if at any time the care of the collection has to be handed over to someone else, and they should never be omitted or scamped. All cultures should be carefully checked every time the collection is sub-cultured, and if there is the slightest suspicion of anything wrong, the particular culture should be plated out, fresh cultures being made from typical colonies.

When the culture collection includes species of several different genera it will be found that chance infection due to imperfect methods of transfer is more readily spotted if the species are stored and sub-cultured in random order. For example, number 1 in the collection may be an *Aspergillus*, 2 a *Mucor*, 3 a *Cladosporium*, 4 a *Penicillium*, and so on throughout, with never more than two or three species of the same genus bearing consecutive numbers. A collection of species of a single genus, particularly one in which there is little colour variation, is far more likely to carry unnoticed contaminants than is a miscellaneous collection.

Mites. These creatures are a most annoying pest to have in the culture laboratory, and are one of the most frequent causes of contamination. A collection may remain free from their depredations for years and then, suddenly, scores or hundreds of cultures are found to be infested, but, unless the mycologist is constantly on the look-out, and unless he knows what to look for, a serious attack may go unnoticed for a long time, and may result in wholesale contamination and loss of many species. Fig. 169 is a photomicrograph of a typical mite in a culture tube. The different species vary somewhat in shape and size, but, when once seen, can never be mis-

FIG. 169.—Mite (adult) in culture tube. × 100. The larvae are somewhat similar in appearance but have only six legs.

FIG. 170.—Eggs of mite as seen in culture tube. × 100.

taken for anything else. An adult mite is usually about one-hundredth of an inch in length, and is thus almost at the limit of unaided vision. When alive and in motion they are, as a matter of fact, readily seen, as small whitish specks, by anyone with normal sight, but, if any cultures are known to be infested, or there is any reason to suspect the presence of mites, *all* cultures should be carefully examined under the microscope, using a 2-inch objective. The presence of eggs should be particularly noted, as these are more difficult to kill than are the adult mites. Like the adults and larvae they are pale coloured, but appear to be dark brown when infested cultures are examined by transmitted light. They are surprisingly large, in comparison with the adults, and so are readily seen, even when lying amongst masses of hyphae and spores (Fig. 170). The evil mites do is twofold. In the first place they eat the cultures and, if left unchecked, may even destroy them entirely. In addition, they crawl from one culture to another with spores adhering to their hairy bodies, spreading infection wherever they go. Petri dishes are readily entered and become contaminated with amazing rapidity if once mites appear in their vicinity. Cultures in tubes are just as readily invaded, since the mites find no difficulty in crawling through the tightest cotton-wool plugs and, according to Thom, have even been known to find their way through paraffined plugs. If plugs are a good fit the mites shed most of the adhering spores *en route* to the cultures, the major portion of the damage being then due to their gastronomic activities, but mixing of cultures does occur and is presumably due to spores which pass unchanged through the digestive tract. The commonest infections introduced by these creatures are species of *Cephalosporium*, and these, with their production of myriads of tiny spores, are exceedingly difficult, and often impossible, to eliminate.

The control of mites is not easy. They are brought in with all kinds of raw materials, they may be introduced with new laboratory fittings and apparatus, or they may be present in cultures sent from other laboratories, and hence hygienic measures sufficiently stringent to prevent their access to the culture room are difficult to realize in practice. Constant vigilance is necessary, with prompt application of suppressive measures at the first indication of an attack. The usual method of killing mites is to expose the cultures to the vapour of some substance which is highly toxic to the mites and relatively non-toxic to moulds. Many fumigants have been suggested but, of these, only two are effective and safe, *p*-dichlorobenzene and crude kerosene. In the author's experience *p*-dichlorobenzene has given consistently satisfactory results, but some workers state that it is toxic to some species of fungi and, in addition, causes some others to grow abnormally. However, it has been used successfully for several years at the Commonwealth Mycological Institute, Kew, with no discernible ill-effects on a wide range of fungi. It is easy to use, infested cultures simply being put in an airtight box with some crystals of the fumigant and left overnight. H. A. Dade, of the Commonwealth Mycological Institute, Kew, states (personal communication) that when ordinary commercial kerosene was first used in his laboratory it was found to be extremely effective, not only for killing mites but also as a repellent. After some time, however, newly purchased kerosene was found to be almost useless. The difference was eventually traced to a new process of refining the oil, in which the toxic material was eliminated. An effective substitute, as a repellent, is "tractor vaporizing oil", obtainable in reasonably small quantities from almost any large garage. A bottle of the oil, fitted with a cork through which passes a lamp-wick, is placed amongst the stored cultures. The oil vapourizes slowly and drives away completely all mites and insects, but the concentration of vapour attained in this way is not high enough to kill mites which are already in the culture tubes. The best way of treating infested cultures is to purify by plating out, the old cultures being then killed off by heat.

An interesting acaricide, described by Curl (1958), is 4:4'-dichloro-α-trichloromethylbenzhydrol, or 1:1-bis(p-chlorophenyl)-2:2:2-trichloro-ethanol. It is sold under the trade name "Kelthane", and is supplied in this country by the Murphy Chemical Co. It has been extensively tested at the Commonwealth Mycological Institute and, unfortunately, found to be variable in its action. Some species of mites are, and remain, susceptible, whilst other species rapidly become completely immune.

Another excellent way of preserving cultures from infestation is to store them in a refrigerator at a temperature not exceeding about 8° C. Mites are not attracted to cold areas and, even if by some chance an infested culture is put into store, the mites, at this low temperature, move very slowly and do not breed, so that there is little danger of the infestation spreading.

Insect pests. Mites are not insects (though commonly regarded as such) but belong to the class Arachnida, order Acarina, and are allied to the spiders and ticks. Though not of so frequent occurrence as mites, true insects do, on occasion, invade cultures in the laboratory. The species most frequently met with are all larger than mites and, whilst they readily enter Petri dishes, are rarely able to make their way into plugged tubes. The insects which are attracted to cultures of fungi include a number of small beetles and weevils, readily recognized by their horny wing cases, but the most common predators are wingless species of primitive insects called Psocids, better known as book-lice. Fig. 171 shows a typical psocid taken from a culture of *Pleospora herbarum*, and showing masses of spores and fragments of perithecia still sticking to it. Some species differ from the one illustrated in having closely-set long hairs on the legs, but in other respects the general appearance is the same.

During the second world war a number of research institutions, engaged in testing the resistance to moulds of equipment for the Far East, installed chambers in which temperature and relative humidity were adjusted to simulate as closely as possible the conditions in tropical jungle. Every one of these tropic chambers became infested with psocids (and mites). No attempt was made to control them, since they helped to spread infection by moulds and since the real jungle abounds in such creatures. All the insects which infest cultures are readily killed by exposure to *p*-dichlorobenzene.

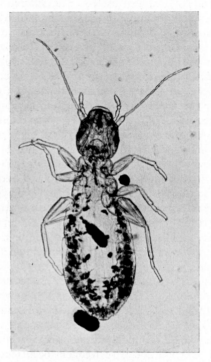

FIG. 171.—Psocid—mounted in lactophenol. × 50.

MAINTAINING TRUENESS TO TYPE

Probably the most difficult problem facing the curator of a culture collection is how to keep all the species in his care, through a long series of subcultures, true to type. Some moulds vary little or not at all, successive transfers coming up in every way, as far as can be seen, identical with the original. Unfortunately, on the other hand, a few fungi seem to be decidedly unstable in artificial culture and defy all efforts to maintain their distinguishing features. Between these two extremes are a large number of species which can be kept in good condition only if special precautions are taken.

The usual type of deterioration is the increased production, in successive sub-cultures, of sterile mycelium at the expense of fruiting structures. For example, species of *Chaetomium* (and of other Pyrenomycetes) tend to lose the power of producing perithecia; many dark coloured Hyphomycetes, such as species of *Alternaria, Stemphylium,* and *Helminthosporium,* often produce masses of floccose, whitish or grey mycelium over a thin basal felt of dark coloured hyphae, and spores are formed very tardily and sparingly or not at all; some species of *Fusarium* deteriorate very rapidly, growing as masses of ropy, slimy mycelium, with complete lack of typical macroconidia; some strains of *Penicillium luteum* and allied species frequently become entirely conidial or completely sterile, lacking both perithecia and conidia; even some of the Aspergilli, normally amongst the least troublesome of all fungi in this respect, occasionally deteriorate and become almost sterile. In many cases deterioration of this kind can be prevented by suitable choice of culture media. Some fungi will produce typical fruiting structures only on media containing low concentrations of nutrients, or, what is often equivalent, a low concentration of readily available carbohydrate. Species of *Phoma,* if kept on corn-meal or oat-meal agar, which contain much unhydrolysed starch but only very small amounts of soluble sugars, will usually continue to produce abundant pycnidia, whereas the same species, grown on sugary media, rapidly become sterile. On the same starchy media *Penicillium luteum* grows slowly but produces ascocarps freely, whereas on Czapek or wort agar it rapidly deteriorates. Most of the dark coloured Hyphomycetes also retain their capacity for sporulation best on media containing carbohydrates which have to be hydrolysed before they can be utilized. In contrast to the foregoing, some members of the *Aspergillus glaucus* series deteriorate on substrates of low concentration, but are rapidly restored to vigour if grown on culture media containing 20–40 per cent of cane sugar. Some fungi will not grow on synthetic media, requiring certain accessory growth factors which are found in many vegetable decoctions, but which are absent from media compounded of pure salts and sugars (see also Chapter XII).

Another cause of deterioration of certain fungi is lack of light. Some species of common moulds appear to grow equally well in light or dark, but in many others sporulation is stimulated by light, and it is always advisable to try the effect of, first, exposure to diffused light, then, if this fails, a short exposure to bright sunlight, on any cultures which are showing undue development of sterile aerial mycelium. A number of species of *Penicillium,* particularly the Biverticillata, respond well to this treatment, even old degenerate strains of *P. luteum* producing typical conidiophores if incubated in the open laboratory instead of in a closed incubator. Many Ascomycetes, and many dark coloured Hyphomycetes, produce fruiting structures sparingly or not at all in the dark but grow typically in diffused daylight. Many of the pathogenic Helminthosporia do not spore at all in artificial culture under ordinary conditions, but Weston (1933) has shown that *H. avenae* can be induced to form normal conidia if cultures in quartz Petri dishes are submitted to one or two short irradiations from a quartz mercury-vapour lamp.

A cause of variation inherent in the fungus itself has been brought to light by Hansen (1932, 1938) and termed the "Dual phenomenon". The spores of many fungi are bi- or multi-nucleate, the nuclei being of two types, one promoting mycelial growth (M), the other production of spores (C). Cultures originating from conidia containing nuclei of the C type only remain typical through a series of sub-cultures and produce abundant fruiting structures. Spores containing only M nuclei give rise to cultures which are predominantly mycelial, with scanty spore production; whilst cultures from single spores containing both M and C nuclei (MC) are of intermediate type. Mixed cultures from several spores, including CC, MM, and MC types, produce, by anastomosis, mycelium containing both kinds of nuclei. The method of obtaining homotypic sporing cultures is to prepare a fairly large number of single-spore cultures. When these reach maturity they will probably show marked differences in degree of freedom of sporulation and amount of sterile mycelium produced. Those which produce spores most freely, and show least tendency to development of floccose mycelium, are selected and from them further series of single-spore cultures are made. This process of selection is continued until a culture is obtained which produces spores of one kind only, the CC type, any of which will develop into a similar, freely sporing colony. The only flaw in this system is that occasionally a CC culture changes into a CM type owing to mutation. Hansen and Snyder (1944) state that the well-known instability of some strains of *Penicillium notatum*, used for the production of penicillin, can be explained as an example of the dual phenomenon, but Baker (1944) maintains that the conidia of this species are predominantly uni-nucleate.

Another type of deterioration is, fortunately, not common, but when it does occur is more serious, in that it usually results in total loss of the culture. Growth in successive transfers remains fairly typical, but becomes less and less vigorous, until the vitality of the fungus seems to dwindle away completely. If the preservation of such a strain is desirable it is necessary to carry out a fairly extensive cultural investigation, trying the effect of different sources of carbon and nitrogen, of the addition of various trace elements and accessory growth substances, of adjusting the medium to different reactions, of growing the fungus at various temperatures, and so on, until a set of conditions can be formulated which will ensure vigorous growth.

Many fungi show to some degree a tendency to spontaneous mutation. It is commonly evidenced by what is known as "sectoring", a more or less circular colony of the mould showing one or more fairly clean-cut sectors, which vary from the rest of the colony in such characters as colour, texture, or degree of sporulation. Sectoring may be enhanced, or may be induced in normally stable species by exposing cultures to abnormal conditions, such as a temperature near to the death point, or by incorporating small amounts of toxic substances in the culture medium. In these cases the new type of growth may either be stable or may revert back to the original on restoring to normal conditions. Some species regularly give rise to true mutants when grown in the normal way and, if the new strain happens to be of more vigorous habit than the parent, the latter may be entirely swamped unless the change is noticed in good time and the two forms separated. In comparatively recent times deliberately induced mutation, by means of X-rays, ultra-violet rays, or toxic chemicals, has been the means of obtaining strains of *Penicillium chrysogenum* giving enormously increased yields of penicillin, and has also provided a useful tool to the geneticist, in a large number of mutants of *Neurospora*.

A final form of variation from type, which the biochemist knows only too well, concerns, not the appearance or structure of the fungus, but alteration in chemical activity, or loss of power to produce some particular metabolic product. Two examples, from amongst the many recorded in the literature, may be cited to illustrate this type of change. Wehmer, in 1918, described a new species of *Aspergillus* which gave good yields of fumaric acid when grown on glucose media, and which was therefore named *A. fumaricus*. In a later paper, 1928, he reported that his fungus no longer produced fumaric acid but gave gluconic acid instead. Birkinshaw and Raistrick (1931) describe the production, by four strains of *Penicillium spinulosum* Thom, of a substituted toluquinone which gave to the culture medium an intense purple colour. All the strains, when freshly isolated, gave appreciable amounts of the quinone, but all gave gradually decreasing yields, and after repeated sub-culturing refused to produce the least trace of the substance. This kind of variation gives rise to serious difficulties when an attempt is made to utilize the metabolic activities of fungi for the large-scale production of substances of industrial importance. Before any such process can become a commercial proposition it may be necessary to carry out much preliminary work in order to ensure that any tendency to loss of activity can be completely controlled. However, some of the methods of preserving cultures, which have been described earlier, are a great help in conserving biochemical properties unchanged for several years at least.

COLLECTING FUNGI

Methods of isolation of fungi from mouldy materials have already been described (p. 244 *et seq.*). The student of mycology whose interests go beyond his immediate industrial problems will wish to become acquainted with a larger variety of microfungi.

Many interesting species of moulds can be isolated by plating out soils from different localities. Various media should be used, and it is advisable to add bacteriostatic substances, and an anti-spreader, such as ox-gall (see Chapter XVI).

A large variety of interesting fungi have been collected, by various workers, on twigs and small pieces of wood, which form part of the ground litter under trees. Not all of these can, at present, be grown on artificial media, but the majority present no difficulty. They include a large number of Ascomycetes and Fungi Imperfecti.

Many species can be found with fair certainty when required if the habitat be known, and most original descriptions of fungi mention the habitat. Many of the commonest moulds are almost omnivorous, and have no special habitat, but others are limited in range of hosts, and quite a number are to be found on only one particular host. It has already been pointed out that a few species of *Penicillium* can usually be found, when wanted, in particular habitats, although not necessarily confined to these. During work on the taxonomy of the genus *Paecilomyces*, *P. elegans* (= *Penicillium elegans* Corda) was found in exactly the situation described by Corda in 1838, on the inside of the loose bark of fallen pine trees. Instances such as this could be multiplied, showing the value of the recording of habitat when drawing up diagnoses of new species.

CULTURE COLLECTIONS

There exist a number of large collections of cultures of microfungi, maintained in what are primarily centres of research, but which are organized for the regular supply of cultures to other institutions, firms, and individual workers. The largest and best known of these is the Centraalbureau voor Schimmelcultures, at Baarn, Holland. A very comprehensive catalogue is published at intervals of a few years, the latest one listing well over 6000 species.

In this country the National Collection of Type Cultures, which used to include all types of micro-organisms, is now restricted to maintaining a collection of pathogenic bacteria. Fungi are supplied by the Commonwealth Mycological Institute, Kew, which keeps its catalogue up to date by new editions every few years, and by the issue of supplements at shorter intervals. The collection is not so large as that at Baarn, but is growing and becoming increasingly useful.

There are also collections in the Northern Utilization Research Branch, U.S. Department of Agriculture, Peoria, Illinois; at the Instituto Oswaldo Cruz, Rio de Janeiro, Brazil; the Nagao Institute, Tokyo; the Institute for Fermentation, Osaka; the Laboratoire de Cryptogamie, Paris. There are also a number of smaller and specialist collections.

In addition to these, a number of institutions and industrial firms maintain their own, sometimes extensive, collections. They do not issue catalogues, nor do they supply cultures on demand, but are often willing to send cultures as a matter of courtesy to other workers.

REFERENCES

Baker, G. E. (1944). Nuclear behavior in relation to culture methods for *Penicillium notatum* Westling. *Science*, **99**, 436.

Birkinshaw, J. H., and Raistrick, H. (1931). On a new methoxydihydroxy-toluquinone produced from glucose by species of *Penicillium* of the *P. spinulosum* series. *Phil. Trans.*, Ser. B, **220**, 245–54.

Curl, E. A. (1958). Chemical exclusion of mites from laboratory fungal cultures. *Plant Dis. Reptr.*, **42**, 1026–9.

Fennell, D. I. (1960). Conservation of fungous cultures. *Bot. Rev.*, **26**, 79–141.

Fennell, D. I., Raper, K. B., and Flickinger, M. H. (1950). Further investigations on the preservation of mold cultures. *Mycologia*, **42**, 135–47.

Hansen, H. N. (1938). The dual phenomenon in imperfect fungi. *Mycologia*, **30**, 442–55.

Hansen, H. N., and Smith, R. E. (1932). The mechanism of variation in imperfect fungi: *Botrytis cinerea*. *Phytopathology*, **22**, 953–64.

Hansen, H. N., and Snyder, W. C. (1944). Relation of dual phenomenon in *Penicillium notatum* to penicillin production. *Science*, **99**, 264.

Martin, S. M. (1963). Ed. *Culture Collections. Perspectives and problems.* Toronto: Univ. Toronto Press. (Proceedings of the specialists' conference on culture collections, Ottawa, Aug. 1962.)

Mehrotra, B. S., and Hesseltine, C. W. (1958). Further evaluation of the lyophil process for the preservation of Aspergilli and Penicillia. *Arch. Microbiol.*, **6**, 179–83.

Raper, K. B., and Alexander, D. F. (1945). Preservation of molds by the lyophil process. *Mycologia*, **37**, 499–525.

Thom, C., and Raper, K. B. (1945). *A manual of the Aspergilli.* Baltimore: The Williams & Wilkins Co.

Wehmer, C. (1918). Über Fumarsäure-Gärung des Zuckers. *Ber. dtsch. chem. Ges.*, **51**, 1663–8.

Wehmer, C. (1928). Abnahme des Säuerungsvermögens und Änderung des Säure bei einem Pilz (Gluconsäure- statt Fumarsäure-Gärung). *Biochem. Z.*, **197**, 418–32.

Weston, W. A. R. Dillon (1933). Sporulation of *Helminthosporium avenae* in artificial culture. *Nature, Lond.*, **131**, 435.

Wickerham, L. J. and Andreasen, A. A. (1942). The lyophil process: its use in the preservation of yeasts. *Wallerstein Labs Commun.*, **5**, 165–9.

Chapter XIV

The Control of Mould Growth

> Probably the prevalent opinion about fungi, held even
> amongst men of science, is that they are responsible for a
> hideous storm of terror. They ravage crops and forests,
> attack animals—birds, fish, insects and men—destroy
> stored products, foods of all kinds, buildings and fabrics,
> and spoil paper and paintings. Indeed they appear to be one
> of the chief enemies of man, his possessions and his culture.
>
> J. Ramsbottom, *Mushrooms and Toadstools*, 1953

The growth of moulds in undesirable places is the cause of almost incalculable loss in industry and of less serious, but still extensive, damage in many households. It is therefore desirable that the available methods of combating and limiting their ravages should be as efficient as possible and generally known.

Much may be done to minimize trouble from moulds in the factory by strict attention to hygiene. Damp and dirty walls, corners of floors and ceilings which are not easily cleaned, or waste organic material left lying about are all liable to develop patches of mould and so to act as reservoirs of infection, the spores being carried to all parts of the factory by air currents. Leaky steampipes or water condensed from process steam on ceilings and parts of machinery may cause drips on to valuable material, creating conditions eminently suitable for the growth of fungi. Suitable precautions to prevent infection from such sources as these are obvious. In some cases, where the neighbourhood is excessively dusty, or where there is a known local source of airborne infection, and where, at the same time, the manufactured product is of sufficient value to justify the expense, it may be advisable to purify the air entering the building.

Even with ideal factory conditions, however, it is seldom possible to prevent entirely the access of mould spores to factory products. Latent infection may be carried in raw materials, complete sterility of the air in a workroom is almost impossible to achieve owing to air currents created by the movements of materials and workpeople, and the latter may introduce spores in large numbers on their clothing. Therefore, even if damage by moulds within the factory be completely prevented, there remains the problem of ensuring that products remain in good condition after they leave the factory and until they eventually reach the consumer. There are four main lines of attack on this problem:

1. To ensure continuous sterility by preventing the access of mould spores to an already sterile or to a sterilized product.
2. To arrange that the material to be protected is kept, or keeps itself, in such physical condition that growth of moulds is severely limited or prevented entirely.
3. To limit or prevent mould growth by means of toxic substances, usually known as antiseptics.
4. To manufacture goods from materials which are highly resistant to attack by moulds.

Sterilization methods. The first method is of very limited application but is usually, and should be always, completely effective in cases where it can be carried out. A detailed account of sterilization procedures, including much of the theoretical background of the subject, is given by Sykes (1965). The best-known application of the method is, of course, the food-canning industry. In the canning of meat, fish, and certain vegetables, the main object of sterilization is to destroy bacteria but, since moulds are less resistant to heat than are the majority of bacteria, any cooking operation which will destroy the latter will also automatically kill any moulds present. In the fruit-canning industry, however, high temperatures and long cooking have to be avoided, as they detract from the appearance of the products. Fortunately the acidity of most fruits is of great assistance in suppressing bacterial growth and processing can therefore be conducted at temperatures below 100° C. As already stated in Chapter V, one fungus, *Byssochlamys fulva*, has caused considerable trouble in the fruit canneries, owing to its ability to withstand a temperature only a little below the maximum attained in processing. This means, of course, that there is only a very small temperature difference between incomplete sterility and overcooking, necessitating accurate temperature control and great care being taken to ensure that all parts of the contents of the cans reach the safe temperature (see Olliver and Rendle, 1934).

One branch of the food industry, the packing of fruit for export, applies the principle of isolation in a special way. The moulds which rot fruits are specialized parasites and thus, if the access of spores of *Penicillium digitatum* and *P. italicum* can be prevented, the main cause of spoilage is removed. A good deal of work has been done on the surface sterilization of citrus fruits by means of antiseptic washes, chiefly solutions of borax or boric acid. A fair measure of control is thus obtained, but a small proportion of fruits remain infected. If packed without further precautions infection may spread and result in serious loss, but, if the fruits are wrapped individually in waxed or chemically treated papers, any infected fruits may be extensively rotted and covered with a thick growth of mould without infecting any other fruits in their vicinity. A good account of the control of citrus moulds is given by Fawcett (1936). Wrapping of apples and pears has similarly diminished to a remarkable extent losses due to *Penicillium expansum* rot. It is not unusual to find, in a case of wrapped apples, one or two fruits which are nothing but a mass of mushy brown pulp, with patches of green coremia of the mould, whilst the rest are quite unaffected.

A method of partial sterilization which has received considerable attention during recent times utilizes the lethal properties of ultraviolet radiation. It has been found that both mould spores and bacteria are rapidly killed by exposure to a source of light rich in ultra-violet rays, the most resistant spores being, as might be expected, those with dark coloured walls (see Conklin, 1944). It is somewhat doubtful how far the effect is due to the direct action of the rays and how far to the action of ionized oxygen produced in the neighbourhood of the source of light. James (1936), for example, claims that direct irradiation is unnecessary, and that air circulation during the treatment is beneficial, indicating that it is ionized oxygen which is the potent factor. In any case the effect is exerted only on the surface of the irradiated material, since neither ultra-violet rays nor ionized gas can penetrate far into most kinds of matter. This is of little consequence for such materials as bakery products, which are usually sterile as they leave the oven and become infected only on the surface during cooling. Such goods, after irradiation and wrapping in irradiated paper, are virtually sterile throughout and remain in good condition until they reach the consumer. The chief applications of the method at present are to various foodstuffs and tobacco, its use for other materials being limited by questions of cost and difficulties of sterile packing. Many papers on the subject have been published, chiefly in the journals devoted to the food industries, but these cannot be adequately summarized here.

Physico-chemical methods. In attempting to limit or prevent mould growth by physico-chemical means the most important factor which must be controlled is the amount of available moisture. Spores of fungi cannot germinate without moisture and, therefore, if the amount of water in any material can be kept below a certain critical point, fungoid growth will be entirely prevented. This critical point varies for different materials and must be determined experimentally in each individual case. The reason for such variation is that the absolute amount of water in any material seldom corresponds with what may be termed the available amount, the latter depending on the chemical constitution and physical state of the particular substance. For example, the "safe" limit of moisture in wool is much higher than the amount allowable in cotton, the difference being accounted for by their very different attractions for water. If dry samples of wool and cotton are exposed to the same atmosphere the wool will take up approximately twice as much moisture as the cotton and, leaving out of account differences due to chemical composition, the two samples will be approximately equally liable to become mildewed, or, if the respective percentages of moisture are below the respective safe limits, will be equally resistant. According to Galloway (1935) the minimum relative humidity permitting growth varies from 75 per cent to 95 per cent for different species of moulds. Adequate protection, therefore, is obtained only if materials are stored in an atmosphere of R.H. less than 75 per cent. Tests on numerous types of materials have shown that a few species of moulds may grow very slowly at 75 per cent R.H., and that the safe limit is the moisture content attained when the material is in equilibrium with an atmosphere of 70 per cent R.H.

The method of obtaining the necessary data for control of mould growth by this means is, first, to determine the relationship between atmospheric relative humidity and moisture content of the particular material over any desired range of temperature, then to expose samples infected with mould spores to atmospheres of predetermined humidities. In this way is found the lowest relative humidity, and hence the lowest percentage of moisture in the material, at which mould growth commences. The infection may be that naturally occurring on any particular raw material, or, preferably, the tests may be carried out with single organisms which are known to cause damage in practice. A simple method of conducting such tests in the laboratory is to suspend small samples over solutions of sulphuric acid of known strengths, contained in stoppered bottles or desiccators. A series of such containers can be accommodated in an incubator in order to obtain accurate control of temperature. The following table will be found useful by those wishing to undertake tests of this kind. It is compiled from the data given by Wilson (1921), and covers only the range which is useful for work on moulds.

R.H. per cent	H_2SO_4 per cent	Density at 25°C
35	50·9	1·400
50	43·4	1·325
65	36·0	1·265
75	30·4	1·218
90	18·5	1·128
100	0·0	0·997

The figures, when plotted, give smooth curves, making it possible to obtain accurate intermediate values. The data refer to a temperature of 25° C, but, for any particular strength of acid, there is very little change in R.H. with variations of temperature of up to $\pm 5°$.

Laboratory tests give useful and necessary information but, in attempting to apply the results in practice, complications may arise, chiefly due to temperature variations. If any hygroscopic substance is freely exposed to the air, the amount of moisture in the substance, when equilibrium has been reached, is a determinable function of the relative humidity of the air, but a great many industrial products are transported, stored, and marketed in packages which are airtight or through which diffusion of air is extremely slow. The great majority of solid substances tend to lose moisture as the temperature rises, whereas the capacity of the air for holding moisture is enhanced with increasing temperature. Two questions then arise, for the answering of which we have far from adequate data at the present time. How does the moisture distribution inside an airtight package, which contains solid material together with a comparatively small weight of air, vary as the temperature is changed? Also, can spores of moulds acquire the moisture necessary for germination from a humid atmosphere, even though the substrate with which the air is in contact is comparatively dry? An interesting paper by Galloway (1935) attempts to answer the second of these questions. He made use of the fact that cellulosic materials, exposed to an atmosphere of definite humidity, contain different percentages of moisture according as they are being dried from a wet state or moistened from the dry state. He was thus able to use substrata which contained different percentages of water when exposed to the same atmosphere, but were otherwise identical, and also materials, otherwise identical, which contained the same percentage of moisture when exposed to atmospheres of different humidities. The experiments showed that atmospheric moisture is more effective than moisture in the substrata for bringing about the germination of mould spores. This fact may possibly supply the explanation of a number of mysterious cases of damage by moulds, when the spoiled material has been apparently too dry to permit of growth, but it is impossible to generalize until we have an answer to the first of the above questions and until we have a means of extending Galloway's experiments over a greater range of the variables. There are promising fields for research along these lines.

Another factor which tends to lower the permissible limit of moisture in stored products is the possibility of local variations of temperature within the package. If one side of an airtight container be exposed to a source of heat, such as direct sunlight, a ship's boilers, or heating pipes in a warehouse, whilst the opposite side is kept comparatively cool, it is reasonable to expect that migration of moisture will occur, and the extra water acquired by the cool portions of the package may be sufficient to cause germination of dormant spores. Against the possibility of damage arising in this way a manufacturer has no certain means of protection. The utmost he can do is to allow as large a safety margin as possible in the total moisture content of his product, and to urge merchants to take suitable precautions during transport and storage of the same. A practice which is finding increasing use for goods shipped to the tropics is to pack the materials in airtight containers along with a suitable receptacle containing a desiccant. The best dehydrating agent for this purpose is undoubtedly silica gel, as it is clean in use and can readily be reactivated by heat.

Another method of preventing mould growth, but one, however, which can be used only in special cases, is to limit the amount of *available* moisture in moist materials by arranging that there is a high concentration of soluble matter, thus ensuring that the water present is in the form of a solution of very high osmotic pressure. The best-known example of this method is the manufacture of fruit preserves. It has been found from experience that the concentration of sugar in the finished article must be close to 65 per cent in order to ensure sterility. This percentage includes, of course, the natural sugar of the fruit, and usually means about 60 per cent of added sugar. With amounts only slightly less than this a few cosmopolitan species of moulds, notably members of the *Aspergillus glaucus* series and *A. candidus* can grow quite well. Pickling in brine is another process relying for its efficiency on the same phenomenon.

A method of protecting certain foodstuffs from bacterial attack, namely, cold storage, is effective in preventing damage by moulds only if the temperature is considerably below freezing-point. A fairly large number of species of moulds can grow at $0°$ C, or even at slightly lower temperatures, and a few will grow at $-6°$ C or even lower (see Chapter XII).

Antiseptics. The control of mould growth by the use of antiseptic substances has been practised, or at least attempted, for some considerable time. It is, however, only during the last thirty years that organized research on the subject has put the practice on a scientific basis, explained many failures of the past, and made available new substances of high antiseptic value and general applicability.

It should be stated at the outset that antiseptics are not necessarily fungicides. Fungus spores are often very resistant to toxic substances, and most substances which will kill them are unavailable for use in industrial products because they are corrosive, give off objectionable odours, or are highly poisonous to human beings. Antiseptics merely inhibit growth, and their action is limited by a number of factors, such as the availability of nutrient present along with the antiseptic, the amount of available moisture, the concentration of antiseptic substance and, in some cases, time. In certain cases, such as wood preservation, it is possible to use powerful antiseptics in high concentration and effect complete protection, but often there are technical objections to the use of mould preventives in really efficacious amounts, as well as questions of cost, and the most that can be achieved is to prevent mould growth under more or less normal conditions of use and storage.

The concentration of antiseptic required to prevent growth usually increases rapidly as the humidity of the atmosphere increases, and there are a number of antiseptics which are effective at normal humidities but which become entirely useless when the relative humidity exceeds 95 per cent. There are many antiseptics which at very low concentrations, actually stimulate the growth of fungi. With somewhat higher concentrations germination of spores is delayed and, with still larger amounts, completely inhibited. A study by Morris (1927) of a large number of antiseptics showed some interesting differences in behaviour when these were incorporated in substrata otherwise favourable to mould growth. There were naturally large differences in the amounts of the various antiseptics required for a definite retardation of growth, but, apart from this, it was found that the curves relating concentration of antiseptic to degree of inhibition were of different types. With some substances the effect was strictly, or almost strictly, proportional to concentration, whilst with others there seemed to be a maximum effect, short of complete inhibition, which could not be exceeded by the use of higher concentrations of the antiseptic.

An important point which is often overlooked is what may be termed the specificity of antiseptics. A substance may give adequate protection so long as the potential infection is limited to certain species of moulds, but may fail completely if spores of another species are introduced. For example, salicylic acid is a very efficient preventive of the growth of most of the dark coloured moulds and of many species of *Penicillium*, but can actually be utilized as a source of carbon by *Aspergillus niger*. Zinc chloride is a popular and normally efficient antiseptic for certain kinds of textiles, but, in the concentrations generally used, has no effect on *Aspergillus terreus*. Morris (1927) found that salts of thallium are of high efficiency against most of the species used in his tests, but very much less effective against *Cladosporium herbarum* and *Alternaria tenuis*. Hexamine, on the other hand, has comparatively low antiseptic value against most fungi, but is particularly effective against the few moulds which can grow in the presence of thallium. It is advisable, therefore, when considering the question of an antiseptic for protecting any particular product, to find first of all what are the usual species of moulds which infect the product, and then test out various antiseptics with these species.

A further kind of specificity, which is often important, is the effect of the substrate on the efficiency of an antiseptic. It has been noticed by many workers that the order of efficiency of a number of antiseptics, when these are applied to a certain product, is often completely changed when they are used to protect a different material. It is therefore unsafe for one industry to place too much reliance on results obtained with the very different materials of another industry. Also to be taken into consideration is the fact that some antiseptics are effective only in hygroscopic materials, that is, substances which normally contain a certain percentage of water which fluctuates with the relative humidity of the air. These antiseptics are quite useless in such materials as oil paints and lacquers.

Another phenomenon, which has useful applications, is known as "synergism", this being, briefly, the enhancement of the inhibiting effect of one antiseptic by the presence of another. Many examples of synergism are known, of which one may be quoted as being of special interest. During the second world war the demand for "Shirlan", for protecting clothing, etc., going out to the jungles of Asia, exceeded the potential supply. A search for substitutes showed that a substance used as an accelerator in the vulcanization of rubber, mercaptobenzthiazole, was approximately as effective as "Shirlan". Further tests showed that if the two antiseptics were mixed in equal proportions, the total amount could be reduced by about 40 per cent without reducing the degree of protection.

The actual testing of antiseptics, so as to give results of value, is by no means a simple matter. A usual method of obtaining a preliminary valuation is to grow a number of species of moulds, in the laboratory, on media containing varying percentages of the antiseptic, and to measure growth rates. The basal medium used may be one of the usual agar culture media or, better, a medium containing the particular material to be protected as sole source of nutriment, the material being used either in the form of a stiff paste made from shredded material, or as a fine suspension in a plain agar gel. To begin with, a number of media are prepared containing, say, 0·01, 0·05, 0·1, 0·5, and 1·0 per cent of the antiseptic, respectively. A series of Petri dishes are prepared with equal quantities of the various media, at least two and preferably more dishes for each combination of species and concentration of antiseptic, and single colonies of the various species are planted centrally in the dishes. Rates of growth are recorded as colony diameters, usually with comments on the type of growth, such as density of mycelium and degree of sporulation. Fuller details of the method are given by Morris (1927). In this way a fair comparison between different antiseptics may be achieved, and some idea obtained of the amounts required to suppress growth. The method, although often the only one available for preliminary tests, is, however, open to the objection that such test media contain a very high percentage of water, and the conditions are thus not comparable with those obtaining in practice. It also ignores the fact, mentioned above, that different materials (in this case the culture media and the material to be protected) may require different antiseptics. It is therefore necessary before the true value of an antiseptic can be assessed, to make further tests under works conditions. In some industries, such as the manufacture of toilet preparations, incorporation of antiseptic with product can be carried out with very small batches of material, and a number of tests can be run at comparatively little expense. Even here, however, the preliminary tests, as outlined above, are advisable, as they give results quickly, whereas experiments with industrial products for export may take months. With such materials as textiles, where the antiseptic has to be added to size or dressing, a single factory experiment may be costly and dislocating to normal work, so that it is absolutely necessary to obtain as much information as possible from laboratory tests.

It remains to give some details of the substances in common use in industry as antiseptics. A more extensive list, though not quite up to date, is given by Frear (1949), a list compiled especially for the leather industries is given by Turner et al. (1948), and the antiseptics particularly suitable for cellulosic fabrics and similar materials are listed by Skelly (1957). A comprehensive review of industrial fungicides available in the U.S.A. has been published by Wessel and Bejuki (1959), and there is a more recent, though not exhaustive, survey of commercial products from a number of Western countries by Hueck-van der Plas (1966). The British Wood Preserving Association prepares a biennial *Wood Preservatives and Fire Retardants Register*, which lists the majority of wood preservatives available in Great Britain.

INORGANIC SUBSTANCES

Zinc chloride was originally suggested for the preservation of wood by Sir W. Burnett in 1838 (E.P. 7747). For this purpose it has been entirely supplanted by creosote preparations and copper compounds, but it has found extended use in the cotton industry as an ingredient added to the size which is pasted on to the warp threads to assist weaving. It has to be used in fairly high concentrations, usually reckoned as 10 per cent on the weight of adhesive substances used in the size, but it is reasonably cheap and has the additional advantage of adding weight to the yarn. It is only moderately effective and, as stated above, may fail entirely when the goods are infected with certain Aspergilli, notably *Aspergillus terreus*. It cannot, of course, be used for fabrics which are to be exposed to the weather, since it is very soluble and hence readily leached out.

Chromium compounds. Chromic oxide, Cr_2O_3, is an excellent antiseptic when used in fairly high concentration. The so-called mineral khaki, extensively used for military fabrics, is composed of ferric and chromic oxides, and materials dyed with this mixture shows a very high resistance to attack by moulds. Chromic oxide can be used alone in cases where a pale blue-green colour is not objectionable, the amount required corresponding to approximately 1 per cent of chromium, calculated on the weight of the material. For fabrics which have to come in contact with the skin it is essential that the impregnation be carried out in such a way that not more than a trace of chromate ion is left in the material, as hexavalent chromium is liable to cause dermatitis.

Chromium fluoride is used for proofing woollen goods. About 1 per cent on the weight of the material confers not only protection against moulds but also a high degree of resistance to attack by moths and other insects.

Copper compounds. Copper sulphate has been suggested in the past for various trades, but was never popular. On the whole it is, weight for weight, more efficient than zinc chloride, but is unreliable in the same way as the latter and is more expensive. A few species of moulds are extremely tolerant of copper ions, having been found growing in the concentrated solutions used for electrotyping.

Copper naphthenate has found extensive use, either alone or mixed with other antiseptics, for the preservation of timber. This and other organic compounds of copper, such as the oleate and ricinoleate, are used for the protection of fabrics which normally come into intimate contact with soil, such as sandbags. These compounds are very efficient antiseptics against soil-borne bacteria and certain soil fungi, but are much less effective against common moulds. If used on materials which are exposed to the weather they are, like the inorganic compounds of copper, gradually leached out, and, in addition, they accelerate the tendering effect of strong sunlight on cellulose.

Copper carbonate and basic carbonate in various forms, such as the well-known Burgundy and Bordeaux mixtures, are widely used for spraying plants as a preventive of mildew. Pasteur was the first to introduce the use of such a mixture at a time when the vineyards of France were threatened with ruin by the rapid spread of the vine mildew.

Mercury salts. These, though quite effective, are little used on account of their well-known poisonous nature, and the fact that the metal is readily deposited from solutions of the salts by other common metals.

Thallium salts are powerful antiseptics but are little used, partly owing to their high cost and partly due to the fact that they are very poisonous. Thallium carbonate was used in the cotton industry for a short time before its use was rendered unnecessary by the advent of the much cheaper and equally effective "Shirlan".

Sodium silicofluoride behaves somewhat erratically on most materials but has been widely recommended in the past, and has reappeared on the market at various times under names which give no hint of its chemical composition. One disadvantage for some purposes is its low solubility in water. It has been found to be one of the best antiseptics for the protection of raw rubber, the material being soaked in a 0·25 per cent solution.

Boric acid and borax. Both substances are of low toxicity to most moulds but find a limited application in some industries where highly poisonous substances are debarred. They are used, either singly or in combination, as a wash for citrus fruits to minimize attack by the common rot organisms, *Penicillium digitatum* and *P. italicum*. A number of papers relating to their use are to be found in the various agricultural journals published by the exporting countries.

ORGANIC SUBSTANCES

The number of known compounds having antiseptic properties is very large and is constantly being added to. Many, however, are too expensive for most purposes, whilst others have very limited applications owing to pronounced colour, odour, volatility, or instability, with the result that there is only a comparatively small number of substances which can be recommended for use in industry.

Creosote finds its main application for the preservation of wood. When merely brushed on it gives partial protection to wood which remains above ground but for more severe conditions, in particular for timber which is to be partially buried in the soil, or which is to be used in contact with the soil, the antiseptic is applied under pressure, sometimes following vacuum, or at a temperature sufficiently high to boil the sap out of the wood. Its use for railway sleepers, pit-props, piles, and telegraph-poles has meant increased safety for large numbers of workers as well as an enormous saving in the cost of replacements. Its powerful and somewhat objectionable odour, even when it is partially purified as in certain proprietary preparations, its dark colour, and its action on the skin prevent its more general use.

There is an interesting fungus, often called the "creosote fungus", which grows on creosoted wood, and which can actually utilize creosote and coal tar as sources of carbon (Marsden, 1954; Christensen *et al.*, 1942; and Hendy, 1964). It is a species of *Cladosporium*, but is sometimes referred to the genus *Hormodendrum*, as *H. resinae* (see pp. 96, 97).

Organomercurials. A large number of organic compounds of mercury are known and all are very potent antiseptics. The alkyl compounds are appreciably volatile at ordinary temperatures and are mainly used in agriculture for the control of seed-borne parasitic fungi. They are applied as dusts containing large amounts of inert fillers, in order to ensure even distribution over the seeds, and are mostly sold under trade names. The aryl compounds are not volatile except at high temperatures and a number of them have been suggested for protection of industrial materials. Colin-Russ (1940) tested 19 antiseptics for their value in protecting leather and shoe-lining materials, the best being "merfenil" (phenyl mercuric nitrate), He states that this is effective at a concentration of 0·00175 per cent, whereas the amount of "Shirlan" required for equal protection was 0·1 per cent. Marsh and Duske (1942) recommend phenyl mercuric acetate as the ideal antiseptic for military fabrics, the procedure being to soak the material in a 1:2000 solution, wring out and dry at room temperature. It is stated that the substance is cheap to apply, that it is not readily washed out, and that there is no danger of dermatological action. It is unusual, however, to carry out large-scale drying of processed fabrics at room temperature and Marsh and Duske give no data regarding the effect on the antiseptic of drying by normal methods. A serious disadvantage of the organic mercurials is that, although they are not ionized, they are readily decomposed, at temperatures above about 40° C, by contact with common metals such as copper, aluminium, and iron. This precludes their use in goods which are dried, after processing, by passage over heated metal cylinders. As regards the question of dermatological action, there is at present insufficient information available to decide whether there is any danger attending the use of these substances in fabrics which come into intimate contact with the skin. The alkyl compounds are known to have a vesicant action and must be handled carefully, but the aryl compounds are much less active in this respect, and the concentrations necessary for suppression of mould growth are extremely low.

When carrying out tests with these substances it is necessary to take one precaution. Traces of the substances cling persistently to glass, making it necessary to decontaminate all glassware which has been used for the tests by boiling in nitric acid.

Organotin compounds. Several organic compounds of tin, notably triphenyltin acetate and tri-*n*-butyltin oxide, have been used as fungicides in textiles, plastics, adhesives and various other materials. In many situations they have proved to be very effective but, like the organomercurials, are suspect on the grounds of mammalian toxicity.

Phenol is probably more efficient for preventing bacterial rather than mould growth. Nevertheless, it shows a fairly high toxicity to most fungi. In an acid medium its odour is distinctly noticeable, and objectionable for some purposes, whilst as alkaline phenate its antiseptic value is lower.

Chlorophenols. The halogen-substituted phenols are very much more efficient antiseptics than phenol itself, the best, and most readily available, being pentachlorophenol and its water-soluble sodium salt. It is used in a variety of products, such as leather, cordage, packing papers, and even in some cosmetics. In recent years it has also come to be widely used in mould-proof paints and varnishes. Unlike "Shirlan", it is quite effective in oily materials, and, although the pure substance is appreciably volatile at ordinary temperatures and has a somewhat pronounced odour, both volatility and odour are markedly reduced when the substance is incorporated into any material which dissolves it. Lauryl pentachlorophenate (LPCP) is also widely used, particularly as a textile preservative. It is considerably less toxic to man than many other chlorophenol derivatives.

Dichlorophen (5:5'-dichloro-2:2'-dihydroxydiphenylmethane). This is a popular fungicide on the grounds of its low human toxicity, and it has found many applications in contexts where this is a necessary property. It is, however, generally not so active as the powerful organometallic compounds, although it has been used in place of organomercurials as a slimicide in the paper industry, whenever the toxicity of the mercury compounds is unacceptable.

8-Hydroxyquinoline. This, in the form of its copper complex, is used for the protection of canvas, rope, nets, etc., which have to withstand the leaching action of water, usually as rain or sea-water. It is soluble in many organic solvents and can be applied to some materials from such solution. Alternatively it is obtainable as an emulsion, which can be diluted with water for application. Treatment of materials should aim at putting on approximately 0·18 per cent of copper as the quinolinolate. In this country a full range of suitable preparations is marketed by Cuprinol Ltd., under the trade name "Cunilates".

Paranitrophenol. This is a valuable antiseptic for use in products in which its colour is no detriment. The free acid has a pale, though definite, yellow colour, whilst the alkaline salts are intensely yellow. It is one of the very best antiseptics for protecting vegetable-tanned leather, a material which, in its normal state, becomes mouldy very readily. If the antiseptic

is added by the tanner it may be incorporated before the leather is greased
or in the grease itself, the amount required for full protection being 0·3 per
cent on the weight of the leather. Goods manufactured from untreated
leather may be sprayed or brushed with a 1 per cent solution of paranitrophenol in 50 per cent aqueous alcohol, with the aim of putting on 0·3 per
cent. The antiseptic is also used for the protection of raw rubber, the
material being given a prolonged soaking in a 0·15–0·3 per cent solution.

Magnesium dehydroacetate. Various metallic salts of dehydroacetic
acid have been suggested as possible biocides, but magnesium dehydroacetate is probably the only one as yet to find a practical use. This almost
insoluble compound can be introduced into composition cork during
manufacture, imparting good resistance to mould attack.

Salicylic acid has been much recommended but has never proved to
be entirely reliable in use. When tested in the laboratory against a repre-
sentative collection of moulds, it has been found to inhibit many species
completely, even in low concentration, whilst a few species, notably the
Aspergillus niger series, are little or not at all affected, and, as stated above,
can actually utilize the acid in a culture medium containing no other
source of carbon.

Shirlan, the anilide of salicylic acid, has been patented as an antiseptic
by the British Cotton Industry Research Association (1928). It was selected
by them, after testing a very large number of organic compounds, as the
most efficient antiseptic for use in the cotton industry for exported goods,
the requirement being for an antiseptic which is colourless, odourless, non-
volatile in steam, and unaffected by other materials used in sizing and
finishing or by contact with common metals. Because of its high efficiency
and reasonable cost it has found increasing use in many industries other
than the cotton industry. As mentioned above, during the second world
war the demand became so great that it completely outstripped potential
supply, and substitutes had to be sought.

It is marketed in two forms, the free acid (Shirlan), which is only slightly
soluble in water, and the readily soluble sodium salt (Shirlan NA). The
amount required for adequate protection of cotton goods exported in the
ordinary way is approximately 0·05 per cent on the weight of the goods.
During the war, when goods had to be protected against "jungle" conditions, the amount was raised to 0·1 per cent. This quantity is also sufficient
for most of the materials in which the antiseptic is used. As already indicated,
Shirlan is not effective in non-hygroscopic substances which have an oily
or greasy base.

Mercaptobenzthiazole. This is the best of several substances which
are used as accelerators in the vulcanization of rubber and which have
antiseptic properties. It is used for the same purposes, and in the same
amounts, as Shirlan.

Diphenyl. This substance cannot be regarded as a satisfactory antiseptic
for general use, since it is effective only against a very limited number of

species of moulds. It does inhibit the species of *Penicillium* which parasitize citrus fruits and, therefore, finds its chief use for impregnating the paper used for wrapping such fruits.

Betanaphthol. This has been recommended by Colin-Russ (1940) for leather. It is not so efficient as paranitrophenol but, if the latter substance is in short supply, may be used, but must not be expected to give the same protection under severe conditions.

(Aryloxythio)trichloromethanes. Fawcett, Spencer and Wain (1958) have tested a number of these interesting compounds, and found that the most useful of them is (2:4:5-trichlorophenoxythio)trichloromethane. It is appreciably volatile and has given promising results in controlling downy mildew of lettuce (*Bremia lactucae*), for protecting apples in storage, and for preventing the growth of *Penicillium italicum* on oranges.

5:6-dichloro-2-benzoxazolinone. This is described by Dahl and Kaplan (1961) as a very efficient antiseptic for leather. Vegetable-tanned leather is adequately proofed by addition of 0·4 per cent of the substance. The chemical is stable, colourless, non-volatile, and relatively resistant to leaching by water.

Captan [N-(trichloromethylthio)-4-cyclohexane-1:2-dicarboximide]. This substance has been used for some considerable time in agriculture. More recently it has been tested by a number of investigators for its possible use in industrial products. At present it cannot be stated whether it has any advantages over the more usual anti-mould agents.

Sorbic acid. This is one of the rare antiseptics which is completely non-toxic to the animal organism. A series of seven papers, published simultaneously in 1954 (Deuhl *et al.*, Melnick *et al.*, Smith and Rollin), discuss all aspects of its use, particularly for impregnating the wrappings of packaged cheese. In experiments with rats and dogs it was shown that, if sorbic acid is ingested with food, it is rapidly metabolized and has no ill effects whatsoever. When cheese is wrapped in impregnated paper there is fairly rapid migration of the antiseptic from paper to cheese. If used in adequate amount, 2·5–5·0 g per 1000 sq. in., there is no loss on storage, but, if smaller quantities are used, mould growth occurs, and this increases rapidly as it destroys the sorbic acid.

ANTISEPTICS FOR FOODSTUFFS

The number of antiseptics allowed in foodstuffs is, rightly, very limited. It has long been recognized that protection by means of antiseptics is no substitute for care and strict hygiene in manufacture, and suitable facilities for low-temperature storage and transport. The official attitude in this country has always been that particular antiseptics which may be used should be specified, and the allowable amounts defined, rather than that a list of forbidden preservatives should be issued.

For many years the only antiseptics allowed were sulphur dioxide and benzoic acid, and these only in a limited number of foods. A Food Standards Committee was set up in 1947 to advise the appropriate government departments, and in 1951 a Preservatives Sub-committee was appointed. The report of the latter was published in 1959, and the recommendations therein have since been implemented. Many foodstuffs are, of course, spoiled by bacterial rather than fungal action, but the number of foods which are attacked by fungi is still considerable.

The list of foods which may contain preservatives is a long one, but the number of permitted antiseptics is still very limited. The most widely used antiseptic is sulphur dioxide, this being allowed in a variety of fruit and vegetable preparations, including both alcoholic and soft drinks. Strict limits are imposed on the concentrations allowed in the foods as consumed, larger amounts being permitted in raw materials which are subsequently processed by cooking. In a number of cases benzoic acid may be substituted for an equivalent amount of SO_2. Methyl p-hydroxybenzoate and propyl p-hydroxybenzoate are allowed as substitutes for benzoic acid in certain products, such as rennet, flavouring syrups, and liquid coffee extracts. Bread may contain up to 0·3 per cent, of the weight of the flour, of propionic acid, usually added as the equivalent amount of sodium or calcium propionate. Sorbic acid, up to 1000 ppm, is allowed in flour for confectionery, cheese, and marzipan. The use of diphenyl, or o-phenylphenate, in wrappings for citrus fruits is limited, since the antiseptic migrates slowly into the fruit. The H.M.S.O. publication should be consulted for full details of the recommendations.

VOLATILE ANTISEPTICS

A number of gases are highly toxic to fungi, formaldehyde, ammonia, and sulphur dioxide being the most important, and one of these may be used when it is required to sterilize a room or other large space.

In addition, a number of liquids and easily sublimed solids show marked antiseptic or fungicidal properties in the vapour phase. During the second world war a number of these were investigated as to their possible use for the protection of sealed optical instruments, such as binoculars and telescopes. The chief of the substances suggested were thymol, Cresatin (metacresol acetate), and the alkyl mercurials. The latter substances were usually incorporated in the black lacquer used to coat the insides of such instruments. Thymol as crystals, or cresatin mixed with three times its weight of ethyl cellulose, was used in small capsules, screwed into the bodies of the instruments in such positions as not to interfere with the optical paths, and fitted on the insides with coarse sintered glass filters, which allowed the passage of vapour but prevented particles of solid becoming detached and falling on to lenses or prisms. The results of tests did not give consistent results, for, although the substances were effective in preventing mould growth on the glass components, they sometimes caused corrosion or filming on the glass, or occasionally attacked the balsam cement between lenses and prisms. However, it is possible that the results obtained may have useful applications in other directions.

The use of resistant materials. Work during the second world war on the mould proofing of military radio sets gradually led to the conclusion that the only satisfactory method is to build sets entirely of highly resistant materials. Many of the materials are, of course, used in various other industries, so that the results obtained during the work on radio sets are of fairly wide application. The following list includes the more important of the resistant materials, together with some comments on some other materials.

Plastics. Thermo-setting plastics are highly resistant. They may develop a thin, sparse growth of *Paecilomyces varioti*, but this is never sufficient in amount to be serious. "Filled" plastic sheets are vulnerable if the filling material is liable to attack by moulds, particularly on the cut ends. Plastic sheet filled with fabric or wood may disintegrate entirely if mould can gain access through exposed filler on the cut ends.

Other plastics vary in resistance, but polythene and most grades of polyvinyl chloride and perspex are satisfactory. However, the degree of resistance shown by these and other plastics depends on the plastisizers used, since some of these support mould growth rather readily (see Berk *et al.*, 1957).

Textile fibres. Cotton, in its natural state, is very liable to attack by fungi. It becomes relatively resistant if scoured by boiling for an hour with 1 per cent soda-ash, but even then should be protected with a suitable fungicide. Acetate rayon is resistant but other rayons require the protection of antiseptic. Pure silk, if completely degummed, is highly resistant.

Recently several commercial processes have been developed for the impregnation of cotton materials with methylolmelamine resins. These processes result in the deposition within the textile fibres of a resistant, insoluble condensate which forms an effective physical barrier to invasion by fungal hyphae. A full review of the method is given by Kempton and Kaplan (1964), who report excellent results from rotproofing trials of properly treated fabrics.

Rubber. Raw rubber readily becomes mouldy but properly vulcanized rubber is very resistant. Most of the synthetic rubbers are also satisfactory.

Bitumen. This does not support mould growth but, if it becomes at all tacky in a hot climate, may attract dust, which will most probably become mouldy.

Waxes. Hydrocarbon waxes are resistant. Beeswax is, even when mixed with preponderating amounts of other waxes, very liable to become mouldy, and its use should be avoided.

Cork. It is used in various ways in ordinary radio sets, chiefly as shock-absorbers, but should be avoided in sets for the tropics. Natural cork is fairly resistant but the usual types of bonded cork support profuse growths of mould. Cork gaskets, where these are essential should, if possible, be bonded with rubber.

Leather. Chrome-tanned leather is usually highly resistant to moulds. Vegetable-tanned leather is one of the most easily attacked of all materials. Semi-chrome leather is intermediate, seldom supporting more than a slight growth of mould, but never completely resistant.

In general, a smooth finish, on any type of material, is better than a rough or tacky surface, since it does not harbour dust.

REFERENCES

Berk, S., Ebert, H., and Teitell, L. (1957). Utilization of plastisizers and related organic compounds by fungi. *Indust. engng Chem.*, **49**, 1115–24.

British Cotton Industry Research Association, Fargher, R. G., Galloway, L. D., and Probert, M. E. (1928). English Patent 323579.

Christensen, C. M., Kaufert, F. H., Schmitz, H., and Allison, J. L. (1942). *Hormodendrum resinae* (Lindau), an inhabitant of wood impregnated with creosote and coal tar. *Amer. J. Bot.*, **29**, 552–8.

Colin-Russ, A. (1940). A contribution to the study and control of mould growth in leather and other materials. *J. int. Soc. Leath. Chem.*, **24**, 395–408.

Conklin, D. B. (1944). Ultra-violet irradiation of spores of certain molds collected from bread. *Proc. Iowa Acad. Sci.*, **51**, 185–9.

Dahl, S., and Kaplan, A. M. (1961). 5, 6-dichloro-2-benzoxazolinone as a leather fungicide. *J. Amer. Leath. Chem. Ass.*, **56**, 686–98.

Deuhl, H. J., Alfin-Slater, R., Weil, C. S., and Smyth, H. F. (1954). Sorbic acid as a fungistatic agent for foods. I. Harmlessness of sorbic acid as a dietary component. *Food Res.*, **19**, 1–12.

Deuhl, H. J., Calbert, C. E., Anisfeld, L., McKechan, H., and Blunden, H. D. (1954). *Ibid.* II. Metabolisms of α, β-unsaturated acids with emphasis on sorbic acid. *Food Res.*, **19**, 13–19.

Fawcett, C. H., Spencer, D. M., and Wain, R. L. (1958). Investigations on fungicides. IV. (Aryloxythio) trichloro-methanes. *Ann. appl. Biol.*, **46**, 651–61.

Fawcett, H. S. (1936). *Citrus diseases and their control.* 2nd Ed. New York: McGraw-Hill.

Food Standards Committee report on preservatives in food. H.M.S.O., 1959.

Frear, D. E. H. (1949). *A catalogue of insecticides and fungicides.* Vol. II. Waltham, Mass.: Chemica Botanica.

Galloway, L. D. (1935). The moisture requirements of mould fungi with special reference to mildew in textiles. *J. Text. Inst.*, **26**, T 123–9.

Greathouse, G. A., and Wessel, C. J. (1954). Ed. *Deterioration of materials. Causes and preventive techniques.* New York: Reinhold Publ. Co.

Hendy, N. I. (1964). Some observations on *Cladosporium resinae* as a fuel contaminant and its possible role in the corrosion of aluminium alloy fuel tanks. *Trans. Brit. mycol. Soc.*, **47**, 467–75.

Hueck-Van der Plas, E. (1966). Survey of commercial products used to protect materials against biological deterioration. *Int. biodetn Bull.*, **2**, 69–120.

James, R. F. (1936). Moulds and bacteria killed by new lamp. *Food Ind.*, June 1936, 295–7.

Kempton, A. G., and Kaplan, A. M. (1964). Evaluation of rot- and weather-resistance of cotton fabrics treated with methylolmelamine resins by wet fixation methods. *U.S. Army Natick Laboratories, Microbiological Deterioration Series*. Report No. 7.

Marsden, D. H. (1954). Studies of the creosote fungus, *Hormodendrum resinae*. *Mycologia*, **46**, 161–83.

Marsh, W. S., and Duske, A. E. (1942). Mildew-proofing military fabrics. *Text. World*, **92**, 58–60.

Melnick, D., and Luckmann, F. H. (1954). Sorbic acid as a fungistatic agent for foods. III. Spectrophotometric determination of sorbic acid in cheese and in cheese wrappers. *Food Res.*, **19**, 20-27.

Ibid. IV. Migration of sorbic acid from wrapper into cheese. *Food Res.*, **19**, 28–32.

Melnick, D., Luckmann, F. H., and Gooding, C. M. (1954). *Ibid*. V. Resistance of sorbic acid in cheese to oxidative deterioration. *Food Res.*, **19**, 33–43.

Ibid. VI. Metabolic degradation of sorbic acid in cheese. *Food Res.*, **19**, 44–58.

Morris, L. E. (1927). Mildew in cotton goods. Antiseptics and the growth of fungi on sizing and finishing materials. *J. Text. Inst.*, **18**, T 99–127.

Olliver, M., and Rendle, T. (1934). A new problem in fruit preservation. Studies on *Byssochlamys fulva* and its effect on the tissues of processed fruits. *J. Soc. chem. Ind., Lond.*, **53**, T 166–72.

Skelly, J. K. (1957). Mildew and rot-proofing. *Text. Cord. Quart.*, **7**, 30–35, 37.

Smith, D. P., and Rollin, N. J. (1954). Sorbic acid as a fungistatic agent for foods. VII. Effectiveness of sorbic acid in protecting cheese. *Food Res.*, **19**, 59–65.

Sykes, G. (1965). *Disinfection and sterilization*. 2nd Ed. London: Spon.

Turner, J. N., Musgrave, A. J., and Rose, C. D. (1948). Fungicides for the leather industry. An annotated list. *J. int. Soc. Leath. Chem.*, **32**, 127–43.

Wessel, C. J., and Bejuki, W. M. (1959). Industrial fungicides. *Industr. engng Chem.*, **51**, 52A–63A.

Wilson, R. F. (1921). Humidity control by means of sulfuric acid solutions with critical compilation of vapor pressure data. *Industr. engng Chem.*, **13**, 326–31.

Chapter XV

Industrial Uses of Fungi

> As we study and compare the fungi, it becomes more and more evident that these organisms were not created for the innocent amusement and recreation of the taxonomic botanist.
>
> C. L. Shear, *Proc. Int. Congr. Plant Sci.*, 1929

As an offset to the incalculable damage caused by fungi there are a number of industrial processes in which the biochemical activities of moulds are turned to good account. Only very brief descriptions of these can be given here, but there is an extensive literature on the subject and to this the reader is referred for fuller information. A few such processes have already been mentioned in previous chapters, but, for convenience of reference, are described again here.

Alcoholic fermentation. There are two industries, brewing and baking, which use processes of great antiquity, both dependent on the fact that yeasts convert sugar into alcohol and carbon dioxide. In the brewing of alcoholic beverages alcohol is the important product, but the carbon dioxide, once allowed to escape as useless, is now a valuable by-product, being collected, solidified, and marketed as "dry ice". In baking, on the other hand, the production of alcohol is incidental and it is the carbon dioxide which is valuable, causing the dough to rise and giving lightness to the bread. Both industries were, until comparatively recent times, almost entirely empirical, but are now, in most countries, on a strictly scientific basis. During the last half-century both have built up extensive and highly specialized literatures of their own, and hence do not need any further mention here.

However, one effect, rather than a use, of a mould in the wine industry deserves mention. In the manufacture of sweet wines in the Bordeaux district and in the Rhineland the grapes are attacked by *Botrytis cinerea*, a fungus which infects many cultivated plants, and, in wet seasons particularly, can cause serious losses of soft fruits in this country. Far from being regarded as a nuisance in the wine-growing areas, the disease is called in France "la pourriture noble", and in Germany "Edelfäule", both terms meaning "the noble rot". The fungus causes much of the water of the grapes to evaporate, thus increasing enormously the percentage of sugar, the amount of the latter utilized by the mould being very small. When the wine is made the fermentation stops when the alcohol content reaches

10–12 per cent, and whilst there is still a considerable amount of residual sugar.

Yeasts secrete the enzyme complex often referred to as "zymase", which effects the conversion of sugar to alcohol, but lack diastase, the enzyme which breaks down starch to sugar. There are, however, a number of fungi which secrete a whole range of enzymes and can ferment complex carbohydrates without a preliminary saccharification. In certain processes for the production of alcoholic beverages from starchy materials moulds alone are used, but in others, and particularly in processes for producing industrial alcohol, moulds effect the saccharification of the starch, after which a yeast is allowed to act on the sugar produced, since, although the mould can complete the conversion to alcohol, the yield is better when yeast is used for the second stage. Fitz in 1873 was the first to show that alcohol is produced by a mould, *Mucor mucedo* (later identified as *M. racemosus*), and other workers have since shown that similar results are obtained with several other species of this and related genera. In the "Amylo" process, which is still used in many countries in more or less modified form, *Mucor rouxii* was first used but was later replaced by various species of *Rhizopus*. Good general accounts of earlier practice are given by Lafar (1903), Wehmer (1907), and Galle (1923), whilst an improved "Amylo" process is described by Erb and Hildebrandt (1946).

Various Mucoraceae are also active agents in the so-called starters, used for initiating alcoholic fermentations and marketed in various Eastern countries, chinese rice being the best-known example. Japanese "Koji" differs from most of the other starters in making use of quite a different group of fungi, strains of the *Aspergillus flavus-oryzae* series. *A. flavus* is also one of the active agents in the production of African native beer. In addition, moulds of the same series are used for the production of soy sauce, this being made by fermentation of soybeans, the starter being produced by growing the fungus on cooked beans, usually mixed with some other starchy material to aid rapid growth.

Miso. This is an important commodity in Japan, made by fermentation of soybeans. It is sold as a pale brownish paste, from which is made a potage type of soup, serving as a regular morning dish. According to Sakaguchi (1961) the annual production of miso in Japan exceeds 500,000 tons. Cooked soybeans, "koji" made from rice, and 10–12 per cent of salt, are ground together along with a little miso from a previous batch. The mixture is packed into closed containers and allowed to ferment at 35–40° C, without aeration, for 3–5 months. It is then allowed to age for a month or more at room temperature before being packaged for sale.

Two interesting papers by Shibasaki and Hesseltine (1960, 1961) describe the use of U.S. soybeans for production of miso. These beans are extensively imported into Japan, but, owing to their varying size and hardness, do not cook so easily or evenly, and do not give such a satisfactory product as the Japanese beans. It was shown that it is possible to make

L

satisfactory miso from U.S. beans, using more or less traditional methods, but that it is more satisfactory to remove the hulls and crack the beans into several pieces (not to grind them into flour). The cracked beans cook easily, and the normal fermentation time can usually be reduced by about 50 per cent, resulting in a superior product.

A number of other fermentation products are made in the Far East, far too many to describe here. A most interesting paper by Hesseltine (1965) gives full details of a large number of these products.

Enzyme preparations. As a result of an intensive study of the enzymes of the *A. flavus-oryzae* series, Takamine introduced into commerce a number of products of high enzymic activity, particularly suitable for the dextrinization of starch and the desizing of textiles. The products have been sold under the names "Takadiastase", "Polyzime", "Digestin", and "Oryzyme". Takamine (1914) summarized his work in a short paper, Takamine and Oshima (1920) described the preparation of "Polyzime", whilst Harada (1931) has given a general account of *A. oryzae* enzymes.

In more recent years considerable attention has been given to the possibility of producing amylases by mould fermentation in submerged culture (Adams *et al.*, 1947; Balankura *et al.*, 1946; Le Mense *et al.*, 1947). The fungus used is a selected strain of *Aspergillus niger*.

Mould-ripened cheese. The manufacture of mould-ripened cheese is another industry of unknown origin. Until the present century such cheeses were associated with particular districts, such as the caves of Roquefort and the town of Stilton, and the methods used were entirely empirical, the distinct flavours of certain brands being dependent on a combination of slight local peculiarities in the quality of the curd and of a natural local infection of a particular strain of the all-important mould. It was not until 1906 that Thom's studies of the cheese moulds led to a proper understanding of the ripening process, and, even at the present day, when countless cheeses are being made by pure culture methods, there are certain local blue-veined cheeses whose production is a happy accident rather than a matter to be determined at choice, the nature and source of the marbling being a complete mystery to those who make the cheese.

There are two distinct types of mould-ripened cheese, the soft cheeses of the Camembert and Brie types and the green- or blue-veined cheese, of which Roquefort, Gorgonzola, and Stilton are the best known. In the first type the moulds concerned are *Penicillium camemberti* Thom and *P. caseicolum* Bainier, the two giving somewhat different flavours. The curd is made into cakes 3 to 4 cm. thick, salted on the surface and either inoculated with spores of the fungus or placed in an infected room. The initial moisture in the cakes is 55 to 60 per cent, and the air in the ripening room is maintained at a temperature of 50–60° F and a relative humidity of about 88 per cent. Freedom from infection by undesirable moulds depends on maintaining both temperature and humidity within fairly narrow limits. The mould grows on the surface of the curd and gradually softens the whole mass, the process requiring about four weeks for completion. The most serious source of infection and spoilage is *Scopulariopsis brevicaulis*, which gives to the cheese an ammoniacal taste and odour.

For production of the marbled cheese the raw curd is pressed so as to leave irregular cracks. In the pure culture method a sterile curd is inoculated, before pressing, from a culture of the mould (usually on bread), and is later aerated from time to time during the ripening process by piercing with wires. In the older process natural infection is relied on, and the success of the method depends on the fact that few moulds other than *P. roqueforti* can grow with the small amount of oxygen contained in the narrow air spaces in the curd, and hence the chance of contamination with an undesirable species is small. Although some mycologists consider that distinct species are concerned in the ripening of the various types of cheese, Thom has shown that, for all practical purposes, all the strains may be regarded as one species. However, in view of wide variations in biochemical activity between different strains of single species of other moulds, it is probable that the peculiar characteristics of any of the well-known types of veined cheese is as much due to the strain of *P. roqueforti* used as to the composition and method of preparation of the curd. As far as can be ascertained there has been as yet no adequate study of the effect of using different strains of the mould under otherwise identical conditions.

Oxalic and citric acids. There are a few modern processes which have been developed as the direct result of purely academic research into the biochemical activities of moulds. The first worker to make substantial progress in this field was C. Wehmer, who, in 1891, showed definitely that oxalic acid is a fermentation product of *Aspergillus niger* and made an extended study of its formation from various sugars. Oxalic acid, however, has never been made commercially by this method, because the more usual chemical methods of preparation are cheaper. In 1893 Wehmer described the production of citric acid by two species of moulds which were made the types of a new genus, *Citromyces* (now included in *Penicillium*). Other investigators have found that citric acid is a fermentation product of many species of *Penicillium*, but in no case is the yield sufficiently good to enable this method of production to compete with the old-established process of extracting the acid from citrus fruits. In 1917, however, Currie showed that in the fermentation of sugar by species of the *Aspergillus niger* group there is a distinct lag between the acidity of the medium due to oxalic acid and total acidity, the difference representing citric acid. Methods were worked out for suppressing the formation of oxalic acid and increasing the yield of citric acid, the chief essentials being a high initial concentration of sugar (about 15 per cent), low concentrations of ammonium nitrate as source of nitrogen, and an acid reaction of the culture medium (pH about 3·5 or less). In addition, of course, it is necessary to use a strain of *A. niger* specially selected for the purpose. However, a number of difficulties have had to be overcome before such a process could be worked commercially.

Sterilization of culture media on a large scale is an expensive operation; an abundant supply of air is required and must be supplied in a sterile condition, or, alternatively, the fermentation must be carried out in shallow layers of liquid with free aeration and suitable protection from infection, requiring expensive plant and expert supervision; and, perhaps most important of all, any organism is liable to be erratic in its behaviour, making the question of yields uncertain. A large number of patents have been taken out in this field, but the actual methods at present in use have not been made public in their entirety. Citric acid is certainly being made successfully, by mould fermentation, in this and several other countries, sufficiently cheaply and in sufficient quantities to make them independent of imported acid made from citrus fruits.

Gluconic acid is formed from sugars by the action of a large number of species of moulds, chiefly species of *Aspergillus* and *Penicillium*, and a considerable amount of work has been carried out in the United States in an attempt to develop its large-scale production by mould fermentation. The first fungus to be used was *Penicillium purpurogenum* var. *rubri-sclerotium*, and the fermentation was carried out in shallow pans of pure aluminium (May *et al.*, 1927; Herrick and May, 1928; May *et al.*, 1929). Later it was found that certain strains of *P. chrysogenum* gave better results, and still later selected strains of *Aspergillus niger* were used (Wells *et al.*, 1937; Gastrock *et al.*, 1938; Porges *et al.*, 1941). In the meantime, in connection with this investigation, Herrick, Hellbach, and May (1935) described a semi-large-scale plant for growing moulds in submerged culture, with forced aeration. It was largely the experience with this plant for gluconic acid production which later led to the rapid development of the production of penicillin by the submerged culture method. Gluconic acid is used, in moderate quantities, chiefly as the calcium salt, in place of calcium lactate in medicinal and food preparations.

D-Lactic acid. Most of the work on the production of this acid has been done by American investigators. The fungus used is *Rhizopus oryzae* (Ward *et al.*, 1936; Lockwood *et al.*, 1936). Following a preliminary study of the physiology of the mould, a rapid process was worked out, using a rotary fermenter with forced aeration (Ward *et al.*, 1938). The time for the fermentation was reduced to 30–35 hours, and yields of 70–75 per cent were claimed.

Gallic acid was obtained by Calmette in 1902 by fermenting a clear extract of tannin by means of an organism which he named *Aspergillus gallomyces* (a strain of *A. niger*), the fungus being grown in a well-aerated and agitated liquid. Modifications of Calmette's process have been used, and probably are still being used, both in Europe and in America, but it is very difficult to obtain details of the methods employed.

Fumaric acid. Rhodes *et al.* (1959) showed that fumaric acid is produced, in yields of upwards of 65 per cent of the sugar consumed, when a strain of *Rhizopus arrhizus* is grown in shake-cultures in flasks. A second paper (1962) describes production in 20-litre fermenters. The sugar concentration was 10–16 per cent, and it was found to be essential to neutralize the acid continuously by means of calcium carbonate. The temperature during the fermentation was maintained at 33° C. Yields were equal to those obtained in flask cultures. The acid is required chiefly for the manufacture of plastics and varnishes.

Itaconic acid. This substance was first obtained as a mould metabolic product by Kinoshita (1929), using a species which he called *Aspergillus itaconicus* (*A. glaucus* group), but the amounts obtained were too small for successful commercial development. Of more interest in this connection was a paper by Calam, Oxford, and Raistrick (1939) describing the production of itaconic acid by a strain of *Aspergillus terreus*. In America selection of strains of *A. terreus*, and careful study of culture media and environmental factors, have resulted in itaconic acid being obtained in reasonable yield. In addition, selected strains of the mould have been irradiated with ultra-violet rays, and mutants obtained which give enhanced yields. As in other mould fermentation processes, attention has been directed to the possibility of obtaining the acid by submerged culture methods. For example, Nelson *et al.* (1952) have obtained yields of 45–54 per cent on the sugar consumed, in 4–6 days, on a medium containing 6 per cent of glucose.

Glycerol. It was first shown by Pasteur (1859) that glycerol is formed in small amounts during the normal alcoholic fermentation of sugar by yeast, the maximum yield being about 3 per cent of the sugar consumed. During the first world war the acute shortage of fats in Germany led to an investigation of the process with a view to increasing the production of glycerol. It was found that the yield may be much improved by carrying out the fermentation in an alkaline medium, and still more so by adding to the culture medium sodium sulphite, which has the additional advantage that it inhibits the growth of bacteria without affecting the activity of the yeast. To obtain the maximum advantage it is added in several instalments during the fermentation. The yield of glycerol obtained was about 25 per cent of the sugar consumed and actually something over 1000 tons per month were produced by this method. After the war, details of the German process were published by Connstein and Lüdecke (1919). Later the technical production of glycerol by fermentation was developed in the United States of America, using a specially selected strain of yeast and adding sodium carbonate to the fermenting liquor up to a final concentration of approximately 5 per cent (Connstein and Lüdecke, 1924).

Fats are synthesized from carbohydrates by a large number of fungi. The first fungus to have more than theoretical interest in this connection was *Endomyces vernalis* Ludwig (= *Trichosporon pullulans* (Lindner) Diddens et Lodder), it having been used for the large-scale production of fat in Germany towards the end of the 1914–18 war. According to Haehn and Kinttof (1924), the essentials for successful working are a well-aerated medium rich in sugar and containing only a small amount of nitrogen. They claim yields up to 25 per cent of the dry weight of the yeast cells, or 30 per cent of the sugar consumed, also that the yield can be increased by the introduction of alcohol vapour in the air supply.

A number of American workers (Lockwood *et al.*, 1934; Ward *et al.*, 1935) have investigated the production of fat from glucose by a large number of species of moulds. The largest amount is given by *Penicillium javanicum* van Beyma, which gives yields of fat up to more than 40 per cent of the weight of the mycelium, the best results being obtained with a medium containing 40 per cent of glucose. The production and isolation of the fat have been carried out with semi-large-scale apparatus to demonstrate the feasibility of production on an industrial scale if such were to become an economic proposition.

An account of a very thorough investigation of the possibilities of large-scale production of fat is given by Fink, Haehn, and Hoerburger (1937). They describe in detail experiments carried out at the Institut für Garungs-gewerbe, both with *Endomyces vernalis* and with a number of other organisms. They find that the best yield of fat (30 per cent of the sugar) is given, under favourable conditions, by *E. vernalis*, and that *P. javanicum*, in spite of the high fat content of the mycelium, is unsuitable owing to the low yields calculated on the starting material. Promising results were obtained with *Geotrichum candidum* (= *Oidium lactis*), which, though giving a lower overall yield than *E. vernalis*, gives a better yield on a time/production basis. *G. candidum* has the further advantage that it is practically proof against infection, and so can be grown on unsterilized materials. However, the general conclusion reached by Fink *et al.* is that none of the processes so far described is suitable for peace-time production of fat, owing to the high running costs of the plant required. A more recent investigation, carried out in Vienna at the end of the second world war and published by Lundin (1948), led to much the same conclusions, that micro-biological production of fat is far too costly for use except in times of serious emergency.

An excellent review by Woodbine (1959) covers both past and more recent work, and describes investigations carried out by T. K. Walker and his associates at Manchester, and by the author at Nottingham. His main conclusion is that, whilst microbial fat cannot compete with animal or vegetable fats on the basis of yields and costs, it may become of importance on account of its content of the so-called "essential" unsaturated fatty acids. He also considers that production of fat and/or protein by means of micro-organisms may be of importance in dealing with industrial waste materials and effluents.

Proteins are synthesized by a number of fungi, usually as cell constituents, more rarely in the substrate. In particular, the yeasts *Saccharomyces cerevisiae* and *Candida utilis* (formerly known as *Torulopsis utilis*) contain very considerable percentages of protein of high nutritive value. Since yeasts can be grown on the large scale with ammonia as the sole source of nitrogen and with cheap carbohydrate, such as molasses, as source of carbon, it is not surprising that, during the second world war, a good deal of interest was taken, both in this country and in Germany, in yeasts as a possible supplement to the normal diet. Neither country had a surplus of suitable by-product carbohydrate material, but in Germany the manufacture of yeast for food was considered of such importance that thousands of tons were made, using sugar obtained by acid hydrolysis of wood.

In England a large amount of experimental work was carried out, but manufacture was not undertaken on account of the difficulty of importing the necessary carbohydrate. On the basis of the English experiments, however, manufacture was started in Jamaica, where molasses and surplus raw sugar are available in large amounts.

The organism selected for the large-scale experiments was *Candida utilis*, as this is easier to grow and gives a much more palatable product than *Saccharomyces cerevisiae*. The yeast was subjected to chemical stimulation, in the hope of obtaining mutants with more desirable characteristics than the parent strain, and, as a result, two new varieties were described. *C. utilis* var. *thermophila* grows well at $39°$ C, a temperature considerably above that suitable for growth of the parent strain, and can be used for manufacture in hot countries (Thaysen, 1943). *C. utilis* var. *major*, described by Thaysen and Morris (1943), was stated to have cells approximately twice the size of those of the parent. However, the latest monograph on the yeasts by Lodder and Kreger-van Rij (1952) does not recognize the variety *major*, on the ground that both this and the parent strain show such variation in size of cells that the two cannot be separated. The variety *thermophila* is not mentioned in the monograph.

The manufactured product, known as "Food Yeast", is a straw-coloured powder, containing about 45 per cent of protein, as well as valuable amounts of the B group of vitamins. It is palatable, with a slightly nutty taste, and often imparts an improved flavour to other foods, such as soups, stews, pastry, biscuits, and bread, when added in *small* amounts. Full details of the Jamaica experiment are given in a booklet published by Colonial Food Yeast, Ltd. (1944). The manufacture was started with the idea of using the yeast as an addition to human food in countries where there is a deficiency of protein and vitamins in the normal diet. It is doubtful whether more than a small fraction of the output is now used in this way, possibly because the taste of the yeast, if used in too large amounts, very soon palls. Most of the yeast now manufactured, not only in Jamaica but also in S. Africa and the U.S.A., is used for adding to cattle fodder and the food of other domesticated animals. The position has been summarized by Thaysen (1953) and in Arima *et al.* (1957).

A useful review of the possible methods of production of protein by micro-organisms is given by Fawns (1943).

Vitamins. One of the best sources of the vitamin B complex is yeast, which is, in fact, one of the very few readily accessible foods containing the majority of the known substances comprising the vitamin B group. The increasing recognition of the importance of adequate supplies of these vitamins in the diet has led food manufacturers to put on the market a number of preparations of high potency, made from dried yeast, yeast extracts, or autolysed yeast.

One of the B group, riboflavin, is now made in pure form by fermentation. Two closely related yeast-like organisms are used, *Nematospora gossypii* Ashby and Nowell and *Eremothecium ashbyi* Guill. An interesting account of the occurrence, systematic position, and uses of *N. gossypii* is given by Pridham and Raper (1950), whilst details of the manufacturing process are given by Pfeifer *et al.* (1950). Yaw (1952) describes the production of riboflavin by *Eremothecium ashbyi* when grown on a synthetic medium from which the vitamin is readily isolated. A more recent review of methods of production and biosynthesis is by Goodwin (1959).

Ergosterol, the precursor of vitamin D, is synthesized by a number of moulds as well as by yeasts (Pruess *et al.*, 1931, 1932; Birkinshaw *et al.*, 1931) and there are a number of manufactured preparations of irradiated ergosterol, mostly made from yeast.

Vitamin C (ascorbic acid) has been reported by Geiger-Huber and Galli (1945) as a metabolic product of a strain of *Aspergillus niger*. This is more a curiosity than a possible means of manufacturing the vitamin by fermentation, since there are other and easier methods of preparation.

Very soon after vitamin B_{12}, the anti-pernicious anaemia vitamin, was first isolated from liver, it was found to be present in the broth on which *Streptomyces griseus* had been grown for the production of streptomycin (Rickes *et al.*, 1948). Since then the vitamin has been found to be widely distributed. Most Actinomycetes and some bacteria produce it, but not moulds; it is also present in sewage wastes from the activated sludge process, produced presumably by micro-organisms. At present most of the vitamin which is manufactured is obtained as by-products from the manufacture of streptomycin by *Streptomyces griseus* and aureomycin by *S. aureofaciens*. It has been found, as in other fermentation processes, that exposure of the organisms to lethal agents has furnished mutants which give increased yields. Yields have also been improved by adding to the broth traces of cobalt, usually in several stages during the fermentation. There is already an extensive literature on the subject, too extensive to cite here. One of the best sources of references is the annual review of fermentation processes published in the September numbers of *Industrial and Engineering Chemistry*.

Antibiotics. The term is used to denote substances, produced by living organisms other than ordinary plants and animals, which will inhibit the growth of certain micro-organisms. Some of the substances covered by the term inhibit only a limited number of species of bacteria; others inhibit many bacteria and a variety of fungi; still others inhibit, in addition, some of the pathogenic viruses. The most important of these substances are produced by moulds, Actinomycetes, or bacteria, but a large number of active extracts have been obtained from various species of the higher fungi. The latter, whilst of interest, are not of any practical importance unless the active substance is also produced when the fungus concerned is grown in artificial culture.

The idea of using one micro-organism to combat the evil effects of another is by no means a new one. Pasteur was probably the first to describe clearly the production of an antibiotic, and, during the intervening years, many claims have been made as to the therapeutic value of bacterial products.

The first antibiotic to be widely used, and to fulfil all the claims made for it by the pioneers who first investigated it, was, of course, penicillin. Since the end of the last war, the literature on penicillin has grown to such an extent that it is neither possible nor necessary to attempt to summarize it here. There are several popular books describing the romantic discovery and development of penicillin; there is an excellent account with special emphasis on the mycological side of the investigations, in the Raper and Thom "Manual of the Penicillia"; and there are sections on penicillin, of various degrees of fullness, in the compendia on antibiotics mentioned below.

The success of penicillin, and also the fact that its range of antibiotic activity is limited, naturally led to the search for new substances, produced by micro-organisms, which would act on pathogenic bacteria and viruses which are not affected by penicillin. Many species of moulds have been found to show some antibiotic activity and, in a number of cases, the active substances have been isolated in the pure state and fully characterized. However, most of these are too toxic for use in medicine. The most useful of the new substances are produced, not by moulds, but by species of *Streptomyces*. Such are streptomycin, aureomycin, chloromycetin (chloramphenicol), and terramycin, to mention only the most widely used. Streptomycin is, so far, the only antibiotic which has been used successfully against some forms of tuberculosis, but it is decidedly more toxic than is desirable in such a substance. Aureomycin, chloromycetin, and terramycin all show similar ranges of activity. They not only inhibit growth of a wide range of bacteria, but have been used successfully in the treatment of various Rickettsia and virus diseases.

Two useful antibiotics are derived from a metabolic product of a species of *Cephalosporium*. This was isolated in Sardinia, from the sea near to a sewage outfall. The preparation of one substance, cephalothin (trade name Keflin), has been developed in the U.S.A., and of the other, cephaloridine (trade name Ceporin), in England. Both are marked by low toxicity. The original metabolic product, cephalosporin C, had a number of disadvantages. A large number of chemically modified derivatives were prepared, and the substances undergoing clinical trials are the best of these. A good account of the cephalosporins is given by Abraham (1964).

There are a number of useful books dealing with antibiotics in general, although, of course, none of them is quite up to date. The most comprehensive of these is by the Oxford workers who played such a large part in the development of penicillin (Florey et al., 1949). A smaller but very useful book is by Baron (1950). This gives all essential details about each substance in almost telegraphic form, and does not give undue space to the few well-known antibiotics. There is also a useful review by Brian (1951) dealing only with antibiotics produced by fungi.

The search for new antibiotics still goes on, and new ones are frequently announced. However, we are still waiting for a real cure for chronic tuberculosis, for a substance which will cure, or at least ameliorate, influenza, and for a reliable treatment for the common cold.

During comparatively recent years work has been in progress on the possibility of curing plant diseases, other than root-borne diseases, by means of antibiotics. Brian et al. (1946) showed that failure of trees to grow satisfactorily in parts of Wareham Heath is due to a toxic effect produced by *Penicillium nigricans* (identified at the time as *P. janczewskii* Zaleski) on the mycorrhizal fungi (see below). The toxic substance was later shown by Grove and McGowan (1947) to be griseo-fulvin, a mould metabolite first isolated from *P. griseo-fulvum* by Oxford et al. (1939). More recently, experiments have been carried out to find whether griseo-fulvin can be taken up by plants from soil, in sufficient quantity to give immunity to the attacks of parasitic fungi, just as the so-called systematic insecticides are now being used in horticulture and agriculture (see Bian et al., 1951; Tanner and Beesch, 1958). However, it is too early as yet to judge whether this method of using the antibiotic will have practical value.

Another use of griseo-fulvin as an anti-fungal agent has been announced by Williams et al. (1958). They show that fungal diseases of the skin, such as ringworm, can be cured by oral administration of griseo-fulvin. If the permanence of the cures can be confirmed, the antibiotic will be of the greatest value to dermatologists, since some of the dermatomycoses have, up to the present, been virtually incurable. Since then much work has been done on the subject, and several papers have appeared in the medical press (Beare and Mackenzie, 1959; Barlow et al., 1959; Cochrane and Tullett, 1959).

Another mould product which is stated to be effective in treating human mycoses is "Variotin", a metabolic product of a strain of *Paecilomyces variotii* (Takeuchi et al., 1959; Abe et al., 1959; Tanaka et al., 1962).

If experience with antibacterial substances be any guide, it will be useful, and probably essential eventually, to have more than one weapon in the armoury for the attack on fungal diseases.

Modification of steroids. Some of the substances collectively known as steroids have, for a long time, been known to be produced by various types of living organisms. Gradually more and more of these compounds were discovered, and the physiological importance of a number of them recognized. More recently steroid therapy has become widespread, with the result that it has become a matter of urgency to develop means of producing the physiologically active substances in large amounts. Most of the latter have a hydroxy group at carbon-11, whereas the steroids which are most readily available in large amounts lack this particular group. The introduction of —OH at carbon-11 is difficult by purely chemical means. However, Peterson and Murray (1952) showed that progesterone is hydroxylated at carbon-11 by the action of *Rhizopus arrhizus*. Since then various workers have shown that many types of transformations of steroids can be effected by micro-organisms (of which the Hyphomycetes are the most active), including hydroxylation at many different points in the molecule, ketone formation, epoxidation, hydrogenation, cleavage of the side-chain, and various mixtures of these reactions.

An adequate summary would be out of place here. However, an excellent survey of the state of knowledge of these transformations available at the time was given by Vischer and Wettstein (1958), and a fuller account of somewhat earlier practice by Eppstein *et al.* (1956). A more recent review is by Peterson (1963).

Miscellaneous products of moulds. During comparatively recent times intensive investigations on the biochemistry of moulds have been carried out in several countries, with results which are of great theoretical interest, and which may become the basis of important industrial processes in the future. In particular, H. Raistrick and co-workers (1931–59) have shown that a number of species of common moulds can produce from glucose a bewildering variety of complex organic compounds, some of which have not yet been synthesized by chemists. It has been shown already, by the industrial production of penicillin and other antibiotics, by the success of the fermentation process for producing citric acid, and by the rapid development of a method of manufacturing itaconic acid, that, once a use is found for a fermentation product, yields can usually be pushed up to the point where commercial production will pay dividends.

The most useful review of the synthetic activities of moulds is by Raistrick (1949). Articles on this and other aspects of mould metabolism are published from time to time in *Annual Reviews of Biochemistry*. A very extensive list of mould products is given in the *Pfizer Book of Microbial Metabolites* (Miller, 1961).

REFERENCES

Abe, S., Takeuchi, S., and Yonchara, H. (1959). Studies on variotin, a new antifungal antibiotic. II. Taxonomical studies on variotin-producing strains. *J. Antibiot.*, Ser. A, **12**, 201–2.

Abraham, E. P. (1964). The cephalosporin story. *New Scient.*, **24**. No. 417. 430–1.

Adams, S. L., Balankura, B., Andreasen, A. A., and Stark, W. H. (1947). Submerged culture of fungal amylase. *Industr. engng Chem.*, **39**, 792–4.

Anslow, W. K., Raistrick, H., and Smith, G. (1943). Anti-fungal substances from moulds. Part I. Patulin (anhydro-3-hydroxymethylene-tetrahydro-1:4-pyrone-2-carboxylic acid), a metabolic product of *Penicillium patulum* Bainier and *Penicillium expansum* (Link) Thom. *J. Soc. chem. Ind., Lond.*, **62**, 236–8.

Arima, K., Nickerson, W. J., Pyke, M., Schanderl, H., Schultz, A. S., Thaysen, A. C., and Thorne, R. S. W. (1957). *Yeasts*. Ed. W. Roman. The Hague: Junk.

Balankura, B., Stewart, F. D., Scalf, R. E., and Smith, A. L. (1946). Submerged culture of molds for amylase production. *J. Bact.*, **51**, 594.

Barlow, A. J. E., Chattaway, F. W., Hargreaves, G. K., and La Touche, C. J. (1959). Griseofulvin in treatment of persistent fungal infections of the skin. *Brit. med. J.*, Nov. 28, No. 5162, 1141–3.

Baron, A. L. (1950). *Handbook of antibiotics*, New York: Reinhold.

Beare, M., and Mackenzie, D. (1959). Griseofulvin in treatment of infections of scalp due to *Microsporum canis*. *Brit. med. J.*, Nov. 28, No. 5162, 1137–40.

Birkinshaw, J. H. (1963). Miscellaneous products of fungal metabolism—in *Biochemistry of industrial micro-organisms* Eds. C. Rainbow and A. H. Rose, pp. 452–88. London: Academic Press.

Birkinshaw, J. H., Callow, R. K., and Fischmann, C. F. (1931). The isolation and characterization of ergosterol from *Penicillium puberulum* Bainier grown on a synthetic medium with glucose as the sole source of carbon. *Biochem. J.*, **25**, 1977–80.

Brian, P. W. (1951). Antibiotics produced by fungi. *Bot. Rev.*, **17**, 357–430.

Brian, P. W., Curtis, P. J., and Hemming, H. G. (1946). A substance causing abnormal development of fungal hyphae produced by *Penicillium janczewskii* Zal. *Trans. Brit. mycol. Soc.*, **29**, 173–81.

Brian, P. W., Wright, J. M., Stubbs, J., and Way, A. M. (1951). Uptake of antibiotic metabolites of soil micro-organisms by plants. *Nature, Lond.*, **167**, 347–9.

Calam, C. T., Oxford, A. E., and Raistrick, H. (1939). Itaconic acid, a metabolic product of a strain of *Aspergillus terreus* Thom. *Biochem. J.*, **33**, 1488–95.

Calmette, A. (1902). German Patent 129164.

Cochrane, T., and Tullett, A. (1959). Griseofulvin treatment of acute cattle ringworm infections in man. *Brit. med. J.*, Aug. 29, No. 5147, 286–7.

Colonial Food Yeast (1944). *Food yeast: a venture in practical nutrition.* London.

Connstein, W., and Lüdecke, K. (1919). Über Glycerin Gewinnung durch Gärung. *Ber. dtsch. chem. Ges.*, **52**, 1385–91.

Connstein, W., and Lüdecke, K. (1924). Process for the manufacture of propantriol from sugar. U.S. Patent 1511754.

Currie, J. N. (1917). The citric acid fermentation of *Aspergillus niger*. *J. biol. Chem.*, **31**, 15–37.

Eppstein, S. H., Meister, P. D., Murray, H. C., and Peterson, D. H. (1956). Microbial transformations of steroids and their applications to the synthesis of hormones. *Vitam. & Horm.*, **14**, 359–432.

Erb, N. M., and Hildebrandt, F. M. (1946). Mold as an adjunct to malt in grain fermentation. *Industr. engng Chem.*, **38**, 792–4.

Fawns, H. T. (1943). Food production by micro-organisms. Part I—Protein production. *Food Manuf.*, **18**, 194–8.

Fink, H., Haehn, H. and Hoerburgher, W. (1937). Über die Versuche zur Fettgewinnung mittels Mikroorganismen mit besonderer Berücksichtigung des Instituts für Gärungsgewerbe. *Chemikerztg*, **61**, 689–93, 723–726, 744–7.

Fitz, A. (1873). Ueber alkoholische Gährung durch *Mucor mucedo*. *Ber. dtsch. chem. Ges.*, **6**, 48–58.

Florey, H. W., Chain, E., Heatley, N. G., Jennings, M. A., Sanders, A. G., Abraham, E. P., and Florey, M. E. (1949). *Antibiotics.* 2 vols. Oxford: Univ. Press.

Galle, E. (1923). Das Amyloverfahren und seine Anwendungsmöglichkeiten. *Zeit. angew. Chem.*, **36**, 17–19.

Gastrock, E. A., Porges, N., Wells, P. A., and Moyer, A. J. (1938). Gluconic acid production on pilot plant scale: effect of variables on production by submerged mold growth. *Industr. engng Chem.*, **30**, 782–9.

Geiger-Huber, M., and Galli, H. (1945). Über den Nachweis der l'Ascorbinsäure als Stoffwechselprodukt von *Aspergillus niger*. *Helv. chim. Acta*, **28**, 248–50.

Goodwin, T. W. (1959). Production and biosynthesis of riboflavin in micro-organisms. *Progr. industr. Microbiol.*, **1**, 139–177.

Grove, J. F., and McGowan, J. C. (1947). Identity of griseofulvin and "curling factor". *Nature, Lond.*, **160**, 574.

Haehn, H., and Kinttof, W. (1924). Beitrag über den chemischen Mechanismus der Fettbildung aus Zucker. *Chem. Zelle*, **12**, 115–56.

Harada, T. (1931). Preparations of *Aspergillus oryzae* enzymes. *Industr. engng Chem.*, **23**, 1424.

Herrick, H. T., and May, O. E. (1928). The production of gluconic acid by the *Penicillium luteum-purpurogenum* group. II. Some optimal conditions for acid formation. *J. biol. Chem.*, **77**, 185–95.

Herrick, H. T., Hellbach, R., and May, O. E. (1935). Apparatus for the application of submerged mold fermentations under pressure. *Industr. engng Chem.*, **27**, 681–3.

Hesseltine, C. W. (1965). A millennium of fungi, food and fermentation. *Mycologia*, **57**, 149–97.

Kinoshita, K. (1929). Formation of itaconic acid and mannitol by a new filamentous fungus. *J. chem. Soc. Japan*, **50**, 583–93.

Lafar, F. (1903). *Technical mycology*. English translation by C. T. C. Salter. Vol. II, Part 1. London: Griffin.

Le Mense, E. H., Gorman, J., van Lanen, J. M., and Langlykke, A. F. (1947). The production of mold amylases in submerged culture. *J. Bact.*, **54**, 149–59.

Lockwood, L. B., Ward, G. E., May, O. E., Herrick, H. T., and O'Neill, H. T. (1934). Production of fat by *Penicillium javanicum* van Beyma. *Zbl. Bakt.*, Abt. II, **90**, 411–25.

Lockwood, L. B., Ward, G. E., and May, O. E. (1936). The physiology of *Rhizopus oryzae. J. agric. Res.*, **53**, 849.

Lodder, J., and Kreger-van Rij, N. J. W. (1952). *The Yeasts. A taxonomic study.* Amsterdam: North-Holland Publ. Co.

Lundin, H. (1948). Über Fettsynthesen durch Mikroorganismen und Möglichkeiten für ihre industrielle Anwendung. *Jb. Hochsch. Bodenk. Wien*, **2**, 410–20.

May, O. E., Herrick, H. T., Thom, C., and Church, M. B. (1927). The production of gluconic acid by the *Penicillium luteum-purpurogenum* group. I. *J. biol. Chem.*, **75**, 417–22.

May, O. E., Herrick, H. T., Moyer, A. J., and Hellbach, R. (1929). Semi-plant scale production of gluconic acid by mold fermentation. *Industr. engng Chem.*, **21**, 1198–1203.

Miller, M. W. (1961). *The Pfizer handbook of microbial metabolites.* New York: McGraw-Hill.

Nelson, G. E. N., Traufler, D. H., Kelley, S. E., and Lockwood, L. B. (1952). Production of itaconic acid by *Aspergillus terreus* in 20-litre fermentors. *Industr. engng Chem.*, **44**, 1166–8.

Oxford, A. E., Raistrick, H., and Simonart, P. (1939). Studies in the biochemistry of micro-organisms. LX. Griseofulvin, $C_{17}H_{17}O_6Cl$, a metabolic product of *Penicillium griseo-fulvum* Dierckx. *Biochem. J.*, **33**, 240–8.

Pasteur, L. (1859). Nouveaux faits concernant la fermentation alcoholique. *C.R. Acad. Sci., Paris*, **48**, 640.

Peterson, D. H. (1963). Microbial transformations of steroids and their application to the preparation of hormones and derivatives. In *Biochemistry of industrial micro-organisms.* Ed. Rainbow, C., and Rose, A. H., pp. 537–606. London: Academic Press.

Peterson, D. H., and Murray, H. C. (1952). Microbial oxidation of steroids at carbon 11. *J. Amer. chem. Soc.*, **74**, 1871–2.

Pfeifer, V. F., Tanner, F. W., Vojnovich, C., and Traufler, D. H. (1950). Riboflavin by fermentation with *Ashbya gossypii. Industr. engng Chem.*, **42**, 1776–81.

Porges, N., Clark, T. F., and Aronovsky, S. I. (1941). Gluconic acid production: repeated recovery and re-use of submerged *Aspergillus niger* by filtration. *Industr. engng Chem.*, **33**, 1065–7.

Pridham, T. G., and Raper, K. B. (1950). *Ashbya gossypii*—its significance in Nature and in the laboratory. *Mycologia*, **42**, 603–23.

Preuss, L. M., Peterson, W. H., Steenbock, H., and Fred, E. B. (1931). Sterol content and antirachitic activatibility of mold mycelia. *J. biol. Chem.*, **90**, 369–84.

Preuss, L. M., Peterson, W. H., and Fred, E. B. (1932). Isolation and identification of ergosterol and mannitol from *Aspergillus fischeri. J. biol. Chem.*, **97**, 483–9.

Preuss, L. M., Gorcica, H. J., Greene, H. C., and Peterson, W. H. (1932). Wachstum und Steringehalt gewisser Schimmelpilze. *Biochem. Z.*, **246**, 401–13.

Rainbow, C., and Rose, A. H. (1963). Eds. *Biochemistry of industrial micro-organisms.* London: Academic Press.

Raistrick, H. *et al.* (1931–59). Studies in the biochemistry of micro-organisms

Parts 1–18. *Phil. Trans.*, Ser. B, **220** (1931). Parts 19–106. *Biochem. J.* (1931–59).

Raistrick, H. (1949). A region of biosynthesis (Bakerian Lecture). *Proc. roy. Soc.*, A, **199**, 141–68.

Rhodes, R. A., Moyer, A. J., Smith, M. L., and Kelley, S. E. (1959). Production of fumaric acid by *Rhizopus arrhizus*. *Appl. Microbiol.*, **7**, 74–80.

Rhodes, R. A., Lagoda, A. A., Misenheimer, T. J., Smith, M. L., Anderson, R. F., and Jackson, R. W. (1962). Production of fumaric acid in 20-liter fermentors. *Appl. Microbiol.*, **10**, 9–15.

Rickes, E. L., Brink, N. G., Koniuszy, F. R., Wood, T. R., and Folkers, K. (1948). Comparative data on vitamin B_{12} from liver and from a new source, *Streptomyces griseus*. *Science*, **108**, 634.

Sakaguchi, K. (1961). Culture and characteristics of Japanese fermentation industries. *Inst. phys.-chem. Res.*, Aug. 1961.

Shibasaki, K., and Hesseltine, C. W. (1960). Miso, I. Preparation of soybeans for fermentation. *J. biochem. microbiol. Tech. Engng*, **3**, 49–50.

Shibasaki, K., and Hesseltine, C. W. (1961). Miso, II. Fermentation. *Dev. industr. Microbiol.*, **2**, 205–14.

Symposium. (1963). The chemistry and biochemistry of fungi and yeasts. *Proc. Symposium, Dublin*, 1963. London: Butterworth & Co.

Takamine, J. (1914). Enzymes of *Aspergillus oryzae* and the application of its amyloclastic enzyme to the fermentation industry. *Chem. News*, **110**, 215–18.

Takamine, J. (Jr.), and Oshima, K. (1920). The properties of a specially prepared extract, Polyzime, comparing its starch liquefying power with malt diastase. *J. Amer. chem. Soc.*, **42**, 1261.

Takeuchi, S., Yonchara, H., and Umezawa, H. (1959). Studies on variotin, a new antifungal antibiotic. I. *J. Antibiot.*, Ser. A, **12**, 195–200.

Tanaka, N., and Umezawa, H. (1962a). Biogenesis of variotin. I. Incorporation of $CH_3{}^{14}COOH$ and $^{14}CH_3COOH$ into variotin. *J. gen. appl. Microbiol.*, **8**, 149–59.

Tanaka, N., and Umezawa, H. (1962b). Biogenesis of variotin. II. Incorporation of $^{14}CH_3SCH_2CH_2CH(NH_2)COOH$ into variotin. *J. gen. appl. Microbiol.*, **8**, 160–4.

Tanner, F. W., and Beesch, S. C. (1958). Antibiotics and plant diseases. *Advanc. Enzymol.*, **20**, 383–406.

Thaysen, A. C. (1943). Value of micro-organisms in nutrition. (Food yeast.) *Nature, Lond.*, **151**, 406–8.

Thaysen, A. C. (1953). Food and the future. Part 3 (B). Food and fodder yeast. *Chem. & Ind. (Rev.)*. May 9, 446–7.

Thaysen, A. C., and Morris, M. (1943). Preparation of a giant strain of *Torulopsis utilis*. *Nature, Lond.*, **152**, 527–8.

Thom, C. (1906). Fungi in cheese ripening. Camembert and Roquefort. *Bull. U.S. Bur. anim. Ind.*, **82**, 1–39.

Vischer, E., and Wettstein, A. (1958). Enzymic transformations of steroids by microorganisms. *Advanc. Enzymol.*, **20**, 237–82.

Ward, G. E., Lockwood, L. B., May, O. E., and Herrick, H. T. (1935). Production of fat from glucose by molds. Cultivation of *Penicillium javanicum* van Beyma in large-scale laboratory apparatus. *Industr. engng Chem.*, **27**, 318–22.

Ward, G. E., Lockwood, L. B., May, O. E., and Herrick, H. T. (1936). Studies in the genus *Rhizopus*. I. The production of dextro-lactic acid. *J. Amer. chem. Soc.*, **58**, 1286.

Ward, G. E., Lockwood, L. B., Tabenkin, B., and Wells, P. A. (1938). Rapid fermentation process for dextro-lactic acid. *Industr. engng Chem.*, **30**, 1233.

Wehmer, C. (1891). Entstehung and physiologische Bedeutung der oxalsäure im Stoffwechsel einiger Pilze. *Bot. Ztg*, **49**, 233–638.

Wehmer, C. (1893). *Beiträge zur Kenntnis einheimischer Pilze*. Hannover and Jena: Hansche Buchhandlung.

Wehmer, C. (1907). Chemische Wirkungen der Mucoreen. Alcoholische Gärung. Lafar's *Handbuch der technischen Mykologie*. Vol. IV, pp. 506–28.

Wells, P. A., Moyer, A. J., Stubbs, J. J., Herrick, H. T., and May, O. E. (1937). Gluconic acid production. Effect of pressure, air flow, and agitation on gluconic acid production by submerged mold growths. *Industr. engng Chem.*, **29**, 777–8.

Williams, D. I., Marten, R. H., and Sarkany, I. (1958). Oral treatment of ringworm with griseofulvin. *Lancet*, Dec. 6, 1958, No. 7058.

Woodbine, M. (1959). Microbial fat: micro-organisms as potential fat-producers. *Progr. industr. Microbiol.*, **1**, 181–245.

Yaw, K. E. (1952). Production of riboflavin by *Eremothecium ashbyi* grown in a synthetic medium. *Mycologia*, **44**, 307–17.

Chapter XVI

Mycology of the Soil

> As universal agents of decay, from which new life continually arises and is nourished, the fungi make a unique contribution to the maintenance of soil fertility.
>
> S. D. Garrett, *Soil fungi and soil fertility*, 1963

The role of fungi in the soil has been studied by numerous workers, and in many countries, from the earliest days of pure culture methods. Many publications have dealt with the species which can be isolated when soils are plated out on nutrient agar media. The question soon arose as to whether micro-fungi actually live in the soil, or are present as dormant spores, carried by air currents from fungus-infected materials, deposited on the soil, and later washed in by rain. Special methods have been developed for detecting living mycelium in soil, some of which are described below, and there is no longer any doubt that fungi play an important part, along with actinomycetes, bacteria, and various forms of animal life, in breaking down organic matter to form humus. The papers and discussions at an International Symposium on the Ecology of Soil Fungi have been edited for publication by Parkinson and Waid (1960). The book covers a wide range of relevant topics.

Plant pathogens. Apart from the saprophytic fungi, which act as scavengers, a number of important plant parasites live in the soil, and attack economically important plants through their roots, or at ground level. A number of plants are attacked by species of *Verticillium*, such as *V. alboatrum* on hops, cucumber, potato and tomato, and *V. dahliae* on raspberry, hops, tomato and *Chrysanthemum*. *Botrytis cinerea* attacks a wide variety of plants, both below ground and above. *Penicillium gladioli* and *P. corymbiferum* infect bulbs and corms of various species.

Perhaps the commonest soil-borne fungal infections are by species of *Pythium*, which cause damping-off of seedlings of all kinds, and are responsible for enormous losses. In consequence, a considerable amount of work has been carried out on the possibility of biological control of *Pythium* species and other soil-borne parasitic fungi. Weindling (1932, 1934, 1937) and Weindling and Emerson (1936) showed that *Trichoderma lignorum* (= *T. viride*) and *Gliocladium fimbriatum*, common inhabitants of wet soils, produce a substance, gliotoxin, which inhibits the growth of *Pythium*. He further showed that heavy inoculations of *Trichoderma* into wet soils can favourably influence the growth of crops, by suppressing damping-off

fungi. Van Luijk (1938) investigated a number of species of moulds, all normal inhabitants of the soil, as to their inhibitory effect on various species of
Pythium. He found that the metabolism solution obtained by growing *Penicillium expansum* on a synthetic medium containing sugar was highly active
in protecting seedlings from attack by *Pythium* spp., although inoculation
of the soil with the *Penicillium* itself had no effect. The active substance was
not isolated by van Luijk, but Anslow, Raistrick and Smith (1943) have

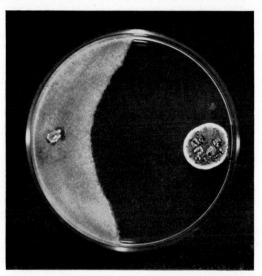

FIG. 172.—Inhibition of *Pythium ultimum* by
Penicillium patulum. ×0·6.

A simple method of testing for anti-fungal activity. In this case the *Penicillium* was sown 3 days before the *Pythium*, since the latter covers a dish
in 2–3 days.

shown that it was probably patulin, isolated by them from a number of
strains of *P. expansum*, and, in better yield, from *P. patulum* (Fig. 172).
Various attempts have been made to establish *P. patulum* in soils, but, even
when the fungus will grow in the soil, it usually does not produce the antibiotic unless carbohydrate material be added at the same time.

Mycorrhiza. This is the term given to a close association of fungus and
plant root. The fungi concerned are mostly, but not invariably, the larger
Basidiomycetes known, in this country, as toadstools. It is generally considered that there are two distinct types of mycorrhiza, termed respectively ectotrophic and endotrophic. In ectotrophic mycorrhiza the fungal
mycelium forms a sheath around the root hairs, and also penetrates between the cells of the root epidermis, and, to some extent, of the cortex.
There is no doubt that the fungus obtains benefit from the association with

the plant. It has been shown that such fungi are unable to utilize complex substances, such as cellulose and lignin, as sources of carbon, and are dependent on the plant for sugar. The benefit to the plant is more problematical, but there is evidence that, in soils which are deficient in nitrogen and/or minerals, the fungus helps to supply the plant with these essentials, phosphate being particularly important. In soils containing adequate supplies of these nutrients the benefit appears to be almost entirely one-sided.

The fungi concerned in the formation of endotrophic mycorrhiza produce very little external mycelium. The hyphae are both intercellular and intracellular, but, as they advance, they are continually being digested by the cortical cells. This type of mycorrhiza is therefore to be regarded as controlled parasitism. However, in the case of some orchids, which have minute and imperfectly differentiated seeds, it appears that, in the early stages of growth of the plant, the orchid is parasitic on the fungus, and the success with which orchid seeds germinate is measured by the completeness with which they dominate the fungus. Many endotrophic mycorrhizal fungi are Phycomycetes, but the fungi associated with orchids are Basidiomycetes, mostly belonging to the genus *Rhizoctonia*, the species of which are imperfect states of species of *Corticium*. It is of interest that the Japanese orchid, *Gastrodia elata*, has as its mycorrhizal associate *Armillaria mellae*, which is one of the most destructive tree-parasites known.

Good accounts of the mycorrhizal association are given by Rayner and Neilson-Jones (1944), Rayner (1945), Ramsbottom (1953), Garrett (1956), and Harley (1959).

The rhizosphere. The term is used to denote an area around the roots of plants which is much richer than the rest of the soil in its microbial population. The phenomenon was first described by Hiltner in 1909, and numerous papers on the subject have appeared since then. It has been generally supposed that the intense stimulation of microbial growth in the vicinity of roots is adequately accounted for by the extra food material available, partly actively excreted by the root, and partly provided by the sloughing-off of dead cells from the root epidermis. However, the results obtained by Jackson, with a species of *Fusarium* (see below under Fungistasis), suggest that other factors may be concerned. The rhizosphere phenomenon is well described by Garrett (1956) and Harley (1959).

Fungistasis in soil. During recent years a number of workers have drawn attention to a remarkable inhibition of germination of mould spores in the soil. Spores of many species lie dormant, even when there is adequate moisture for germination. Dobbs and Hinson (1953) and Dobbs and Bywater (1957) found that such spores will germinate if glucose be added to the soil, but the fungistatic effect is restored when such added carbohydrate has been used up. The fungistasis is to some extent seasonal, reaching a maximum in July to September, and a minimum in January. Autoclaving removes the effect from most soils, but a few highly calcarious soils are unaffected by this treatment.

Jackson (1957) found that fungal spores, introduced into the soil on glass slides coated with agar containing 0·5 per cent peptone, are unable to germinate, but germinate readily when seedling roots grow in the vicinity. A peculiar effect was observed with a species of *Fusarium*, in that the spores germinated in ordinary soil, but the short germ-tubes immediately produced chlamydospores, after which there was no further growth. If now the slide was placed near to a seedling root, the chlamydospores put forth new germ-tubes, which grew towards the root. It is suggested that this may partly explain the relative abundance of fungi in the rhizosphere, compared with the rest of the soil. In a later paper (1958*a*) Jackson recorded a widespread fungistasis in Nigerian soils, removed by autoclaving, and partially and temporarily removed by adding glucose. The test method was to use small agar discs on filter-paper, the latter being pressed into contact with the soil. The agar discs were inoculated with spores of the test fungi. An important observation was that not all the micro-fungi which were tested were inhibited. A third paper (1958*b*) describes the results obtained with a variety of different types of soil at the Experimental Station, Rothamsted. It was shown that the buried slide technique is more sensitive than the agar disc method, but the results run parallel. The degree of inhibition was found to vary with the pH of the soils tested, decreasing with increasing acidity. At pH 4·0 only *Acrostalagmus cinnabarinus* (= *Verticillium lateritium*) was inhibited, and it was shown that this is purely a pH effect. In the most inhibitory soils (alkaline) autoclaving removed the inhibition entirely. Jackson also found that the inhibitory effect is most marked on spores which have just germinated, established mycelium being little affected.

Park (1954, 1955, 1956) has shown that, when spores of fungi, which are not normal inhabitants of the particular soil, are introduced into the soil, they are not only prevented from germinating, but are actually destroyed.

Fungi and the soil fauna. Drechsler (1937) was the first to draw attention to a most interesting activity of some soil fungi, their capture and destruction of eel-worms (nematodes). Since then, Drechsler has published numerous papers (mostly in *Mycologia*), describing many new genera and species of these specialized fungi. For a number of years Duddington has made a particular study of predacious fungi in Britain. Apart from a number of papers describing new species, he has published a useful account of the methods employed in isolating and handling these organisms (1955*a*). He has also published several general reviews (1955*b*, 1957*a* and *b*).

Taxonomically the predacious fungi fall into two groups. The Zoopagales, which are Phycomycetes, forming zygospores as well as asexual conidia, prey mostly on amoebae. The second group are Hyphomycetes and these usually attack nematodes. Many of the species are difficult to grow on ordinary culture media, but thrive if both bacteria and nematodes are present, the eel-worms feeding on the bacteria, and the fungus on the eel-worms.

Several different methods are used by these fungi for capturing their prey. The first, used by the common and widespread *Arthrobotrys oligospora*, is by means of sticky hyphal loops. These are formed in abundance only when eel-worms are present. When one of these creatures attempts to pass through a loop, it is held by the sticky secretion. Then the fungal hyphae pierce the cuticle of the nematode, grow inside it and gradually digest it. A slightly different, though similar, method is used by *Dactylella cionopaga* (Gr. *kion*, a movement; *page*, a trap), in which short, sticky, erect branches are formed on the aerial hyphae. Another variant is the formation of spherical sticky knobs on the hyphae, the best-known fungus of this group being *Dactylella ellipsospora*. The Phycomycete *Stylopaga hadra* produces hyphae which are sticky all over, and is an obligate parasite.

The species of the genus *Dactylaria* catch the nematodes by means of passive rings, which are simply hyphal loops of such aperture that an eel-worm trying to force its way through becomes tightly wedged. As soon as the prey is held, special branches from the hyphae penetrate the cuticle, and the animal is consumed in the usual manner. The commonest species of the genus, *D. candida*, usually forms sticky knobs as well as rings.

The final, and most spectacular way of catching eel-worms, used by *Dactylella bombicodes* and *D. doedycoides*, is by contracting loops. Each loop consists of three cells, of approximately the same diameter as the parent hypha, when at rest. If an eel-worm tries to pass through, the cells swell suddenly, and form a kind of garotte by reducing the aperture to a small fraction of what it was originally. In spite of much investigation, and many ingenious theories, we still do not know how the sudden expansion of the rings is brought about.

A number of experiments have been carried out with a view to increasing the number and voracity of the predacious fungi in soils heavily infested with parasitic nematodes. Whilst much work remains to be done before a sure and inexpensive method becomes generally available, enough has been accomplished to show that healthy plants can be grown on land which is too heavily infested for growing the particular crop without the special treatment. Duddington (1957a) gives a good account of these experiments up to the date of publication.

Methods for isolation and study of soil fungi. The most widely used method, employed in many laboratories for over half a century, is the soil dilution-plate method, in which small samples of soil, as such or after fractionation by sedimentation, are distributed in tubes of molten agar medium, and a number of dilutions made from each tube, each dilution being poured into a Petri dish and incubated for a suitable period. A study of the numerous records of such investigations shows that, whilst new species are often found in the soil, the lists of genera all show a remarkable similarity. This does not necessarily mean that the mycological floras of all soils in all countries are similar. It merely shows that the fungi recorded are those which will grow rapidly, and form spores freely, on the culture media used. Such species of fungi are of world-wide distribution, largely because the types of material which they colonize in the soil are similar in all soils, and partly because they grow on a wide range of the raw materials and manufactured products of industry, and are carried backwards and forwards between all parts of the world. It is a remarkable fact that Basidiomycetes, whose sporophores are such a characteristic feature of our woods and fields in Autumn, and whose mycelium spreads widely in the soil, are very seldom isolated by the usual technique, although many of them grow well on common laboratory media. Also it is seldom that Oomycetes are isolated without using special methods, although certain genera of the Zygomycetes are to be found in almost every list of soil fungi.

It is not surprising that much attention has been given, particularly in comparatively recent times, to devising improved methods of studying the mycological flora of the soil. It has been realized that any fungus which readily forms spores in the soil, and which is capable of growing on laboratory media, will show up as numerous colonies on the usual dilution plates, whereas a widespread and active mycelium would produce very few colonies, or fail to appear at all.

The first to attempt to prove that the soil contains active fungal mycelium was Waksman (1916). He argued that, if a sample of soil be placed on an agar medium, any living mycelium which it contains will grow outwards before any dormant spores have time to germinate. Small crumbs of soil are distributed over the surface of a suitable medium and incubated not longer than 24 hours. Examination under low magnification will usually show that visible mycelium is issuing from some of the particles, this being taken as proof of the presence of living mycelium in the particular soil. The limitation of the method lies in the fact that it favours those organisms which can best utilize the nutrients provided, and those which have a spreading habit of growth.

Rossi (1928) described a method, which was slightly modified by Cholodny (1930), and, as the so-called "Rossi-Cholodny contact slide method", has been widely used by later soil microbiologists. It consists in pressing a sterile microscope slide against a freshly-cut surface of soil, and leaving in contact for some time. Organisms which become attached to the slide are fixed, stained, and examined by the usual methods. This technique has been criticized because a film of moisture deposits on the slide, as on any solid surface in contact with the soil, and usually promotes the proliferation of bacteria. Also the slide affects the aeration of the soil in contact with it.

Brown (1958) used a modified Cholodny slide, coated with an adhesive nitrocellulose in amyl acetate plus 5 per cent castor oil. A very thin, even layer of this mixture is brushed on to the slide, and this is immediately pressed very lightly against the soil profile and held for 20 seconds. After shaking off gross particles which are not properly stuck, the film is stained for one hour with phenolic aniline blue, rinsed and dried. It is examined dry with an 8 mm. metallurgical objective (i.e. corrected for no cover-glass) and a ×25 eyepiece. In this way fungi were detected which did not appear at all in soil dilution plates.

Chesters (1940, 1946) used what he called "immersion tubes". These are 6 by $\frac{3}{4}$ inch hard glass tubes with the lower ends tapered. A number of holes, usually 6, extend internally and upwards in the form of short, fine capillaries. These are made between $1\frac{1}{2}$ and 3 inches from the bottom. For use, melted agar is run into the tubes in sufficient quantity to cover the holes. Special methods are used for filling, and for ensuring sterility, for details of which the original papers should be consulted. To insert a tube into soil, a core is first taken out with a sterile borer of slightly less diameter, then the immersion tube is pushed into the hole, with the plug protected by a glass cup, and left for about 7 days. Active mycelium in the soil grows in through the capillaries, and can be transferred to ordinary culture tubes or plates for examination.

Thornton (1952) described an improved method, using what he termed the "screened immersion plate". This consists of a microscope slide, coated with agar medium, and contained in a shallow, rectangular Perspex box, the box being covered with a Perspex screen in which ten holes are drilled. This lid is not in contact with the agar, but is separated from it by a very small amount, determined by the depth of the box. The "plate" is sunk in the soil and left as long as required. Mycelium which is active in the soil grows through the holes in the screen to reach the agar. The time of exposure should not be so long that mycelia from adjacent holes inter-mingle. The plate is taken back to the laboratory, where sub-cultures are made with the usual precautions.

Chesters and Thornton (1956) carried out an investigation in which various methods of isolating soil fungi were directly compared with the screened plate method. These were (a) immersions tubes; (b) Waksman's direct inoculation method; (c) a modified Rossi-Cholodny slide; (d) dilution plates; (e) Warcup's soil plates (see below). It was shown that, in all cases, the screened plate isolated the most species, and there were usually less numbers of the heavily sporing species than were obtained by the other methods.

Kubiena (1938), in a most interesting and stimulating book, has de-scribed a number of methods of studying soil fungi. One interesting tech-nique, which I have not seen referred to elsewhere, is a method of setting up a microscope to examine a soil profile *in situ* in the field.

In laboratories where the aim is to isolate new species, and new strains of known species, in the search for antibiotics, the dilution plate method is still widely used, and has considerable value. It is advisable to use an anti-bacterial agent in the culture medium (see Chapter XI), in order to avoid inhibition of the fungi by some of the soil bacteria which grow readily on mycological media. Warcup (1950, 1951a and b) has described two methods which result in the isolation of a greater variety of species than by means of the usual dilution plates. The first of these is known as the "soil plate" method. A small amount of soil, 0·005–0·015 g for surface soils, and rather more for samples from lower strata, is placed in an empty Petri dish, and any aggregates are broken up with a miniature spatula. A quantity of agar medium, 8–10 ml., and not hotter than 45° C, is poured into the dish, and the soil dispersed by shaking and rotating before the agar sets. If the soil is very dry or clayey, a drop of water is added before pouring in the agar. The medium recommended is Czapek plus 0·5 per cent yeast extract, acidified with phosphoric acid to pH 4·0. The second method, designed particularly for the isolation of Ascomycetes, consists of partial steam sterilization of the soil. The soil samples, in glass tumblers, are steamed at 100° C for a period long enough to reduce the number of live spores, but not long enough for complete sterilization. The period is usually 3–4 minutes. By this means most of the conidia of Hyphomycetes are killed, but ascospores, which are in general more resistant to heat, usually survive. Many of the species which can be isolated in this way are slow-growing, and are completely swamped when soils are plated out in the usual way.

Interrelationships of soil organisms. In spite of the immense amount of work already accomplished in investigating the microbiology of the soil, it must be admitted that no one has yet been able to give a clear picture of all the life in the soil, and of all the interactions of the different groups of living things. No one group of organisms lives in isolation, but is affected, i.e. consumed, parasitized, inhibited, or stimulated, by numerous other types. To take a simple example, we know that some bacteria attack some fungi, that eel-worms consume bacteria, and that the cycle is completed by the destruction of the eel-worms by predacious fungi. There are, of course, numerous other kinds of living things in the soil, protozoa, mites and spiders, beetles, larvae of these and other insects, centipedes and millipedes, earthworms, snails and slugs, to say nothing of larger creatures, such as rodents and moles. The difficulties of obtaining an overall picture of life in the soil are immense, chiefly due to the fact that the soil is opaque, and any attempt to make the activities of soil organisms visible invariably disturbs the soil.

A symposium held in Holland in 1962 (Doeksen and van der Drift, 1963) brought together zoologists, bacteriologists, and mycologists, in an attempt to obtain a more unified general picture of biological activity in the soil, but the published proceedings show that there is still far to go. A little book by Garrett (1963) is mainly mycological, but does attempt to relate fungal activity to the activities of other types of organisms.

It appears that we shall have to wait for the inspiration of some entirely new technique before any real advance can be made.

REFERENCES

Anslow, W. K., Raistrick, H., and Smith, G. (1943). Anti-fungal substances from moulds. Part I. Patulin (anhydro-3-hydroxymethylene-tetrahydro-

1:4-pyrone-2-carboxylic acid), a metabolic product of *Penicillium patulum* Bainier and *Penicillium expansum* (Link) Thom. *J. Soc. chem. Ind., Lond.*, **62**, 236–8.

Brown, J. C. (1958). Fungal mycelium in dune soils estimated by a modified impression slide technique. *Trans. Brit. mycol. Soc.*, **41**, 81–88.

Chesters, C. G. C. (1940). A method of isolating soil fungi. *Trans. Brit. mycol. Soc.*, **24**, 352–5.

Chesters, C. G. C. (1946). A contribution to the study of fungi in the soil. *Trans. Brit. mycol. Soc.*, **30**, 100–117.

Chesters, C. G. C., and Thornton, R. H. (1956). A comparison of techniques for isolating soil fungi. *Trans. Brit. mycol. Soc.*, **39**, 301–13.

Cholodny, N. (1930). Über eine neue Methode zur Untersuchung der Bodenmikroflora. *Arch. Mikrobiol.*, **1**, 620–52.

Dobbs, C. G., and Bywater, J. (1957). Studies in soil mycology. I. *Rep. For. Res., Lond.*, (1957), 92–4.

Dobbs, C. G., and Bywater, J. (1959). *Ibid.* II. *Rep. For. Res., Lond.*, (1959), 98–104.

Dobbs, C. G., and Hinson, W. H. (1953). A widespread fungistasis in soil. *Nature, Lond.*, **172**, 197.

Doeksen, J., and van der Drift, J. (1963). Eds. *Soil organisms*. Amsterdam: N. Holland Publ. Co.

Drechsler, C. (1937). Some Hyphomycetes that prey on free-living terricolous Nematodes. *Mycologia*, **29**, 447–52.

Duddington, C. L. (1955a). Notes on the technique of handling predacious fungi. *Trans. Brit. mycol. Soc.*, **38**, 97–103.

Duddington, C. L. (1955b). Fungi that attack microscopic animals. *Bot. Rev.*, **21**, 377–439.

Duddington, C. L. (1957a). *The friendly fungi*. London: Faber & Faber.

Duddington, C. L. (1957b). The predacious fungi and their place in microbial ecology. *Microbial Ecology. Seventh Symp. Soc. gen. Microbiol.* Camb. Univ. Press.

Garrett, S. D. (1956). *The biology of root-infecting fungi*. Camb. Univ. Press.

Garrett, S. D. (1963). *Soil Fungi and Soil Fertility*. London: Pergamon Press.

Harley, J. L. (1959). *The biology of Mycorrhiza*. London: Leonard Hill (Books) Ltd.

Hiltner, L. (1909). Über neuere Erfahrungen und Probleme auf dem Bebiet der Bodenbakteriologie, und unter besonderer Berücksichtigung der Gründüngung und Brache. *Arb. dtsch. Landw. Ges.*, **98**, 59–78.

Jackson, R. M. (1957). Fungistasis as a factor in the rhizosphere phenomenon. *Nature, Lond.*, **180**, 96–7.

Jackson, R. M. (1958a). An investigation of fungistasis in Nigerian soils. *J. gen. Microbiol.*, **18**, 248–58.

Jackson, R. H. (1958b). Some aspects of soil fungistasis. *J. gen. Microbiol.*, **19**, 390–401.

Kubiena, W. L. (1938). *Micropedology*. Ames, Iowa: College Press.

Park, D. (1954). An indirect method for the study of fungi in soil. *Trans. Brit. mycol. Soc.*, **37**, 405–11.

Park, D. (1955). Experimental studies on the ecology of fungi in soil. *Trans. Brit. mycol. Soc.*, **38**, 130–42.

Park, D. (1956). Effect of substrate on a microbial antagonism, with reference to soil conditions. *Trans. Brit. mycol. Soc.*, **39**, 239–59.

Parkinson, D., and Waid, J. S. (1960) Eds. *The ecology of soil fungi. An international symposium.* Liverpool Univ. Press.

Ramsbottom, J. (1953). *Mushrooms and toadstools.* London: Collins (New Nat. No. 7).

Rayner, M. C. (1945). *Trees and toadstools.* London: Faber & Faber.

Rayner, M. C., and Neilson-Jones, W. (1944). *Problems in tree nutrition: an account of researches concerned primarily with the mycorrhizal habit in relation to forestry and with some biological aspects of soil fertility.* London: Faber & Faber.

Rossi, G. M. (1928). Il terreno agrario nella teoria e nella realta. *Ital. agric.*, No. 4 (1928).

Thornton, R. H. (1952). The screened immersion plate. A method of isolating soil micro-organisms. *Research, Lond.*, **5**, 190–1.

Van Luijk, A. (1938). Antagonisms between various micro-organisms and different species of the genus *Pythium*, parasitizing upon grasses and lucerne. *Meded. phytopath. Lab. Scholten*, **14**, 45–83.

Waksman, S. A. (1916). Do fungi live and produce mycelium in the soil? *Science*, **44**, 320–2.

Warcup, J. H. (1950). The soil-plate method for isolation of fungi from soil. *Nature, Lond.*, **166**, 117–18.

Warcup, J. H. (1951a). Studies on the growth of Basidiomycetes in soil. *Ann. Bot., Lond.*, N.S. **15**, 305–17.

Warcup, J. H. (1951b). Soil-steaming: a selective method for the isolation of Ascomycetes from soil. *Trans. Brit. mycol. Soc.*, **34**, 515–8.

Weindling, R. (1932). *Trichoderma lignorum* as a parasite of other soil fungi. *Phytopathology*, **22**, 837–45.

Weindling, R. (1934). Studies on a lethal principle effective in the parasitic action of *Trichoderma lignorum* on *Rhizoctonia solani* and other soil fungi. *Phytopathology*, **24**, 1153–79.

Weindling, R. (1937). Isolation of toxic substances from the culture filtrates of *Trichoderma* and *Gliocladium*. *Phytopathology*, **27**, 1175–7.

Weindling, R., and Emerson, O. H. (1936). The isolation of a toxic substance from the culture filtrate of *Trichoderma*. *Phytopathology*, **26**, 1068–70.

Chapter XVII

Mycological Literature

And out of olde books, in good feith, cometh all this newe
science that men lere.

Chaucer, *The Parlement of Foules*

The following list of publications is not intended to be in any sense a
complete bibliography. Apart from a number of works used for identification of fungi, the literature which is cited includes only such as is considered
to be useful to a student whose interest extends to the more general aspects
of mycology and who wishes to keep abreast of modern knowledge. Anyone
who desires an exhaustive list of publications relating to any particular
branch of the subject will find full bibliographies in the various monographs
mentioned below.

Unfortunately a number of the most important mycological publications
are to be found in only a few British libraries. Some of these may be obtained on loan through the various public libraries, but a few works of
reference cannot be borrowed. These may, however, be consulted, by
special permission, at the libraries of the British Museum (Natural History), the Royal Botanic Herbarium, Kew, the Commonwealth Mycological
Institute, Kew, and the various universities (if available there). Some
libraries have a photostat service, providing copies of rare papers at low
cost, and the C.M.I., Kew, lists a number of microfilms of important
mycological literature, copies of which are available. Some important
mycological classics are also available as sets of microcards or microfiches
(prints on transparent film). They are produced by Micro Methods Ltd.,
East Ardley, Wakefield, and new issues are regularly added to the list.
Those possessing a binocular dissecting microscope will find the microfiches quite easy to read, without any special equipment.

I. GENERAL WORKS USED FOR IDENTIFICATION OF FUNGI

BARNETT, H. L. (1960). *Illustrated Genera of Imperfect Fungi*. 2nd Ed.
Minneapolis: Burgess Publ. Co.
The first edition (1955) contained drawings and brief descriptions, with
literature references, of 302 genera, mostly Fungi Imperfecti, but including
some Phycomycetes. The new edition describes 462 genera. Many of the
illustrations are copies from original publications, the remainder being
original.
This is a very useful book, since it includes, not only most of the

common moulds, but also a number of genera which are met with only occasionally, and which are otherwise difficult to identify. The book, however, must be used with circumspection, since the nomenclature is not always up to date, and a number of inadequately described genera, or genera which ought never to have been erected, are taken at their face value. The book is reproduced from typescript and has spiral binding, so usually cannot be borrowed from a library.

CLEMENTS, F. E., and SHEAR, C. L. (1931). *Genera of Fungi*. New York: The H. W. Wilson Co. Reprinted 1954.

This is a second edition, much enlarged, of a book published by Clements in 1909, and intended primarily as a guide to Saccardo's Sylloge. Keys are given to the known genera (i.e. known up to the time of publication), with references to Saccardo, but there are no separate generic diagnoses, and, in consequence, the book is not so easy to use as Lindau's (see below). Although written in English, a fair knowledge of Latin is essential to anyone using the book, as the keys are liberally besprinkled with anglicized Latin terms. There is, however, a fairly extensive glossary of Latin and English terms. The classification is, in several respects, unorthodox and must be followed with caution by the beginner. A number of plates of illustrations are included, but these cover only a small selection of genera.

CORDA, A. C. I. (1837–54). *Icones fungorum hucusque cognitorum*. 6 vols Prague.

The original edition is a very rare work, containing descriptions and beautiful illustrations of a large number of fungi, not always accurate in the light of present-day knowledge, owing to the comparative crudity of microscopes in Corda's time, but nevertheless extremely valuable. Actually, the quality of the illustrations improves markedly after Vol. I, probably because, just about this time, better microscopes were coming on to the market. In Vol. VI the illustrations are by Corda, but, owing to the latter's untimely death at sea, the then unwritten text was completed by Zobel, who also includes a short biography of Corda and a complete list of his publications.

Corda described for the first time a large number of genera and species, including many Hyphomycetales. For those doing serious taxonomic work it was, until recently, unfortunate that there are very few copies of the original edition in the country, and that none of these could be borrowed. However, a microfilm is now available at reasonable cost, produced by Micro Methods Ltd., and a reprint, somewhat reduced, has been published in Germany.

COSTANTIN, J. (1888). *Les Mucédinées Simples*. Paris: Klincksieck.

In spite of its age this is still a very useful book, describing 234 genera mostly of Hyphomycetes, and with line illustrations of about 75 per cent of them. The classification is quite different from that of Saccardo, and foreshadows that of Vuillemin.

ENGLER, A., and PRANTL, K. (1897–1900). *Die natürlichen Pflanzenfamilien*. I Teil, Abteilungen I and I*. Leipzig.

The work is accessible in most botanical libraries and is much used by taxonomists. It gives descriptions and illustrations of most genera, but no separate descriptions of species.

GILMAN, J. C. (1957). *A Manual of Soil Fungi*. 2nd Ed. Iowa: Iowa State College Press.

The first edition was used by most students of soil fungi. It gives keys to, and descriptions of, most of the fungi which have been reported as occurring in soil, and is illustrated by numerous line drawings. The new edition includes many more genera, and the illustrations include 85 photomicrographs.

LINDAU, G. (1922). *Kryptogamenflora für Anfänger*. Band II. Die mikroskopischen Pilze. 2nd Ed. Berlin: Julius Springer.

Published in two parts—Part I including Myxomycetes, Phycomycetes, and Ascomycetes; Part II including Rusts, Smuts, and Fungi Imperfecti. This is a most useful book, but, unfortunately, is out of print and is now difficult to obtain. It gives keys to all the families and most of the genera, with separate generic diagnoses and descriptions of several species in most cases. The lists of species are, necessarily in a book of this size, incomplete, and do not always include the most common forms, but for tracking down a fungus to its genus the book is extremely good. The illustrations are not so good. They are all from line drawings, small, often very sketchy, and not infrequently inadequate. No legends are given and magnifications are not stated, so that often a fair knowledge of structure is required in order to know what they are supposed to represent. They are, nevertheless, at times very useful.

RABENHORST, L. (1884–1921). *Kryptogamenflora von Deutschland, Oesterreich und der Schweiz*. Leipzig: Eduard Kummer.

A series of many volumes, illustrated, including not only Fungi but other Cryptogams. Some portions have not been completed. Band I, Die Pilze, is in ten volumes, of which the most useful to a student of moulds are Vol. 4, Phycomycetes, by A. Fischer, 1892, and Vols. 8 and 9, Hyphomycetes, by G. Lindau, 1904–10.

"Rabenhorst" is to be found in a fair number of libraries and is a very useful work of reference. The lists of species are fairly complete up to the dates of publication, and include most of those commonly occurring, but many of the descriptions are scanty.

A reprint of "Rabenhorst" has been published in America, but is very expensive. Much cheaper are the sets of microcards and microfiches of the volumes on fungi, which are now available.

SACCARDO, P. A. (1882–1931). *Sylloge fungorum omnium hucusque cognitorum*. 25 volumes. Pavia, Italy. Reprinted 1945, 1967.

This unique work was originally planned to give descriptions, in Latin, of all known species of fungi. Begun in 1882, Saccardo had by 1889 covered all the classes of fungi in 8 volumes. By this time, however, the literature of descriptive mycology was growing to such an extent that a series of supplementary volumes was issued, at intervals of a few years. The last 2 volumes, issued after the death of Saccardo in 1920, are edited by A. Trotter.

"Saccardo" is to the mycologist very much what "Beilstein" is to the chemist, indispensable but always, of necessity, several years out of date. Since no further supplements have been published since 1931, the amount of leeway to make up is enormous, and it seems very doubtful whether the work will ever be brought up to date.

From the point of view of handy reference it is unfortunate, but unavoidable, that the species of a single genus are often scattered through several volumes. The work is, however, well indexed and, with the requisite knowledge of Latin, not difficult to use. The main weakness is that many of the descriptions are, through no fault of Saccardo's, incomplete or merely fragmentary. Even at the time when Saccardo began the compilation it was impossible for one man to have first-hand knowledge of all the thousands of species of fungi. To a large extent, therefore, the treatment is non-critical, many of the diagnoses being merely transcriptions into Latin of original descriptions which were totally inadequate.

SACCARDO, P. A. (1877–86). *Fungi Italici autographice delineati.* Pavia, Italy. A collection of illustrations, mostly coloured, of a very large number of fungi, including most of the common moulds.

VERONA, O. AND BENEDEK, T. (1959–). *Iconographia mycologica.* Suppl. to Mycopathologia et Mycologia Applicata, Vol. XI–.

Each part consists of about 50 plates, each of which is a separate once-folded sheet, with a brief description or comment on the page facing the drawing of the particular species. A few of the drawings are borrowed, but the majority are by Verona. They are classified into 4 groups: A, Fungi Imperfecti; B, Phycomycetes; C, Ascomycetes; D, Protomycetes (= the parasitic micro-Basidiomycetes, rusts and smuts).

INDEX OF FUNGI. Kew: Commonwealth Mycological Institute.

Saccardo's Sylloge is a reasonably complete extract of taxonomic literature up to 1920. The continued production of a compendium on this scale has ceased to be a practical proposition, but its place, as a work of reference, has been partially filled by the *Index of Fungi.* The *Index* for the years 1920–39 was compiled by Petrak in Vienna, and is supplied in this country by the C.M.I., Kew. From 1940 the *Index* has been continued by C.M.I., being issued in 2 parts per year.

The *Index* lists names of new genera and species of fungi, with authorities, literature references, and hosts, but does not supply diagnoses. It is invaluable to anyone making a study of any group of fungi.

There are two works, not quite in the same category as the preceding but, nevertheless, of importance to the student, since they constitute the starting-points of nomenclature of fungi. Both are now available as photolithographic reprints.

PERSOON, C. H. (1801). Synopsis methodica fungorum, sistens enumerationem omnium huc usque detectarum specierum cum brevibus descriptionibus nec non synonymis et observationibus selectis. 2 partis, Goettingae.

FRIES, ELIAS MAGNUS (1821–32). Systema mycologicum sistens fungorum ordines, genera et species huc usque cognitas, quas ad normam methodi naturalis determinavit, disposuit atque descripsit. 3 volumes and Index volume, Gryphiswalde.

(1828). Elenchus fungorum, sistens commentarium in systema mycologicum. 2 volumes, Gryphiswalde.

(The original 6 volumes are reprinted in 4 volumes.)

2. MONOGRAPHS AND IMPORTANT PAPERS DEALING WITH PARTICULAR GENERA OF MOULDS

Alternaria

NEERGAARD, P. (1945). *Danish species of* Alternaria *and* Stemphylium. Copenhagen: Einar Munksgaard. London: Oxford Univ. Press.

Although published in Denmark this monograph is written in English. The main emphasis is on the role of these fungi as parasites, but the common saprophytic species are adequately described.

Aspergillus

RAPER, K. B., and FENNELL, D. I. (1965). *The Genus* Aspergillus. Baltimore: The Williams and Wilkins Co.

This is a complete revision of the genus, and entirely supersedes the 1945 Manual of Thom and Raper. All the new species which have been described during the intervening 20 years are carefully considered before acceptance of relegation to synonymy. There are five new groups, and a few species have been reclassified in different groups. A few of the groups have been somewhat excessively split up, distinctions between some of the species accepted being trivial, but this is a comparatively minor fault. There are numerous illustrations, from drawings and photomicrographs, but the colour photographs of the 1945 Manual, which were not really satisfactory, have been omitted. As in the earlier Manual, the activities of the Aspergilli receive extensive treatment.

Chaetomium

SKOLKO, A. J., and GROVES, J. W. (1948). *Notes on seed-borne fungi. V.* Chaetomium *species with dichotomously branched hairs. Canad. J. Res., Sect. C,* **26**, 269–80.

—— (1953). *Notes on seed-borne fungi. VII. Chaetomium. Canad. J. Bot.,* **31**, 779–809.

These two papers constitute a complete monograph of the genus, as known at the time of publication. A key to all the species is given in the second paper. A few species have been described since 1953, these being most readily traced by means of the *Index of Fungi*.

AMES, L. M. (1961). *A Monograph of the Chaetomiaceae.* U.S. Army Res. and Devel. Series No. 2.

This work is not entirely satisfactory, since the key is very difficult to use. It is certainly not so useful as the Skolko and Groves monograph.

Cladosporium

DE VRIES, G. A. (1952). *Contribution to the Knowledge of the Genus* Cladosporium *Link ex Fr. Baarn*: Uitgeverij & Drukkerij Hollandia.

Undoubtedly the best taxonomic study up to date of this difficult and important genus. However, de Vries does not give a check list, accounting for all the names which have been used in the genus, an essential for a real monograph.

Coelomycetes

GROVE, W. B. (1935–7). *British Stem- and Leaf-Fungi.* (Coelomycetes.) 2 vols. Camb. Univ. Press.

This is still the standard taxonomic work on this group of fungi, which includes the Sphaeropsidales and Melanconiales of Saccardo.

Doratomyces

MORTON, F. J., and SMITH, G. (1962). *The genera* Scopulariopsis *Bainier,* Microascus *Zukal and* Doratomyces *Corda. Mycol. Pap.,* No. 85. Commonwealth Mycological Institute.

The Authors show that *Doratomyces* is the correct name for the species previously assigned to *Stysanus.* The known species are described and illustrated, and there is a check list of published names.

Fusarium

WOLLENWEBER, H. W., and REINKING, O. A. (1935). *Die Fusarien.* Berlin: Paul Parey.

Wollenweber's monograph of the genus was published in several parts, in different journals, and is not readily accessible. The present book is undoubtedly easier to use, and with its aid it is possible to identify a fair proportion of *Fusarium* isolated from sources other than diseased plants. Species occurring as parasites are, of course, much easier to identify, since only very few species are likely to be found on any particular host plant.

WOLLENWEBER, H. W., SHERBAKOFF, C. D., REINKING, O. A., JOHANN, H., and BAILEY, A. A. (1925). *Fundamentals for taxonomic studies of* Fusarium. *Jour. Agric. Res.,* **30**, 833–43.

This paper still remains of importance to anyone working with Fusaria. It gives explanations of terms and criteria, and details of culture media, a good deal of the information not being included in Die Fusarien.

Mucorales

ZYCHA, H. (1935). *Kryptogamenflora der Mark Brandenburg.* Band VIa, Pilze II, Mucorineae. Leipzig: Gebrüder Borntraeger. Reprinted 1963.

Mucorineae is equivalent to Mucorales of the more usual classification. This book is easily the best treatment of the whole Order which has yet appeared.

HESSELTINE, C. W. *et al.* (1952–64). A number of papers, mostly published in *Mycologia,* treating various genera, or sections of genera, in a much fuller way than Zycha. Unfortunately, the largest and most commonly occurring genera are apparently the last to be treated, and it will be some considerable time before we can hope for a complete monograph (see Chapter IV for details and references).

Penicillium

RAPER, K. B., and THOM, C. (1949). *Manual of the Penicillia.* Baltimore: The Williams and Wilkins Co. London: Baillière, Tindall & Cox.

This entirely supersedes Thom's 1930 monograph as a working manual. The classification is more satisfactory, the keys are better, there are direct comparisons between the species of individual series, there are very numerous illustrations from both line drawings and photomicrographs, and Thom's conception of the "group species" has now been worked out completely. The book is expensive but is essential to anyone contemplating serious work on species of this genus.

Scopulariopsis

MORTON, F. J., and SMITH, G. (1962). *The genera* Scopulariopsis *Bainier,* Microascus *Zukal and* Doratomyces *Corda. Mycol. Pap.*, No. 85. Commonwealth Mycological Institute.

Describes and illustrates the known species of the three related genera, and includes a check-list of published names.

Yeasts

LODDER, J., and KREGER-VAN RIJ, N. J. W. (1952). *The Yeasts. A taxonomic study.* Amsterdam: North-Holland Publishing Co. Reprinted 1967.

The original monograph by the Dutch workers was published in 3 parts, which appeared in 1931, 1934, and 1942 respectively. The new book includes all the yeasts and yeast-like fungi, both ascosporic and non-ascosporic. Everything possible has been done to simplify the taxonomy of this very difficult group of fungi, and, unlike the previous monograph, the book is written in English.

3. BOOKS ON GENERAL AND SYSTEMATIC MYCOLOGY

AINSWORTH, G. C. and SUSSMAN, A. S. (1965–68). Ed. *The Fungi. An advanced treatise.* 3 vols. Vol. 1 (1965) *The fungal cell,* Vol. 2 (1966) *The fungal organism,* Vol. 3 (1968). London: Academic Press.

This important book reviews many aspects of fungi. Industrial mycology is touched on in Vol. 3.

DE BARY, A. (1887). *Comparative Morphology and Biology of the Fungi, Mycetozoa and Bacteria.* English translation. Oxford Univ. Press.

Although written more than half a century ago, this is still one of the best general accounts of the fungi.

BESSEY, E. A. (1950). *Morphology and Taxonomy of Fungi.* Philadelphia: P. Blakiston, Son & Co.

This is really a second edition of a book published in 1935 as "A Textbook of Mycology". It is the most modern of the books on general mycology and is to be recommended. A very valuable feature is an extensive bibliography under the title "Guide to the literature for the identification of fungi".

FITZPATRICK, H. M. (1930). *The Lower Fungi—Phycomycetes.* New York: The McGraw-Hill Book Co.

A detailed account of the morphology of the fungi of this class.

GÄUMANN, E. A. (1949). *Die Pilze. Grundzüge ihrer Entwickelungsgeschichte und Morphologie.* Basel: Birkhäuser.

A much revised edition of Gäumann's earlier book, of which Gäumann and Dodge (1928) *Comparative Morphology of Fungi* is a translation. The latter is out of date and, unfortunately, there is no English version of the newer edition.

GWYNNE-VAUGHAN, H. C. I., and BARNES, B. (1937). *The Fungi.* 2nd Ed. Cambridge Univ. Press.

M

For the beginner this is a much more readable book than Gäumann and Dodge. The classification adopted differs somewhat from that of Gäumann, and also from that of Bessey.

4. BOOKS ON APPLIED MYCOLOGY

GALLOWAY, L. D., and BURGESS, R. (1950). *Applied Mycology and Bacteriology*. 3rd Ed. London: Leonard Hill, Ltd.

A useful introduction to the study of microbiology. Gives a brief account of micro-organisms, including methods of study, and includes a number of chapters on the importance of moulds and bacteria in particular industries.

HENRICI, A. T. (1947). *Henrici's Molds, Yeasts and Actinomycetes*. 2nd Ed. by C. E. Skinner, C. W. Emmons, and H. M. Tsuchiya. New York: John Wiley & Sons. London: Chapman & Hall.

The first edition, published in 1930, was deservedly popular, especially with students of medical mycology. Henrici was in process of revising for the second edition when he died, and the work was completed by the three authors mentioned. The book has a distinct medical bias.

LAFAR, F. (1904–14). *Handbuch des technischen Mykologie. Zweite Aufl.* *Jena*: Gustav Fischer.

A series of five volumes dealing with the role of micro-organisms in industry. Vol. 1 (1904–7) deals with general morphology and physiology of organisms used in fermentations; Vol. 2 (1905–8) with the food industries; Vol. 3 (1904–6) with water, soil, and manure; Vol. 4 (1905–7) with special morphology and physiology of yeasts and mould fungi; Vol. 5 (1905–14) with the mycology of beer, spirits, wines, fruits, vinegar, leather, and tobacco. Although in some respects out of date, this is still a useful work of reference.

LAFAR, F. (1898–1910). *Technical Mycology*. English translation of the first German edition, by C. T. C. Salter. London: C. Griffin & Co.

Vol. 1 (1898) deals with bacterial fermentations; Vol. 2, Part 1 (1903), Part 2 (1910), with fermentations brought about by fungi. Whilst not so full or up to date as the later German edition, this gives a good account of the history of the fermentation industries and of traditional methods.

PRESCOTT, S. C., and DUNN, C. G. (1959). *Industrial Microbiology*. 3rd Ed. New York: The McGraw-Hill Book Co.

A detailed account of the uses of yeasts, moulds, and bacteria in manufacturing processes, together with brief sections on the morphology and classification of the three groups of organisms. The descriptions of mould fermentations, including both those of present practical importance and those which are still of theoretical interest only, are the fullest yet published in collected form. Methods of testing antiseptics and disinfectants, and the disposal of microbiological wastes, are described in appendices. The third edition describes a number of fermentations which were scarcely, if at all, developed when the second edition was published in 1949, such as transformations of steroids, production of vitamins, and the artificial culture of mushroom mycelium.

SIU, R. G. H. (1951). *Microbial Decomposition of Cellulose*. New York: Reinhold Publishing Corporation.

A very comprehensive book, attempting to cover all aspects of the action of micro-organisms on cellulosic materials. Contains excellent bibliographies.

THAYSEN, A. C., and BUNKER, H. J. (1927). *The Microbiology of Cellulose, Hemicelluloses, Pectin and Gums.* Oxford Univ. Press.

THAYSEN, A. C., and GALLOWAY, L. D. (1930). *The Microbiology of Starch and Sugars.* Oxford Univ. Press.

These two books give descriptions of a large number of organisms, including bacteria, Actinomycetes, and moulds, and summarize the results of researches on the role of these organisms in the decomposition of the various complex carbohydrates.

5. MISCELLANEOUS

AINSWORTH, G. C. (1961). *Ainsworth and Bisby's Dictionary of the Fungi.* 5th Ed. Commonwealth Mycological Institute, Kew.

Owing to the untimely death of Dr. G. R. Bisby in 1958, the final revision of the material for the present edition has been carried out by Dr. Ainsworth alone. An invaluable book, which lists the generic names of fungi, with synonyms, distribution, and number of known species for each genus. There are explanations of mycological terms, common names of fungi, short biographies of well-known mycologists, and many other interesting details concerning important groups of fungi.

ALEXOPOULOS, C. J. (1952). *Introductory Mycology.* London: Chapman & Hall.

A general account of the fungi, including Bacteria and Myxomycetes, illustrated by many drawings and some photographs. The author's nomenclature is not always in accordance with European ideas. A good point is that the etymology of all technical terms is given. The standard of the illustrations is very variable.

CHRISTENSEN, C. M. (1951). *Molds and Man.* Minneapolis: Univ. of Minnesota Press. London: Oxford Univ. Press (1952).

A semi-popular and very readable book. It gives a good account of the importance and ubiquity of micro-fungi.

CIFERRI, R. (1957–61). *Thesaurus litteraturae mycologicae et lichenologicae.* Supplementum 1911–1930. 4 vols. Pavia: Cortina.

This is a most welcome addition to the original Thesaurus (see Lindau and Sydow below). It is reproduced from typescript by photolithography.

DADE, H. A. (1943). *Colour Terminology in Biology. Mycol. Pap.* No. 6. Commonwealth Mycological Institute.

The accurate description of colours has always been a difficulty to mycologists, particularly in the preparation of diagnoses of new species. Dade's paper lists all the useful terms (and many which should be avoided), in both Latin and English, and gives precise definitions.

GARRETT, S. D. (1956). *Biology of Root-infecting Fungi.* Camb. Univ. Press. Deals with all aspects of the relationship between soil fungi and the roots of plants. The author has made this subject his special study over a long period.

GREATHOUSE, G. A., and WESSEL, C. J. (1954). Ed. *Deterioration of Materials. Causes and Prevention Techniques.* New York: Reinhold Publ. Corp.

A very comprehensive account of the causes of deterioration, including chemical, physico-chemical, and biological factors, and of the methods of protection.

GREGORY, P. H. (1961). *The Microbiology of the Atmosphere*. London: Leonard Hill (Books) Ltd.

A fascinating account of the organisms which occur in, and are distributed by, the atmosphere. It deals, to a considerable extent mathematically, with the movement of micro-organisms in moving air, and the patterns of their deposition from suspension.

HARLEY, J. L. (1959). *Mycorrhiza*. London: Leonard Hill (Books) Ltd.

The best and most complete account of this fascinating subject.

HAWKER, L. E., LINTON, A. A., FOLKES, B. F., and CARLILE, M. J. (1960). *An Introduction to the Biology of Micro-organisms*. London: Edward Arnold (Publ.) Ltd.

This is the first attempt in English to give a balanced account of all types of micro-organisms. During recent years bacteriologists have endeavoured to appropriate the term microbiology to their own discipline. This book uses the term in its correct sense, giving an interesting account of fungi, bacteria, viruses, protozoa, actinomycetes, etc., and their interrelationships.

INGOLD, C. T. (1965). *Spore liberation*. Oxford: Clarendon Press.

This interesting book is concerned with the various methods of spore discharge. All chapters except the last (which considers the Bryophyta) deal with mechanisms developed by the fungi. The book is well illustrated.

INGRAM, M. (1955). *An Introduction to the Biology of Yeasts*. London: Pitman.

An interesting book dealing with yeasts from various points of view apart from their taxonomy and uses.

LANGERON, M. (1945). *Précis de Mycologie*. Paris: Masson et Cie.

LANGERON, M., and VANBREUSEGHEM, R. (1952). *Précis de Mycologie*. 2nd Ed. Paris: Masson et Cie. [English translation from the French by J. Wilkinson (1958 and 1965). London: Pitman.]

The first edition is a unique and stimulating book. There is no attempt to describe the various Orders of the Fungi *seriatim*. Instead, a number of topics are discussed, such as: General principles of mycology; General morphology; Anastomosis; Cytoplasmic currents; Reproduction and propagative organs; Liberation and dispersion of spores, etc.

Unfortunately Langeron died before the manuscript for the second edition was completed. It was finished and published by his pupil Vanbreuseghem. The section on Medical Mycology has been enlarged. An English translation has been produced.

LARGE, E. C. (1940). *The Advance of the Fungi*. London: Jonathan Cape. Reprinted 1950.

A stimulating, and at times exciting, account of the importance of fungi in human affairs.

LINDAU, G., and SYDOW, P. (1908–17). *Thesaurus litteraturae mycologicae et lichenologicae*. 5 vols. Leipzig: Borntraeger. Photolitho reprint New York: Johnson Reprint Corporation.

A most useful compendium to anyone doing serious taxonomic work. In vols. I and II authors are listed alphabetically, and for each author there is a complete list of his mycological publications, with full references. Vol. III is a supplement, extending the period covered to 1910. In vols. IV and V the entries contained in the first three volumes are classified according to subjects. Here the only references given are the serial numbers of the entries in vols. I-III.

Plant Pathologist's Pocketbook (1968).
Compiled by Commonwealth Mycological Institute, Kew, England.
This contains much useful information on many aspects of fungi.
ROMAN, W. (1957). Editor. *Yeasts*. The Hague: Junk.
An imposing, and expensive, compilation, by a number of specialists, concerned chiefly with the uses of yeasts. The topics dealt with are: Bakers' yeast; Brewers' yeasts; Wine and fruit yeasts; Saké and similar yeasts; Food and fodder yeasts; and Yeast preparations.

6. JOURNALS, PERIODICALS AND ANNUAL REVIEWS

The following list does not include all the journals devoted primarily to mycology, but only those which fairly regularly contain papers of interest to the student of moulds, and which are reasonably accessible.

Annales Mycologici. Berlin.
Contains original papers, mostly on taxonomy of all classes of fungi, and lists of new literature, but not abstracts. It ceased publication in 1944 (but see *Sydowia*, below). Fairly recently the firm of Stechert-Hafner Inc. of New York have reprinted Vols. 1–42, 1903–44.
Biological Abstracts. Philadelphia, Pa., U.S.A.
"Published under the auspices of the Union of American Biological Societies, with the co-operation of biologists generally." This is a continuation of *Botanical Abstracts*, issued 1918–26. It is published monthly, and covers all branches of biological science. Abstracts are classified, and many of them are written by the authors of the papers abstracted. Starting in January 1967, abstracts of mycological papers appearing in *Biological Abstracts* have, in addition, been published separately, in monthly volumes entitled *Abstracts of Mycology*.
Bulletin Trimestriel de la Société Mycologique de France. Paris.
A quarterly journal with a preponderance of papers on taxonomy. Includes abstracts.
Developments in Industrial Microbiology. New York: Plenum Press.
Published by the American Society for Industrial Microbiology. The first volume appeared in 1960. The series covers a very wide range of subjects, a number of them of interest to mycologists, others concerned exclusively with the activities of bacteria.
Environmental Effects on Materials and Equipment (formerly *Prevention of Deterioration Abstracts*)
A series of abstracts, of which Series A covered biological subjects. This periodical was first issued in 1946 by the National Academy of Sciences, Washington, D.C., but ceased publication in 1965.
Hedwigia. Dresden.
This, unfortunately, ceased in 1944. It was issued monthly, and contained both original papers, relating to Cryptogamic Botany in general and to plant pathology, and excellent abstracts. Half-yearly supplements contained lists of titles of new literature and book reviews.

International Biodeterioration Bulletin. Biodeterioration Information Centre, University of Aston, Birmingham.

First published in 1965, this appears twice yearly and is free of charge. Contains mainly short papers on techniques and current work, together with some review articles. The Centre also publishes a quarterly *Reference Index Supplement*, for a modest charge, containing classified references to papers on most aspects of biological deterioration; it is invaluable to all serious workers in this field.

Materials and Organisms. Berlin.

First issued in 1965 and published quarterly. It contains original papers on biological deterioration of materials.

Mycologia. Lancaster, Pa.

This was originally, 1885–1908, *The Journal of Mycology.* It is published bi-monthly for the New York Botanical Garden, and has been since 1933 the official organ of the Mycological Society of America. It covers all branches of mycology with a preponderance of papers on taxonomy. There are no abstracts.

Nova Hedwigia. Weinheim/Bergstr.

A new journal, on the same lines as the original *Hedwigia*, but not a continuation of the latter. Besides being a vehicle for the publication of original papers, it aims at including references to all newly described taxa. It is planned to publish four to six numbers per year, the first issue being dated May 1959.

Progress in Industrial Microbiology. Ed. D. J. D. Hockenhull. London: Heywood and Co., Ltd.

Vol. I was published in 1959, and subsequent vols. have appeared yearly since then. The reviews cover a wide range of topics, many of which are of interest to the industrial mycologist.

Review of Applied Mycology. Commonwealth Mycological Institute, Kew.

Published monthly. A journal of abstracts only, covering the world literature on plant disease.

Transactions of the British Mycological Society. London.

Four parts per year are issued. Contains original papers on all aspects of mycology. Includes occasional book reviews but no abstracts.

Zentralblatt für Bakteriologie, Parasitenkunde und Infektionskrankheiten, Zweite Abt. Jena.

Issued monthly, with two volumes per year. Contains original papers in German and English, including a fair number on taxonomy of moulds and industrial mycology; also classified abstracts in German. Stopped publication with Vol. 106 in 1943, but has now resumed with Vol. 107 for 1952. (*Erste Abt., Originale* is concerned with medical and veterinary bacteriology.)

Sydowia. Horn, near Vienna.

This is described as *Annales Mycologici*, new series, and started publication in 1947.

Taxon. Utrecht. Published by the International Bureau for Plant Taxonomy and Nomenclature.

There are 6 numbers per year. The journal is concerned mainly with problems of nomenclature, and is the recognized organ for the publication of criticisms of the International Code and suggestions and proposals for its modification. Included also is a news section, and list of new taxonomic literature.

Apart from the journals devoted to mycology, papers on fungi are frequently published in the various botanical journals and in journals devoted to particular industries, whilst articles on applied mycology are to be found in numerous trade periodicals. All of these, however, are reasonably adequately covered by the journals which publish abstracts.

Appendix

Microscopy

> By the help of microscopes, there is nothing so small as to
> escape our inquiry; hence there is a new visible World
> discovered to the understanding.
>
> In natural forms there are some so small, and so curious,
> and their design'd business so far remov'd beyond the
> reach of our sight, that the more we magnify the object, the
> more excellencies and mysteries do appear.
>
> Robert Hooke, *Micrographia*, 1665

The microscope is the most important tool used in the study of micro-fungi. At the same time it is probably the most misused piece of apparatus in the average laboratory, owing to lack of knowledge, on the part of the user, of the basic principles underlying the formation of a trustworthy image. Without a proper understanding of these principles correct interpretation of fine structure, as seen through the microscope, is impossible.

The function of a microscope is to assist the eye to see objects, or details of the structure of objects, which are too small to be seen by the unaided eye. A person with good eyesight can just distinguish the dual nature of two objects whose distance apart subtends an angle of one minute at the optical centre of the eye. This means that, at a distance of 10 inches (the normal near point of distinct vision) the eye can just separate two objects which are $\frac{1}{350}$ inch apart. It is generally assumed that the virtual image seen through the microscope, by anyone with normal sight, lies at 10 inches from the eye. Hence, unless the images, formed by the microscope, of two points or lines in the object are at least $\frac{1}{350}$ inch apart, the eye will see the two points or lines as one. Actually, many people are unable to distinguish detail as fine as this, and in any case, observation at the limit of the eye's capability is very fatiguing. For most practical purposes it may be taken that objects, or images of objects, at 10 inches from the eye must be separated by at least $\frac{1}{100}$ inch (0·25 mm.) to be seen without strain.

Resolution. It will have been noticed that, so far, the emphasis has been on the ability of the eye to separate objects lying close together, rather than on the least dimensions of individual objects which can be seen. It is possible for the eye to see isolated objects of much smaller angular size than one minute of arc, provided that there is sufficient contrast between object and background, or that the eye receives sufficient light from the object to affect the retina. Several thousand stars are visible to the naked eye, but not one of them subtends any measurable angle. Viewed through a tele-scope, many stars are seen to be double, yet these pairs appear to the naked eye as single points of light because their angular separations are too small. The function of the telescope is not to magnify the individual stars, for even in a large telescope a star has no apparent size, but to form separate images of points of light which are too close together to be separated by the unaided eye, and to present these images to the eye sufficiently separ-ated to be readily seen. This function is termed "resolution".

The purpose of a microscope is similar. The first essential is the forma-tion of an image in which all the details of the object are accurately depicted. In addition, this image must be magnified sufficiently for the eye to see all the details. It cannot be too strongly emphasized that magnification without corresponding resolution is useless.

The microscope consists of three systems of lenses: the sub-stage con-denser, the objective, and the eyepiece. Both the objective and the eyepiece magnify but, of these, only the objective is concerned with resolution. The eyepiece merely magnifies the image formed by the objective and cannot reveal anything which is not already depicted in that image. Consequently, the highest power of eyepiece which can be used advantageously depends entirely on the quality of the image formed by the objective.

Objectives. The magnifying powers of objectives are inversely propor-tional to their focal lengths. With any particular objective the magnifica-tion also depends on the distance from the lens at which the image is formed, that is, on the tube-length of the microscope. However, since an objective can be made to give the best possible image at only one particular distance (strictly for only one pair of conjugate foci), the image distance, and consequently the magnification given in conjunction with any parti-cular eyepiece, are fixed by the maker. Different kinds of work demand a great range of magnification, from 10, or even less, to about 2000 diameters, and, from the practical point of view, it is impossible to make one objective capable of giving this range of magnification (in conjunction with eyepieces of various powers) and, at the same time, to give adequate resolution throughout the range. It is usual, therefore, to use at least three objectives, of long, medium, and short focal lengths.

The resolution given by an objective depends on its aperture, which really means the angular size of the cone of rays which enters the objective from any point in the object. Resolution is not strictly proportional to this angle but to what is termed "numerical aperture", or simply "N.A."

$$\text{N.A.} = \mu \sin \frac{\theta}{2}$$

where θ = the angle of the cone of rays,

μ = the refractive index of the medium between object and lens.

The maximum possible angle of the cone of rays is 180°, and sin 90° = 1.

If the medium between the cover-glass and the front lens of the objective is air, refractive index (μ) = 1, then the theoretical maximum N.A. is also 1. In practice, the limiting N.A. for so-called "dry lenses" is about 0·9. Some manufacturers produce 4-mm. objectives with N.A. as high as 0·95, but it is doubtful whether there is any advantage to be gained from taking the aperture as high as this. In general, the shorter the focal length of an objective, and hence the greater its magnifying power, the greater must be its N.A. Otherwise high powers would reveal no more than low powers, and would merely give what is termed "empty magnification". The average figures for N.A. of achromatic objectives (the usual type) are:

Focal length	N.A.
2 inch	0·08–0·15
$1\frac{1}{2}$,,	0·12–0·17
$\frac{2}{3}$,,	0·28
$\frac{1}{3}$,,	0·40
$\frac{1}{6}$,,	0·65–0·70

Objectives of very short focal length must obviously be given the greatest possible N.A. The only way to increase the N.A. above 1 is to increase the refractive index of the medium between cover-glass and lens. The practical way of achieving this is to put a drop of immersion oil, whose refractive index is the same as that of glass (1·515), between lens and cover-glass. The theoretical limit of N.A. under these conditions is 1·5, and the practical limit about 1·4. The majority of so-called $\frac{1}{12}$-inch oil immersion objectives (most of them are really $\frac{1}{14}$ inch) have N.A. about 1·3.

What do these figures mean in terms of useful magnification? The chief limit to useful magnification is the fact that the image of a point, produced by a lens or system of lenses, is not a point but a diffraction disc of finite size. Two adjacent points in an object are not resolved by a lens unless the centres of the two diffraction discs, which form the image, are separated by at least the sum of the radii of the discs. At any given magnification, the higher the N.A. of the objective the smaller are the diffraction discs. Hence, the higher the N.A. of an objective the better is the resolution, and the higher is the useful magnification. It can be shown that N.A. and M (maximum useful magnification) are related by the formula

$$d = \frac{\lambda M}{2 \text{ N.A.}} \quad \text{or} \quad M = \frac{d \times 2 \text{ N.A.}}{\lambda}$$

where λ = wave-length of the light passing through the microscope,
     d = the separation of two points or lines which are just resolvable
         when at 10 inches from the eye.

Taking the figures already given for d, $\frac{1}{350}$ inch (0·07 mm.) for good eyesight, and $\frac{1}{100}$ inch (0·25 mm.) for comfortable working, and using blue-green light of λ = 0·0005 mm., we get

$$(1) \quad M = \frac{0 \cdot 14 \times \text{N.A.}}{0 \cdot 0005} = 280 \text{ N.A.}$$

$$(2) \quad M = \frac{0 \cdot 5 \times \text{N.A.}}{0 \cdot 0005} = 1000 \text{ N.A.}$$

This means that there is seldom any advantage to be gained by working at a magnification greater than N.A. $\times$ 1000, and that, in fact, anyone with good sight can see all that the objective will reveal at a considerably lower magnification. It also means that, other things being equal, the best resolution is obtained by using light of the shortest possible wave-length. The following table gives the maximum useful magnifications for various objectives and the approximate magnifications actually obtained in conjunction with a $\times$ 10 eyepiece. The latter figures are based on a tube-length of 160 mm., for which most objectives are corrected.

Objective	N.A. $\times$ 1000	Actual magnification
$1\frac{1}{2}$ inch	170	40
$\frac{2}{3}$,,	280	100
$\frac{1}{6}$,,	650	450
$\frac{1}{12}$ inch ($\frac{1}{14}$ inch)	1300	960

It is obvious that the low-power objectives have more than sufficient N.A. for their focal lengths, and it would appear that even with the higher powers there is ample margin. However, the figures for maximum useful magnification assume the fulfilment of a number of conditions which are seldom completely satisfied in practice: that the objectives are of first-class quality; that the illumination of the object is such as to utilize to the full the capabilities of the objective; that the microscopist knows how to get the best out of his microscope; and, often the most serious limiting factor, that the object can be seen at all under conditions which should give maximum resolution.

Corrections of objectives. Simple lenses have two major faults which preclude the formation of a well-defined image, chromatic aberration and spherical aberration. Explanations of these terms are given in books on optics or microscopy, but the best account of spherical aberration and its correction is given by Dade (1958). The important point here is that the perfection of a microscope objective depends on the extent to which these aberrations have been eliminated by the use of suitable combinations of lenses. There are two main classes of objectives, achromatic and apochromatic. In achromatic objectives, the kind most widely used, chromatic aberration is only partially corrected and spherical aberration is strictly corrected for only one selected wave-length. Apochromats have almost perfect corrections for both spherical and chromatic aberrations. In consequence, they give cleaner images and can be used with light of any colour, a property which is valuable in visual work and still more valuable in photomicrography. Also, an apochromat, on account of its superior corrections, can be given an N.A. higher than that of an achromat of the same focal length, and hence can be used to give a wider range of useful magnification. Apochromats are, however, much more expensive than achromats and, in addition, have to be used in conjunction with special "compensating" eyepieces.

One other feature of high-quality objectives deserves mention, since it is a cause of disappointment to many who use apochromats for the first time. It is impossible at present to make an objective of high N.A. which will give a flat field, and the higher the N.A. the more pronounced is the curvature of the field. This means that only a small portion of the field of view is in focus at one time. For visual work this is not really a serious matter, since the fine adjustment of the microscope is in constant use when examining specimens, and different zones can be brought into focus as required. In any case, a particular structure which requires special examination can always be brought into the centre of the field, where resolution is at a maximum with any objective. In photomicrography curvature of the field is more serious and, with ordinary equipment, it is not possible to take a picture showing more than a small portion of the field of view in sharp focus. There are on the market special eyepieces which give flat projected images, but they are of use for photomicrography only and cannot be used at all for visual work because the eyepoints are inside the eyepieces.

Substage condensers. In microscopy the great majority of specimens are examined by transmitted light. Under these conditions the full resolving power of an objective is realized only when the object is illuminated by a solid cone of light, strictly aligned with the axis of the objective and of N.A. equal to that of the objective. The substage condenser is a compound lens of short focal length designed to give such a solid axial cone of illumination. In practice, however, it is found that only under special conditions can the full aperture of an objective be utilized. Also, under any conditions, a cone of illumination of larger aperture than that of the objective produces excess of glare, which obliterates fine structure. For these reasons it is necessary to have some means of regulating the N.A. of the condenser, and this is provided by the iris diaphragm fitted just below the back lens.

Since a solid axial cone is obtained when the source of light is focused on the object, some means has to be provided for moving the condenser up and down. On the cheaper microscopes this is a quick-acting screw, or even a simple sleeve through which the condenser slides, but, on the better models, the movement is effected by rack and pinion. In order that the cone of light from the condenser may be aligned with the axis of the objective there should be centring screws fitted to the substage. Assuming then that the condenser is appropriately fitted, there are a few elementary rules which must be observed when setting up the microscope for use.

(1) Only the flat mirror must be used with a condenser. The lens system is computed to work with a parallel or slightly divergent beam of light and is quite unsatisfactory with the convergent beam from the concave mirror.

(2) The source of light must be centred. If no bull's-eye condenser is used (see below) the objective, eyepiece, and condenser should be removed and the mirror adjusted until the source of light appears to be on the axis of the tube. If a bull's-eye is required this should now be inserted so that an image of the light source is thrown on to the centre of the mirror, and the bull's-eye itself appears to be central when viewed down the micro-scope tube.

(3) The substage condenser must be centred. Place in position a low-power eyepiece, a $\frac{2}{3}$- or $\frac{1}{3}$-inch objective, and the substage condenser. Put a slide on the stage and focus it, then rack up the condenser until the light source, or the surface of the bull's eye if one is used, is in focus also. Now close the substage iris as far as it will go and focus this by carefully racking up the microscope tube. When the image of the iris is as sharp as possible bring it to the centre of the field of view by means of the centring screws on the substage, and make sure that the iris opens and closes concentrically with the field. Open the iris somewhat and refocus the specimen on the stage. If these operations have been correctly performed the object is now illuminated with a solid axial cone of light, the angle of which can be con-trolled by the substage iris. If the source of light is not quite uniform, and its image interferes with observation, the condenser may be moved, up or down, *very slightly* in order to even out the light. After this the position of the condenser should not be altered whatever objective is used to examine that particular slide. It will probably need refocusing when the slide is changed for another one, because slides vary in thickness. The common practice of racking the condenser down when using a low-power objective, and up for high powers, is entirely wrong. The iris should always be used for controlling the angle of the cone of rays to suit objectives of varying N.A.

With some condensers, particularly the cheaper kinds, the iris cannot be focused in the manner described. In such a case focus the slide, focus the source of light by means of the condenser, remove the eyepiece and look down on the back lens of the objective. If the iris is now opened and closed it will be seen that an image of it appears to lie on the back lens. Adjust the centring screws until this image appears to be central.

(4) Adjust the N.A. of the condenser. Remove the eyepiece and look down the tube at the back lens of the objective. This should appear to be a perfectly even disc of light, or should show a central disc of light sur-rounded by a dark annulus. On opening or closing the substage iris this disc should increase or decrease in size but should always be of uniform brightness. Set the iris so that the diameter of the disc of light is $\frac{2}{3}$ to $\frac{3}{4}$ that

of the back lens of the objective. The object is now illuminated with a cone of light of N.A. $\frac{2}{3}$ to $\frac{3}{4}$ that of the objective. Good objectives work best under these conditions on the majority of subjects. With colourless and very faint objects it may be necessary to close the iris further in order to see anything at all, but inferences from observations are then not always to be trusted, particularly as regards dimensions, since much of what the eye sees consists of diffraction rings around the exceedingly faint images.

It has been stated that only rarely can the full aperture of an objective be utilized. This is because, with a full cone of illumination, there is nearly always a certain amount of glare which tends to create a kind of fog over the image. Only a first-class objective will, under any conditions, stand a cone of illumination corresponding to more than $\frac{3}{4}$ of its N.A. (A full analysis of the question of glare is given by Beck in *The Microscope*.) In consequence, the maximum theoretical resolving power of objectives is seldom realized. However, what may be termed the "working N.A." of an objective is not cut down to that of the cone of illumination, the general rule being that resolution is proportional to the mean of the N.A.s of objective and condenser. In order to know the working N.A. at any time it is useful to have a scale fitted to the condenser mount which will indicate the N.A. of the condenser for any setting of the iris.

There are several different kinds of condensers, varying in focal length and perfection of corrections. The type more used than any other is known as the Abbe condenser. It is very imperfectly corrected for spherical aberration and gives a cone of light which is by no means uniform in intensity. Nevertheless, good work can be done with these condensers if they are used properly, and they are inexpensive. They are less sensitive to errors in centration than the more highly corrected types and should be chosen for preference if the substage is not fitted with centring screws. Achromatic condensers are comparable with objectives in their corrections for chromatic and spherical aberrations. There are three types on the market, two of them designed to work dry, that is, with air between condenser and slide, and the third made to work with the top lens oiled to the slide, just as an oil-immersion objective is oiled to the cover-glass. The two dry condensers differ in focal length and size of back lens, but either type will give a perfectly even cone of light of maximum N.A. between 0·9 and 1·0, and so will cover the full aperture of any dry objective and about $\frac{3}{4}$ of that of a $\frac{1}{12}$-inch oil-immersion. The large type of longer focal length is the easier to use and is the more satisfactory for use with medium-power objectives. The top lens can be removed by unscrewing, to give a condenser of greater focal length and lower N.A., suitable for low powers. The increased focal length means that the size of the illuminated area is increased, whilst the reduced N.A. (about 0·4) is ample for the $\frac{2}{3}$-inch and lower-power objectives.

The N.A. of an oil-immersion condenser is usually 1·3, or somewhat higher, so that it will permit the use of a $\frac{1}{12}$-inch objective at full aperture.

Such a condenser can also be used dry, when the maximum N.A. is 1·0, and the top lens can be removed to give a condenser of increased focal length and lower N.A. These condensers are thus adequate for almost all purposes, but they are more expensive than dry achromatic condensers, and it is seldom that a condenser of N.A. greater than 1·0 is required.

McClure (1958) claims that the substage condenser is unnecessary if a thin piece of flashed opal glass be placed immediately below the slide. Attempts to obtain thin opal glass having been unsuccessful. I have tested the method using instead a piece of thin, but optically fairly dense, opal perspex. There is no doubt that the method gives adequate resolution, but it has two disadvantages. For high magnifications, which demand illumination of high intensity, it is impossible, without a lamp of very high candle-power, to obtain a sufficiently bright image. Also there is no means of controlling the aperture. Unless work is confined to the use of fairly low magnifications, it is probably cheaper to buy an achromatic condenser than to obtain a very powerful and reliable illuminant (see also p. 360).

Correction for thick slides. Most achromatic condensers have one limitation which is well known, although the method of overcoming it is not. Condensers are made to focus through slides up to about 1·2 mm. thick, with a lamp distance of 8 inches. Slides vary much in thickness and not infrequently one is called on to examine one which may be as much as 1·5 mm. thick. Also, cavity slides are much used in some laboratories, and these are usually 1·5–2·0 mm. thick. It is not surprising that the majority of photographs from hanging-drop preparations show poor resolution and excessive diffraction effects. The trouble is aggravated when the distance from condenser to lamp is more than 8 inches, which is very frequently the case, particularly in photomicrographic apparatus. The method of dealing with thick slides is to treat the condenser as if it were short-sighted, as in effect it is. Any optician will provide, for a few shillings, a negative spherical spectacle lens, ground to a diameter of $1\frac{3}{8}$ inches, so that it will fit into the swing-out ring below the condenser. A lens of $2\frac{1}{2}$ diopters minus will enable the condenser to focus through all but the thickest cavity slides, and a second lens of minus $3\frac{1}{2}$ diopters may be reserved for these.

Correction for cover-glass thickness. After the microscope has been set up, and the condenser and illuminant arranged to give a solid axial cone of light of the correct N.A., there is usually one further adjustment which must be made in order to obtain the maximum resolution. A dry objective gives a correct image only at one specified tube-length and with one definite thickness of cover-glass. Any departure from correctness of either dimension introduces spherical aberration. Tube-length is, of course, controllable, but appreciable variation is found in even the best quality coverglasses. In addition, it is often necessary, when mounting moulds in fluid, to leave a fair amount of medium under the cover-glass, because the application of pressure would break up delicate structures. Any fluid between a part of the specimen under examination and the cover-glass has a similar effect to an increase in thickness of the glass.

The spherical aberration introduced by using a cover-glass of the wrong thickness may very largely be corrected by an alteration in the tube-length of the microscope. The method of doing this will be described in detail because it is not given in most of the books on microscopy.

It is much easier to learn the method by observation of a bright object on a dark ground than of an opaque object on a light ground. A very suitable object, provided a powerful source of light is available, can be prepared by lightly silvering a No. 1 cover-glass and mounting it on a slide, silver side down, in a drop of balsam. On examination under the microscope the film will be found to contain many holes, some of them very small. Bring a minute hole to the centre of the field and, with the substage iris wide open, examine with a $\frac{1}{3}$- or $\frac{1}{6}$-inch objective. At the correct focus the image of the hole will be a bright spot, probably with a faint light ring around it. Now put the object *slightly* out of focus, by focusing first down then up. If the tube-length is correct the appearances on both sides of correct focus will be alike, a somewhat ill-defined ring of light. If the tube-length is incorrect, or the cover-glass of the wrong thickness, the images will be different. When the tube-length is too short, or the cover-glass too thin, the appearance on focusing down is a more definite ring, and on focusing up a blurred spot brightest in the centre. A tube too long or a cover-glass too thick gives the opposite effects. It is an easy matter to adjust the tube-length until the appearances on both sides of the focus are alike, or very nearly alike, and with a little practice, anyone can make successive determinations of correct tube-length which agree within 1–2 mm.

The method of conducting the so-called "star test", and the information to be obtained from it, are discussed in more detail by Slater (1957).

Having become accustomed to making the adjustment with a bright object, take any ordinary slide and find a minute speck of dust (it will be a most unusual slide if none can be found), bring it to the centre of the field and examine it by throwing it slightly out of focus first in one direction and then the other. The appearances are analogous to those described above, but the out-of-focus images are much fainter. If the tube-length is too short for the particular cover-glass the speck expands into a faint ring on focusing down, and becomes fuzzy on focusing up.

High-power dry lenses are very sensitive to variations in cover-glass thickness. With a $\frac{1}{6}$-inch objective a variation of 0·01 mm. will involve altering the tube-length by about 10 mm. This means, of course, that with a cover-glass only a little too thick, or with a thick layer of mounting medium, the tube of an ordinary microscope cannot be shortened sufficiently to effect the correction. On the other hand, it is quite impossible to obtain even a tolerable image of an uncovered specimen, as this would involve extending the tube far beyond its limit. It is not surprising that many bacteriologists, working regularly with uncovered slides, use a $\frac{1}{6}$-inch objective only for locating a suitable field, and aver that the majority of bacteria cannot be seen clearly with a lower power than a $\frac{1}{12}$-inch. Actually the smallest bacteria can be clearly seen with a $\frac{1}{3}$-inch objective, provided that the tube-length is adjusted correctly. Metallurgists get over the difficulty by having all their objectives specially corrected to work with no cover-glass.

The $\frac{1}{3}$-inch objective is much less sensitive than the $\frac{1}{6}$-inch to variation in cover-glass thickness, but is, on the other hand, more sensitive to deviations from the correct tube-length. A lens corrected for 160 mm., in conjunction with a cover-glass 0·17 mm. thick, can be corrected for use on uncovered specimens by extending the tube to about 200 mm.

The $\frac{2}{3}$-inch or 16-mm. objective is little affected by moderate variations in cover-glass thickness. Hence it is a good plan, when purchasing such a lens for mycological work, to have it corrected for 200 mm. Most manufacturers supply, at no extra cost, objectives corrected for any specified tube-length. The $\frac{2}{3}$-inch gives a somewhat higher magnification at 200 mm. than at 160 mm. and can, by shortening the tube, be corrected to give a good image of a mould in a culture tube with walls up to about 2 mm. thick.

Oil-immersion objectives are not affected by variations in cover-glass thickness, provided that the oil used is, as it should be, of the same refractive index as the glass. The correct tube-length is usually marked on the objective mount, but it is advisable to check this by the method described above. A thick layer of mounting fluid, if its refractive index differs from that of glass, will slightly affect the performance of an oil-immersion lens. However, most objectives of this type have very short working distances, with the result that a layer of fluid, between specimen and cover-glass, sufficiently thick to necessitate tube-length correction, would probably make it impossible to focus the specimen.

The oil used for this purpose until comparatively recent times was thickened cedar-wood oil. It is a semi-drying oil, and, if left on slides or the front lenses of objectives, sets to a varnish-like consistency and is not easy to remove. The oil usually supplied with O. I. objectives at the present day, and also readily obtainable in bulk, is a synthetic fluid, which is non-drying. Although it is always best to remove the oil from both objective and slide immediately after examination is complete, no harm is done if this is omitted, and the oil left *in situ* for several days. The main disadvantage of this is that the tacky surface attracts dust.

To remove the oil from an objective the front lens should first be gently wiped with a very soft cloth, or, better, a piece of the special lens tissue made for the purpose, then wiped again with a piece of tissue moistened with a drop of pure xylene. Alcohol should on no account be used. For cleaning slides either unsealed or ringed with shellac cement, the most convenient method is to use an acetone wash-bottle with a fine jet. This has the advantage that no wiping, with consequent danger of disturbing the specimen, is necessary. Washing an unringed slide removes also any mounting medium around the cover-glass, and leaves the slide ready for ringing. Slides sealed with nail varnish cannot be cleaned with acetone, since this is a solvent for the cement. The method recommended by the Commonwealth Mycological Institute is to wipe off most of the oil with a soft cloth, then remove the residual smear by adding a drop of saliva and wiping this off. There is, of course, always a danger of moving the specimen, any portions logged for reference being moved. The oil can also be removed by xylene without the cement being affected. It is best to absorb as much of the oil as possible by carefully applying the edge of a piece of thick filter-paper, then add a drop of xylene and absorb this in the same way. If this is carefully done there should be no disturbance of the specimen.

Illuminants. Daylight is satisfactory only for very low powers. For examination of cultures laid across the microscope stage it is one of the best illuminants, because the condenser cannot be focused through the tube and a diffuse source of light gives the most satisfactory results. For medium and high powers, with the condenser properly focused, diffuse light gives rise to excessive glare, the best illuminant being a fairly small source of uniform and high intrinsic intensity. For visual work the Welch lamp, the thorium disc lamp, and the flat-wick oil-lamp are all suitable. The Welch lamp uses a coiled filament electric lamp, the unevenness in the beam of light being smoothed out by multiple reflections in a thick glass rod placed end-on to the filament. In the thorium lamp a small cylinder of composition similar to that of gas mantles is made incandescent by means of a flat, non-luminous gas-flame. For very high powers, particularly when colour screens are used, and for most photomicrographic work, a more intense source of light is desirable. The two most satisfactory kinds of lamp are the enclosed tungsten arc, the best-known type of which is the "Pointolite", and the high intensity low voltage lamps. There are two forms of the latter. One, known as the "solid source lamp", has a filament composed of a number of close coils in parallel, placed very near together, the type most suitable for microscopic work being rated at 6 volts, 8 amps. The other type uses a ribbon of tungsten, which gives a rectangular light source of absolutely even intensity, and is rated at 6 volts, 18 amps. Both types of lamp are run from the A.C. mains, using a transformer to reduce the voltage. The great advantage over the "Pointolite" is that a rheostat or stepped transformer may be used to regulate the intensity to suit objectives of different powers.

Bull's-eye condensers. When the source of light is very small, as is the case with the "Pointolite" lamp and, to a lesser degree, the thorium lamp, its image formed by the substage condenser is also very small. This image may be big enough to fill the field of a $\frac{1}{12}$-inch objective but covers only a small portion of the field of a low-power objective. The size of the image, i.e. the area illuminated, may be increased by interposing a suitable lens, usually termed a bull's-eye condenser, between the lamp and the mirror. The lens is placed, with its optical axis aligned with that of the rest of the system, at such a distance that an image of the illuminant is formed in the back focal plane of the substage condenser. As a first approximation it may be placed so that the image falls on the mirror, and a final adjustment be made later. The bull's-eye is now treated as the source of light, its image being focused on the specimen. Finally the position of the bull's-eye is adjusted until its image, as seen through the microscope, is an even disc of light.

The cheapest kind of bull's-eye is a single plano-convex lens of $2\frac{1}{2}$–3 inches diameter and 3-4 inches focal length, but this is at best a makeshift since it forms a very poor image, due to excessive chromatic and spherical aberrations. A popular type with photomicrographers is that designed by the late E. M. Nelson, and known by his name. It consists of two lenses and gives a comparatively good image. Its main fault is that, owing to its long focal length, the image of the illuminant which it forms is not sufficiently large to cover the back lens of the substage condenser, thus cutting down the N.A. of the latter, unless the lamp is about 2 feet from the mirror. At this distance, the image of the bull's-eye is not usually large enough to cover a sufficient proportion of the field of a $\frac{2}{3}$-inch objective, even with the top lens of the substage condenser removed. At a distance of about 10 inches the bull's-eye will throw an image big enough to utilize about 0·6 of the N.A. of the condenser, and this is sufficient for most dry objectives. When using an oil-immersion objective, provided that a bull's-eye is required at all, the alternatives are to move the lamp further from the mirror, or to substitute a bull's-eye of shorter focal length. A suitable short-focus lens is the field lens (the larger of the two) of a ×6 eyepiece. It is a single lens but is comparatively thin and fulfils the purpose admirably. The author's outfit includes such a lens, mounted so that it is readily interchangeable with a Nelson condenser.

An ingenious and useful lamp is described by Barer and Weinstein (1953). There are three auxiliary lenses, A, B, and C, and these are used as A, AB, or AC, giving a range of focal lengths. Without moving the lamp as a whole it is possible to obtain a beam of adequate N.A. for objectives from 16 mm. to 2 mm., and simultaneously to fill the fields of view, with any of the usual types of dry condenser, and without the necessity for unscrewing the top lens. A "solid source lamp" is used, rated at 6 volts, 8 amps, with the intensity controlled by a multiple-contact switch and tapped transformer. The lamp is made by the Singer Instrument Co., Reading.

The apparatus is very satisfactory in use, the intensity of illumination being sufficient for photomicrography at high magnifications. A slight inconvenience lies in the fact that for low powers, ×100 or less, the light at its minimum intensity is still too bright for visual observations. However, this is readily remedied by the use of neutral light filters. The front of the lamphouse is totally enclosed, so that practically no strong light

issues in the direction of the microscope. Also, since in the Köhler system
of illumination, it is an image of the bull's-eye, or of the iris in front of it,
which is focused on the specimen, it is a distinct advantage to have the
position of the bull's-eye fixed, and to focus the light by moving the lamp-
bulb. In the majority of other types of lamps the bulb is fixed and the
bull's-eye moves for focusing.

Another set-up for microscopic illumination is described by Baker and
Llowarch (1959). The illuminant is the Philip's "mirror-condenser lamp",
originally designed for use in cinematograph projectors. The bulb, 8 volts,
6·25 amps, has an ellipsoidal back and hemispherical front, the whole of
the glass surface being silvered, except for a small circular window in the
centre of the front. The filament is a tight coil placed at the primary focus
of the ellipsoid, the reflected light being concentrated at the secondary
focus, outside the bulb. Light striking the silvered portion of the hemispherical front is twice reflected, and is concentrated at the same focal
point. In order to even out the image of the filament, a glass rod, 10 cm by
12·5 mm., with one end ground flat and the other left plain, is placed
axially, with the ground end in or near the image of the lamp filament.
The light is evened out by multiple reflections as it traverses the rod. A
bull's-eye, preferably the spherically corrected type, fitted with iris diaphragm and colour-filter holder, transmits the beam to the substage condenser. The illumination of the slide is very intense, but may be moderated
as required by means of a rheostat. It is emphasized that, to ensure long
life, the lamp should be started with low current. If the full current be used
for starting, the lamp burns out fairly quickly. A lamphouse is not needed,
since the only light issuing from the bulb must pass through the
small circular window. Free exposure to the air at least ensures adequate
cooling.

An interesting and useful paper on illuminants and principles of illumination for microscopy is by Ockenden (1946). The illuminants which
were available at the time are described and evaluated.

The bull's-eye, contrary to a somewhat general opinion, does not increase the intensity of illumination. It increases the amount of light gath-
ered from the lamp, but this light is spread over a larger area. The real
purpose of the bull's-eye is to increase the area of illumination on the
specimen under examination.

In photomicrography particularly it is advisable to be able to control the
size of this illuminated area, since otherwise much unwanted light enters
the camera and may fog the image. Even in visual work it is found that
resolution is improved by restricting the area of illumination to about half
the diameter of the field of view. Control is effected by means of an iris
diaphragm, mounted as close as possible to the bull's-eye and on the side
of it away from the lamp. In such a set-up there are thus two diaphragms.
The one below the substage condenser controls the N.A. of the cone of
light entering the objective, and has no effect on the size of the illuminated

area. The bull's-eye diaphragm has exactly the opposite effect. It controls the proportion of the field of view which is illuminated and has no effect on the effective N.A. of the substage condenser.

Colour filters. Filters, or screens as they are often called, are used for a number of different purposes. The first of these is the use of neutral screens to moderate a light which is too intense. These filters are absolutely necessary when one of the more intense sources of light is used with low-power objectives. They can be obtained in a range of different densities, each transmitting a definite fraction of the incident light without appreciably altering its colour.

The second use is for the control of contrast. A filter of colour complementary to that of the specimen will increase contrast. For example, many of the photographs in this book were taken from specimens stained a faint yellow with picric acid, a blue filter being used to obtain as much contrast as possible. A filter of nearly the same colour as the specimen will show up details of structure, or reduce excessive contrast. Many of the photographs of dark coloured moulds were taken with orange light in order to show details, since brown is merely orange plus grey.

Another use of colour filters is to increase resolution. Other things being equal, resolution is inversely proportional to the wavelength of the light forming the image. Hence maximum resolution is attained by using deep blue light (violet light being very unpleasant to work with), provided that the objective is fully corrected for blue light. Only apochromatic objectives and specially corrected achromats can be used to achieve better resolution by this means. Ordinary achromats give maximum resolution with a bright green filter, since they are strictly corrected for a wave-length in the yellow-green portion of the spectrum.

Colour filters may be obtained as circles to fit the substage ring, or as 2-inch squares for use between lamp and mirror. The filters which give most precise control over the wave-length of the light entering the microscope are made of dyed gelatine, cemented between thin sheets of carefully selected glass. A wide range of these is available, making it possible to select a narrow band from any part of the spectrum. They should be handled with care, must not be exposed to heat, and should preferably be kept in the dark when not actually in use. Filters of solid glass, in which the colouring matter is incorporated during manufacture, are also available. They are more robust than the gelatine filters, are more resistant to heat, and, whilst they do not show quite such clean transmission bands, are capable of meeting most of the requirements of the microscopist.

Fleming (1966) has described an interesting colour-filter which transmits a very narrow band of green light, particularly suitable for use with achromatic and fluorite objectives. It is composed of three filters, plus a heat-absorbing filter between colour screen and lamp. Four different combinations of filters are described, three of these using screens of American origin. The fourth uses Chance-Pilkington solid glass filters as follows: ON 16, 6 mm. thick; OY 3, 2 mm.; OGR 1, 2·65 mm. The heat-absorbing glass is HA 1, 2 mm.

The use of opal glass to increase contrast. An interesting method of increasing the contrast of fine coloured structures is described by Kurozumi and Shibata (1960). A piece of thin opal glass is placed immediately below the slide, the substage condenser being focused in the usual way. It is recommended that a cell to accommodate the opal glass be made by cementing two one-inch squares of thin glass (portions of a thin slide) on to the ends of a slide. The preparation to be examined rests on the two squares, leaving a pocket into which a piece of thin opal glass, one inch wide, can be inserted as required and readily removed when not needed. Naturally, with such a combined thickness of glass below the specimen, the substage condenser will require its spectacle lens (p. 357).

Thin opal glass, i.e. not more than 1 mm. thick, is far from easy to obtain, and thicker glass cuts off too much light. A substitute is very thin opal Perspex sheet (see also p. 357).

Choice of equipment. The choice of a microscope stand is often a question of personal fancy. It may also be a question of cost. Good work can be done with a comparatively inexpensive microscope of reliable make, but, of course, extra money spent on the stand means, or should mean, ease of working and a great saving of time. Of late years a number of manufacturers have departed from conventional design and their products have, to modern eyes, a very attractive appearance. "Modern lines", however, should not blind the purchaser to the importance of sound construction. Particularly to be noted are the smoothness of the sliding motions, the robustness and simplicity of the fine adjustment, and the provision of means for taking up slack in bearings, since the best machine will show signs of wear with long use. An interesting paper by Dade (1956) discusses, amongst many other subjects of moment, the question of simple versus complex equipment.

A properly fitted substage, with rackwork focusing adjustment and centring screws, is a good investment at the time of buying the stand. It can be fitted subsequently but is then seldom so satisfactory. A mechanical stage is sooner or later a necessity if much microscopic work is done. The built-in type is the better job, but some stages which can be attached to existing microscopes are quite adequate and need not be purchased until actually required.

As stated above, three objectives are normally required. Most of the complete outfits sold include $\frac{2}{3}$-, $\frac{1}{6}$-, and $\frac{1}{12}$-inch objectives but, for mycological work, this is not the best choice. A lower power, 2- or $1\frac{1}{2}$-inch, is required for examination of living cultures and, if only three objectives are bought, the $\frac{2}{3}$- and $\frac{1}{6}$-inch are best replaced by a $\frac{1}{3}$-inch. Oil-immersion lenses of $\frac{1}{7}$-inch and $\frac{1}{8}$-inch focal lengths are easier to use than the so-called

$\frac{1}{12}$-inch objectives, and give sufficiently high magnification for most purposes. However, a long series of measurements can be very fatiguing unless the magnification is higher than is actually necessary, and here the $\frac{1}{12}$-inch has the advantage. If only three objectives are to be used the best choice is then $1\frac{1}{2}$-, $\frac{1}{3}$- and $\frac{1}{12}$-inch. It may be found to be slightly cheaper to add a $1\frac{1}{2}$-inch to the standard outfit of $\frac{2}{3}$-, $\frac{1}{6}$-, and $\frac{1}{12}$-inch but, unless the microscope is provided with a quadruple nosepiece (or objective changers) instead of the usual triple nosepiece, there is bound to be frequent replacement of the $1\frac{1}{2}$-inch by the $\frac{1}{12}$-inch, and *vice versa*. If a $\frac{1}{3}$-inch or 8-mm. objective is purchased, an apochromat is a sound investment, particularly if photomicrography is contemplated. Its N.A., 0·65, is as high as that of many $\frac{1}{6}$-inch achromats, but it has a longer working distance and gives a brighter image. In addition, as stated above, it is much less sensitive than the $\frac{1}{6}$-inch objective to variations in cover-glass thickness. It will easily stand a ×20 eyepiece to give a magnification of about 400 diameters, and is capable of giving almost perfect images, with light of any wave-length, of both covered and uncovered specimens.

Two eyepieces are required with the objectives recommended and these should be ×8 or ×10 and ×15, instead of the usual ×6 and ×10. Better still, if an 8-mm. objective is used, is to have three eyepieces of powers ×8, ×12, and ×20.

Micrometry. For taking measurements of parts of fungi, and, of course, of other specimens, two pieces of apparatus are required, a stage micrometer and an eyepiece micrometer. If much measuring is to be done it is best to acquire an extra eyepiece to accommodate the disc micrometer permanently. The power of the micrometer eyepiece is a matter of individual preference, some workers finding a low power suitable, whilst others obtain the best results with a ×12 or ×15. It should be fitted with a glass disc micrometer, this being accurately calibrated, in conjunction with each of the objectives, against a stage micrometer. It will usually be found that the eyepiece micrometer is in focus when resting on the diaphragm, but, if it is slightly out, it is not difficult to move the diaphragm the necessary amount. Most manufacturers list one or more special micrometer eyepieces, fitted with focusing eye-lens. Until comparatively recently these were always of low power, but now a fairly wide selection is available.

A source of some inconvenience is that the value of one division of the eyepiece micrometer is seldom a whole number, and this means that all measurements have to be converted to real values. By altering the tube-length of the microscope it is possible to arrange that one division of the eyepiece scale = 10μ with the 16 mm. objective, 5μ with the 8 mm., and 1μ with the 2 mm. oil-immersion, but, of course, the quality of the image is diminished just when it should be as good as possible. The solution adopted by the author was first, to determine the exact tube length which gives 1 div. = 1μ with a 2 mm. objective, using a ×15 eyepiece. An objective was then bought, adjusted by the manufacturer to work exactly at this tube-length. This has proved to be highly satisfactory.

It is interesting that a Ramsden type ocular gives the desired values at a normal tube-length, because the diaphragm, and hence the micrometer, are placed outside the field lens, and not within the eyepiece. Unfortunately this type of eyepiece is not easy to obtain.

The stereoscopic binocular microscope. This type of microscope, of which the Greenough Binocular is the prototype, consists essentially of two complete microscopes inclined to one another at such an angle that true stereoscopic vision is attained. The image, unlike that in the ordinary microscope, is right side up, since each tube of the instrument contains an erecting prism. The objectives and eyepieces are carefully paired, to ensure that the magnifications of the two microscopes are identical. There is provision for adjusting for interocular distance, and one eyepiece has a focusing adjustment. The range of magnification is usually about $\times 6$ to $\times 140$.

This type of instrument is widely used by botanists and zoologists for dissections. It is also a useful addition to the equipment of the mycologist for the examination of mouldy materials, and for the isolation of moulds. With the aid of a fine needle it is not difficult to pick off a few spores from a single spore-head. The microscope is particularly valuable in taxonomic work for the examination of herbarium specimens.

The phase-contrast microscope. The technique of phase-contrast has come into prominence during comparatively recent years. Its value lies in the examination of material which cannot be stained, and some wonderful photographs and films have been produced, illustrating such processes as cell-division. Contrast is obtained by differences in alteration of phase of the incident light by different types of structure.

So far this method of examination has found little application in the study of moulds. It does not give increased resolution, but rather the reverse, since an annular diaphragm is placed in the path of the incident light.

PHOTOMICROGRAPHY OF MOULDS

Photomicrography is treated fully in a number of excellent books, and the following notes are intended to assist the mycologist, who may be interested in keeping a pictorial record, to overcome some of the difficulties particularly associated with the photography of moulds.

Slides. The main difficulty is the preparation of suitable slides. Given a satisfactory preparation, the photography usually presents few difficulties, provided that the worker understands his microscope and knows something of the technique of ordinary photography. During visual examination of a specimen the fine adjustment of the microscope is in constant use, giving a sense of depth of focus, but, when the image is projected on to a photographic plate, only one plane can be in focus. Hence, for photography, it is necessary to have a slide which shows all the essential features of the mould virtually in one plane. There is no general method of securing perfectly flat mounted preparations. The specimen should be teased out as much as possible without breaking up sporing structures, and the cover-glass should be pressed well down, but often, many slides have to be prepared in order to obtain one that is satisfactory. On the other hand, it frequently happens that slides not specially prepared for photography, and made with the minimum of manipulation, are almost perfect. In such cases it is an advantage if a set-up is available, like the one described below, with which a photograph can be obtained quickly and with very little trouble. For the photography of moulds which break up very easily it is usually necessary to use slide cultures, or cultures on cellulose sheet (see Chapter XI).

Finders. It is often desirable or necessary to mark the position of a particular structure on a slide, so that it can be found again when required. This is usually accomplished by the use of the verniers which are fitted to most mechanical stages. The position of the object can then be recorded on the slide-label. Sometimes it is necessary to examine a valuable slide with a different microscope. In this case it is possible to determine the relationship between the two sets of verniers, but this is seldom really satisfactory in practice.

There are several devices for marking the slide, of which the best is the diamond marker. This is shaped more or less like an objective, but, in place of the front lens there is a small diamond chip, mounted on a rotating collar, and with a screw for decentring the diamond. The fitting is balanced against a delicate spring, so that, when the diamond point is lowered on to the cover-glass, it presses only lightly. Rotation of the collar causes the diamond to mark a ring on the cover-glass, the size of the ring being determined by the amount the diamond is decentred. It is easy to find the ring on any other microscope, and, with a little search, to locate even a very small object within it.

Another method of recording position is the "England Finder", a modification of a device which has been known for many years as the "Maltwood Finder" (Maltwood, 1858). Both depend on having some means of locating a slide accurately on the stage of the microscope. Most mechanical stages are already provided with such a means. If no mechanical stage is fitted to the microscope, the simplest means of accurate location is by means of three pins in the stage.

The Maltwood Finder consists of a slide completely covered with small engraved squares, all numbered. The England Finder is similarly engraved, but each square is divided as shown in Fig. 173, the squares being consecutively numbered horizontally and lettered vertically. To use the finder the object is centred in the microscopic field. The slide is then carefully removed and the finder substituted. The number and letter of the square, and the position of the centre of the field within the square are recorded. To find the object on another microscope the finder is examined first and the logged position found, then the slide is substituted. A simple method of recording the position of an object which is visible or recognizable only under a high power, is to draw a sketch of the relevant square, or portion of it, on the slide label, or on a separate label on the other end of the slide, and mark with a dot the exact location within the square.

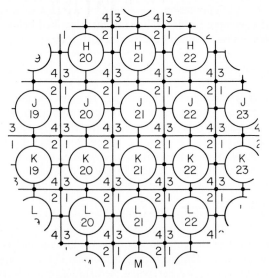

FIG. 173.—Portion of England finder. × 15.

Diffraction and contrast. One of the most frequent troubles is to obtain sufficient contrast in a photomicrograph without introducing diffraction effects, manifested as a kind of halo round the images. In the formation of an optical image there is always a certain amount of diffraction, but this should not be sufficient to be noticeable in a photograph or to blur outlines. Unfortunately a photographic negative will often show diffraction rings when these have been quite unnoticed during visual examination of the specimen, or even when viewing the image on the ground-glass screen of the camera.

The cause of excessive diffraction is illumination of the specimen by a too narrow cone of light, that is, by closing the substage iris too far in an effort to increase contrast. If a slide is examined with a $\frac{2}{3}$-inch objective, the iris of the condenser being correctly set, and then a $\frac{1}{3}$-inch or $\frac{1}{6}$-inch is swung into position, it is noticed at once that there is marked contrast, but that all the outlines are surrounded by broad haloes. As the iris is opened these disappear and the image gains in clarity but, at the same time, contrast is reduced. When photographing subjects of very low contrast a contrasty negative is a sure sign that illumination was incorrect. The aim should be to use the largest cone of illumination which will allow of the image being focused. With some subjects the use of a $\frac{2}{3}$ cone is impossible, as the image under these conditions is almost or quite invisible. Here the choice is between no photograph at all or one showing some diffraction effects. Still, the majority of moulds are either self-coloured or can be stained to give adequate contrast with correct, or nearly correct, illumination.

In photographing deeply coloured or well-stained specimens contrast can be controlled over a wide range by the use of colour screens. The general principles governing the use of these have been given above, but two additional precautions are necessary. First, when using achromatic objectives, the image must be focused on the ground glass with the colour screen in position, because these objectives do not bring all colours to a focus at the same point. Secondly, the plate or film must, of course, be sensitive to light of the wave-length used.

Other important factors which influence contrast are the time of development of the negative and the type of printing paper used, but these are purely photographic matters.

Use of infra-red radiation. In order to show detail in photographs of very dark-coloured structures, when the use of a deep red colour filter and panchromatic film proves to be inadequate, recourse must be had to radiation of longer wave-length and greater penetrating power.

As sources of the range of infra-red radiation which can be used for photography, all the usual types of lamps used for visible light photomicrography are satisfactory. The only special requirements are an infrared filter and a supply of films or plates which are sensitive to this portion of the spectrum. Both are readily obtainable from manufacturers of photographic materials.

The main difficulty is focusing, since no appreciable amount of visible light is passed by the I.R. filter. The best method is one of direct calibration. A suitable test object is focused first in green light, then in red light, and the amount by which the fine-adjustment screw-head has to be rotated is noted. Obviously the screw-head will have to be turned further in the same direction in order to bring the infra-red into focus. Various formulae have been published for calculating the amount of displacement required, but the amount varies for different objectives, and it is more satisfactory to make a series of exposures, with the I.R. filter in position, noting carefully for each exposure the amount the fine-adjustment head is turned. The amount which gives the sharpest negative is recorded, and will always be the same for that particular combination of objective and eyepiece.

One other thing needs to be noted. Infra-red radiation readily penetrates thin layers of wood, and some varieties of hard rubber. Hence, if dark-slides are used, they should have metal shutters to prevent fogging of the film.

A good account of photography by infra-red radiation is by Clark (1946). Another useful paper is by Armitage (1938). Needham's book will also be found useful.

The author's outfit. There is a wide selection of photomicrographic cameras on the market: horizontal, vertical, combined horizontal and vertical, and small cameras which rest on top of the draw-tube. Horizontal cameras are not really suitable when most of the subjects to be photographed are mounted in fluid, since movement is liable to occur as soon as the slide is brought into a vertical plane. The small eyepiece cameras allow of only a limited range of magnification and require very careful manipulation to avoid vibration. The chief drawback to most of the vertical outfits, when the same microscope has to be used for visual work as well as photography, is the amount of time required for alignment of the illuminating system with the microscope every time a changeover is made. The author's outfit has, it is believed, certain advantages which make for ease of working, and which may be of interest to others.

The microscope is fitted with concentric rotating stage (a luxury but useful), centring objective changers, centring substage with dry achromatic condenser, and an aluminized surface mirror, the latter to avoid the multiple reflections which are always given by an ordinary glass mirror. Objectives include a 35-mm. micro-anastigmat for very low powers, an achromatic $1\frac{1}{2}$-inch, a semi-apochromatic 16-mm. corrected for a tube-length of 200 mm., an 8-mm. apochromatic, and a 3-mm. oil-immersion apochromatic.

The "Pointolite" lamp and fitting, described in previous editions of this book, has been discarded in favour of the Singer lamp, described on p. 360. The latter has been found to be very satisfactory in use.

The microscope and lamp, carefully aligned with one another, are both permanently fixed to a stout board, covered with velvet on the underside so that it will slide easily over the work-table. When not in use the microscope is covered by a bell-glass, resting in a circular groove cut in the board. The light is always centred and the microscope is ready for use as soon as the lamp is switched on. The work-table itself projects from a wall and is sufficiently low to allow of the microscope being used in the vertical position without strain.

For supporting the camera a 20-inch length of 1-inch diameter steel shafting is clamped vertically between two cast-iron brackets projecting about 4 inches from the wall, at a suitable height over the work-table. Two ring fittings, about 4 inches apart, and provided with clamping screws, are fastened to the camera back and slide on the steel pillar. The camera has thus no direct connection with the microscope or the table and is very securely supported. The camera is an old half-plate field camera, the dark slides having adaptors for quarter-plates. The use of a camera of larger size than the plates normally used means that a circular image large enough to cover the plate can be projected without any danger of reflections from the inside of the bellows. The front of the camera has been adapted to take inter-changeable sliding panels. One panel carries a short cardboard tube for making a light-tight connection with the microscope. A second panel carries a length of 3-inch diameter cardboard tube, terminating in a short piece of narrower tube as on the first panel, for increasing the range of extension of the camera. A third panel is fitted with a 6-inch anastigmat, for photo-graphing Petri dishes, or other large objects, laid on the table.

A pointer is fastened to the back of the camera and moves over a half-metre scale fixed vertically on the wall. By projecting a stage micrometer and measuring the image on the ground glass, the magnifications given by various combinations of objectives and eyepieces have been plotted against camera extension (the figures for the latter being purely arbitrary and not actual distances of plate from eyepoint of microscope), so that any desired magnification can be obtained by reading off from the graph the appropriate setting of the pointer on the scale. When anything is seen of which a pictorial record is required it is a simple matter, requiring only a few minutes, to slide the microscope board under the camera, and set the latter to give exactly the magnification decided on.

The whole apparatus is in a dark-room, since thus it is much easier to control stray light than when working in a brightly lit room.

The apparatus has been used for many years with quarter-plates as the sensitive material, but, during recent years, the cost of plates and cut films has become so high that the camera has been adapted to use the much cheaper 35 mm. film. Dade (1953) describes an apparatus which differs in a number of ways from that described here, but also describes how an old half-plate camera can readily be adapted. One of the most interesting of his suggestions is to use the film cut up into 10 cm. lengths (sufficient for two exposures), so that negatives can be processed at once, and fresh ex-posures made if the first are unsatisfactory, instead of having to wait until 36 exposures have been made, as in the conventional types of apparatus.

The reversing back of the camera has been replaced by a panel of $\frac{1}{4}$ inch wood, in which a rectangular hole, 10 cm. by 35 mm. is cut, with half the hole occupying the centre of the panel. A shallow well is formed by cover-ing the under side of the hole with a brass plate, in which an aperture 36×24 mm. is cut, the plate being screwed to the wood so that this aperture is exactly in the centre of the panel, and this centred in the optical axis. A piece of stout brass sheet, 10 cm. by 35 mm., and fitted with a central knob for lifting, serves to press the film flat in the well. During exposures the back of the camera is covered with a piece of black cloth, to ensure that no stray light reaches the film. With a simple outfit of this type it is virtually essential to have it fitted up in a dark-room, so that the film can be inserted, turned round between the exposures, and removed, with all lights turned off.

The magnifications at which most photomicrographs are taken are exactly one-third those desired in the final print, the negatives being, of course, printed through an enlarger to give the precise magnification required. Apart from the shorter camera extensions required, which is advantageous to anyone starting the construction of such an apparatus, the lowering of the initial magnification, and printing by enlargement, give improved depth of focus. Dade (1952) has shown that, at any particular working N.A., the depth of focus is inversely proportional to the square of the magnification, whereas, when a negative is enlarged, the decrease in depth of focus is directly proportional to the degree of enlargement, since the film has no appreciable thickness. Dade gives some striking examples, comprising photographs of an artificial test-object, which compare different combinations of initial magnification and degree of enlargement.

Dade (1953) also describes a device, easily made from Perspex sheet, for holding one or two 10 cm. lengths of film, and keeping the film flat, so that development can be carried out in a quarter-plate dish.

In previous editions of this book it was stated that the eye is more sensitive, and tires less easily, when the microscope is used in a dark-room than when used in brightly lit surroundings, *vide* Coles (1921). That great microscopist, E. Eliot-Merlin, used to say "I find that a darkened room and a well-rested eye are essential for all delicate objects". Hartridge (1954), however, in a paper dealing particularly with the function of the eye in microscopy, states that the practice of using a microscope in a darkened room is wrong. The general illumination should more or less match the illumination of the microscopic object. For maximum resolution a high intensity of illumination is required, but this causes glare unless the surroundings are adequately lit.

REFERENCES

Armitage, F. D. (1938). Infrared photomicrography. *Microscope*, June, 1938, **11**, 123–5.

Baker, J. R., and Llowarch, W. (1959). The use of the Philips "Mirror condenser lamp" in microscopy. *Quart. J. micr. Sci.*, **100**, 321–4.

Clark, W. (1946). *Photography by infrared. Its principles and applications.* 2nd Ed. New York: Wiley.

Dade, H. A. (1952). Depth of focus in low power photomicrography. *J. Quekett microsc. Club*, Ser. 4, **3**, 367–79.

Dade, H. A. (1953). A photomicrographic apparatus for 35 mm film. *J. Quekett microsc. Club*, Ser. 4, **3**, 463–79.

Dade, H. A. (1956). On common-sense in microscopy. *J. Quekett microsc. Club*, Ser. 4, **4**, 275–87.

Dade, H. A. (1958). On spherical aberration in the microscope. *J. Quekett microsc. Club*, Ser. 4, **5**, 3–25.

Fleming, W. D. (1966). A narrow band-pass colour filter for microscopy. *J. Quekett microsc. Club*, **30**, 105–8.

Hartridge, H. (1954). The optimal conditions for visual microscopy. *J. Quekett microsc. Club*, Ser. 4, **4**, 57–88.

Kurozumi, T., and Shibata, K. (1960). New technique for the microscopic observation of colored particles with opal glass. *Cytologia*, **25**, 229–32.

McClure, A. E. (1958). A cheap substitute for the substage condenser. *J. Quekett microsc. Club*, Ser. 4, **5**, 102–3, 199.

Maltwood, T. (1858). On a finder for registering the position of microscopic objects. *Quart. J. micr. Sci.*, N.S. **6** (Trans.), 59–62.

Ockenden, F. E. J. (1946). Illuminants and illumination for microscopical work. *J. Quekett microsc. Club*, Ser. 4, **2**, 103–26.

Slater, P. N. (1957). The star test—Its interpretation and value. *J. Quekett microsc. Club*, Ser. 4, **4**, 415–22.

BOOKS ON MICROSCOPY AND PHOTOMICROGRAPHY

The literature of microscopy is extensive and the books mentioned below are only a selection from the many which have been published. However, they cover between them most, if not all, of the requirements of the mycologist. Readers who wish to know more about recent advances in microscopy should consult the *Journal of the Royal Microscopical Society*.

ALLEN, R. M. (1958). *Photomicrography*. 2nd Ed. New York: Van Nostrand.

A useful book, including, besides standard photomicrographic technique, the construction of home-made apparatus and also the method of making "microphotographs".

BARER, R. (1956). *Lecture Notes on the Use of the Microscope*. 2nd Ed. Oxford: Blackwell.

An excellent introduction to the correct use of the microscope.

BARNARD, J. E., and WELCH, F. V. (1936). *Practical Photomicrography*. 3rd Ed. London: Edward Arnold & Co.

A standard work on the subject. Includes a full description of photomicrography with ultra-violet light, a technique largely developed by the authors.

BARRON, A. L. E. (1965). *Using the Microscope*. London: Chapman & Hall, Ltd.

This book is really a third edition of *The Intelligent Use of the Microscope*, by C. W. Olliver, the second edition of which was published in 1951. The new edition brings the text up to date, gives more attention to elementary optical theory, describes the interference microscope, and deals with photomicrography in a much fuller way. Many of the original illustrations are deleted, often, but not always, justifiably. The book still remains a useful one, especially for beginners, since it deals with many matters which are often taken for granted in other publications.

BECK, C. (1938). *The Microscope*. 2nd Ed. London: R. & J. Beck, Ltd.

The first edition was in two volumes, Elementary and Advanced. For the second one-volume edition much elementary descriptive matter has been omitted. It discusses the practical applications of modern theories of microscopical vision and deals fully with a number of topics which are barely mentioned in most other books. It is stimulating reading for those who wish to get the very best out of their instruments.

BIRCHON, D. (1961). *Optical Microscope Technique.* London: George Newnes, Ltd.

A very practical and useful book covering all aspects of optical microscopy. There is also a short section on photomicrography and cinematography. There are many references to papers dealing more fully with particular subjects. The book includes rather too much description of methods of metallurgical microscopy.

COLES, A. C. (1921). *Critical Microscopy.* London: J. & A. Churchill.

A very personal and unconventional book which passes on some of the teaching of that great microscopist E. M. Nelson.

HIND, H. L., and RANDLES, W. B. (1927). *Handbook of Photomicrography.* 2nd Ed. London: George Routledge & Sons, Ltd.

Another standard book on the subject. It contains some beautiful examples of photomicrographic practice, of wide range, and gives full details of the apparatus and methods used.

LANGERON, M. (1942). *Précis de Microscopie.* 6th Ed. Paris: Masson et Cie. A very comprehensive treatise on microscopy and well worth study. It includes methods of preparation of slides of all kinds of subjects.

LAWSON, D. F. (1960). *The Technique of Photomicrography.* London: George Newnes, Ltd.

A very complete account of the practical side of photomicrography. Whilst in no way neglecting high-magnification work, it deals much more fully than most other books with low-power photography. Many of the illustrations are useful in showing what can be done in this line. There are also descriptions of a number of modern illuminants, and of the use of flash in photographing moving objects.

NEEDHAM, G. H. (1958). *The Practical Use of the Microscope, Including Photomicrography.* Springfield, Illinois: Charles C. Thomas.

This is an excellent book, dealing with both elementary and advanced microscopy and photomicrography. There are sections on fluorescent, ultra-violet, infra-red, reflecting, phase-contrast, and electron microscopes, also chapters on the polarizing microscope, chemical microscopy (i.e. elementary crystallography), microprojection and drawing, and the testing of objectives. The Author compares directly alternative pieces of equipment, such as different kinds of substages, objective changers, and types of binocular microscopes. There are also comparisons between different available methods, such as critical illumination and the Köhler system. There are 173 figures, including many illustrations of microscopes and set-ups, but all are to the point.

The book is expensive, but there is no other which covers such a wide range of microscopical techniques.

POLYCARD, A., BESSIS, M., and LOCQUIN, M. (1956). *Traité de Microscopie, Instruments et Technique.* Paris: Masson et Cie.

A very comprehensive and valuable book for those who read French with sufficient ease.

SHILLABER, C. P. (1944). *Photomicrography in Theory and Practice.* New York: Wiley. London: Chapman & Hall (1945).

This is probably the most comprehensive book on photomicrography in existence. It does not, of course, describe the more modern advances in microscopy, such as phase contrast, interference microscopy, and the reflecting microscope, but, within the limits of what may be termed "classical microscopy", it is first class. The author discusses scores of points which are either ignored or inadequately treated in most other books on the subject. The numerous illustrations are all to the point, there being none of the pretty type of photomicrograph commonly included to demonstrate an author's skill. The special techniques required for the examination of a great variety of industrial materials are discussed.

Author Index

Subject Index

Page numbers in italics refer to diagnoses of genera and species, those in heavy type to illustrations